Generalized Fractional Order Differential Equations Arising in Physical Models

Generalized Fractional Order Differential Equations Arising in Physical Models

Santanu Saha Ray
Subhadarshan Sahoo

CRC Press
Taylor & Francis Group
Boca Raton London New York

CRC Press is an imprint of the
Taylor & Francis Group, an **informa** business

CRC Press
Taylor & Francis Group
6000 Broken Sound Parkway NW, Suite 300
Boca Raton, FL 33487-2742

© 2019 by Taylor & Francis Group, LLC
CRC Press is an imprint of Taylor & Francis Group, an Informa business

No claim to original U.S. Government works

Printed on acid-free paper

International Standard Book Number-13: 978-1-138-36681-7 (Hardback)

This book contains information obtained from authentic and highly regarded sources. Reasonable efforts have been made to publish reliable data and information, but the author and publisher cannot assume responsibility for the validity of all materials or the consequences of their use. The authors and publishers have attempted to trace the copyright holders of all material reproduced in this publication and apologize to copyright holders if permission to publish in this form has not been obtained. If any copyright material has not been acknowledged please write and let us know so we may rectify in any future reprint.

Except as permitted under U.S. Copyright Law, no part of this book may be reprinted, reproduced, transmitted, or utilized in any form by any electronic, mechanical, or other means, now known or hereafter invented, including photocopying, microfilming, and recording, or in any information storage or retrieval system, without written permission from the publishers.

For permission to photocopy or use material electronically from this work, please access www.copyright.com (http://www.copyright.com/) or contact the Copyright Clearance Center, Inc. (CCC), 222 Rosewood Drive, Danvers, MA 01923, 978-750-8400. CCC is a not-for-profit organization that provides licenses and registration for a variety of users. For organizations that have been granted a photocopy license by the CCC, a separate system of payment has been arranged.

Trademark Notice: Product or corporate names may be trademarks or registered trademarks, and are used only for identification and explanation without intent to infringe.

Visit the Taylor & Francis Web site at
http://www.taylorandfrancis.com

and the CRC Press Web site at
http://www.crcpress.com

This work is dedicated

to

my parents

Contents

List of Figures xv
List of Tables xxvii
Preface xxix
Acknowledgments xxxiii
Authors xxxv

1. Introduction and Preliminaries of Fractional Calculus 1
    1.1   Introduction 1
    1.2   Preliminaries of Fractional Calculus 1
          1.2.1   Useful Mathematical Functions 2
                1.2.1.1   Gamma Function 2
                1.2.1.2   Beta Function 4
                1.2.1.3   Mittag–Leffler Function 4
                1.2.1.4   Mellin–Ross Function 6
                1.2.1.5   Wright Function 6
                1.2.1.6   Error Function 6
                1.2.1.7   Hypergeometric Functions 7
                1.2.1.8   *H*-Function 9
          1.2.2   Riemann–Liouville Integral and Derivative 10
          1.2.3   Caputo Fractional Derivative 11
          1.2.4   Grünwald–Letnikov Fractional Derivative 12
          1.2.5   Riesz Fractional Integral and Derivative 13
          1.2.6   Modified Riemann–Liouville Derivative 18
          1.2.7   Local Fractional Derivative 18
                 1.2.7.1   Local Fractional Continuity of a Function 18
                1.2.7.2   Local Fractional Derivative 19
    1.3   Conclusion 21

2. Overview of Numerous Analytical Methods 23
    2.1   Introduction 23
    2.2   Homotopy Perturbation Method with Adomian Polynomial 23
          2.2.1   Adomian Polynomial 24
    2.3   Homotopy Perturbation Transform Method 25
    2.4   Modified Homotopy Analysis Method 26
    2.5   Modified Homotopy Analysis Method with Fourier Transform 27
    2.6   Fractional Sub-Equation Method 29
    2.7   Improved Fractional Sub-Equation Method 31
    2.8 (G'/G)-Expansion Method 32
    2.9   Improved (G'/G)-Expansion Method 33
    2.10   Proposed Tanh Method for FDE 36
    2.11   Modified Kudryashov Method 37
    2.12   Proposed New Method 39

vii

viii   *Contents*

2.13	Jacobi Elliptic Function Method	41
2.14	Proposed Successive Recursion Method	43
2.15	Conclusion	44

3. New Analytical Approximate Solutions of Fractional Differential Equations 47

3.1	Introduction	47
3.2	Outline of Present Study	47

     3.2.1   Time-Fractional Predator–Prey Models 48

          3.2.1.1   Time-Fractional Lotka–Volterra Model for an Interacting Species Model (Case I) 48

          3.2.1.2   Time-Fractional Simple Two-Species Lotka–Volterra Competition Model (Case II) 48

          3.2.1.3   Time-Fractional Lotka–Volterra Competition Model with Limit Cycle Periodic Behavior (Case III) 49

     3.2.2   Time-Fractional Coupled Non-Linear K–G–S Equations 49

     3.2.3   Time-Fractional Coupled Non-Linear K–G–Z Equations 50

     3.2.4   Time-Fractional Order Non-Linear-Coupled Sine-Gordon Equations 51

   3.3   Application of HPM Methods for Solutions of Fractional Predator–Prey Model 51

     3.3.1   Application of HPM for Analytical Approximate Solutions for Time-Fractional Lotka–Volterra Model for an Interacting Species Model (Case I) 52

          3.3.1.1   Numerical Results and Discussion for Time-Fractional Lotka–Volterra Model for an Interacting Species (Case I) 53

          3.3.1.2   Stability Analysis for Time-Fractional Lotka–Volterra Model for an Interacting Species Model (Case I) 54

     3.3.2   Application of HPM for Analytical Approximate Solutions for Time-Fractional Simple Two-Species Lotka–Volterra Competition Model (Case II) 55

          3.3.2.1   Numerical Results and Discussion for Time-Fractional Dimple Two-Species Lotka–Volterra Competition Model (Case II) 57

          3.3.2.2   Stability Analysis for Time-Fractional Simple Two-Species Lotka–Volterra Competition Model (Case II) 57

     3.3.3   Application of HPM for Analytical Approximate Solutions for Time-Fractional Lotka–Volterra Competition Model with Limit Cycle Periodic Behavior (Case III) 59

          3.3.3.1   Numerical Results and Discussion for Time-Fractional Lotka–Volterra Competition Model with Limit Cycle Periodic Behavior (Case III) 62

          3.3.3.2   Stability Analysis for Time-Fractional Lotka–Volterra Competition Model with Limit Cycle Periodic Behavior (Case III) 63

   3.4   Implementation of the HPM and HPTM for the Solution of Fractional Coupled K–G–S Equations 65

     3.4.1   Implementation of the HPM for the Solution of Fractional Coupled K–G–S Equations 65

Contents

3.4.2 Implementation of the HPTM for the Solution of Fractional Coupled K–G–S Equations 68

3.4.3 Numerical Results and Discussion 72

3.4.3.1 The Numerical Simulations for HPTM Method 72

3.4.3.2 The Numerical Simulations for Absolute Errors in HPM and HPTM Solutions 77

3.4.3.3 Comparison of HPM and HPTM Solutions with Regard to Exact Solutions 78

3.5 Application of HPTM and MHAM Methods for the Solutions of Fractional Coupled K–G–Z Equations 79

3.5.1 Implementation of HPTM Method for the Solutions of Fractional Coupled K–G–Z Equations 80

3.5.2 Implementation of the MHAM for Approximate Solutions to the K–G–Z Equations 83

3.5.3 Numerical Results and Discussion 87

3.5.3.1 The Numerical Simulations for MHAM 87

3.5.3.2 The $\hbar$ Graph and the Numerical Simulations for $u(x, t)$ and $v(x, t)$ in MHAM 90

3.5.3.3 To Analogize the Solutions of K–G–Z Equations by MHAM with Exact Solution for Different Values of $\hbar$ 93

3.5.3.4 Comparison of Absolute Errors for MHAM and HPTM Solutions 93

3.6 Application of the MHAM and HPTM for Approximate Solutions of Fractional Coupled S-G Equations 96

3.6.1 Implementation of the MHAM for Approximate Solutions of Fractional Coupled S-G Equations 96

3.6.2 Implementation of the HPTM for Approximate Solutions of Fractional Coupled S-G Equations 100

3.6.3 Numerical Results and Discussion 102

3.6.3.1 The $\hbar$ Graph and the Numerical Simulations for $u(x, t)$ and $v(x, t)$ in MHAM 102

3.6.3.2 The Numerical Simulations for MHAM 103

3.6.3.3 Comparison of Absolute Errors for MHAM and HPTM Solutions 105

3.7 Conclusion 106

4. New Analytical Approximate Solutions of Riesz Fractional Differential Equations 109

4.1 Introduction 109

4.2 Outline of Present Study 109

4.2.1 Riesz Fractional Diffusion Equation and Riesz Fractional Advection–Dispersion Equation 109

4.2.2 Camassa–Holm Equation 110

4.3 Implementation of the MHAM-FT for Approximate Solution of Riesz Fractional Diffusion Equation 111

4.3.1 The $\hbar$ Graph and Numerical Simulations for MHAM-FT and Discussions 113

4.4	Implementation of the MHAM-FT for Approximate Solution of Riesz Fractional Advection–Dispersion Equation	115
	4.4.1 The $\hbar$ Graph and Numerical Simulations for MHAM-FT Method and Discussions	117
4.5	Implementation of the MHAM for Approximate Solution of Riesz Time-Fractional Camassa–Holm Equation	118
	4.5.1 The $\hbar$ Graph and Numerical Simulations for MHAM	121
	4.5.2 Comparison of Present MHAM Solution with Regard to VIM Solution	123
4.6	Conclusion	124

5. New Exact Solutions of Fractional Differential Equations by Fractional Sub-Equation and Improved Fractional Sub-Equation Method 125

5.1	Introduction	125
5.2	Outline of Present Study	125
	5.2.1 Space-Time Fractional ZK Equation	125
	5.2.2 Space-Time Fractional Modified ZK Equation	126
	5.2.3 (3 + 1)-Dimensional Time-Fractional KdV–ZK Equation	126
	5.2.4 (3 + 1)-Dimensional Space-Time Fractional Modified KdV–ZK	126
5.3	Application of the Fractional Sub-Equation Method for the Solution of Space-Time Fractional ZK Equation	127
	5.3.1 Numerical Simulations of Space-Time Fractional ZK Equation	130
	5.3.2 Influence of Fractional Order Derivative on Solitary Wave Solutions of Fractional ZK Equation	131
5.4	Implementation of Fractional Sub-Equation Method to the Space-Time Fractional mZK Equation	132
	5.4.1 Numerical Simulations of Space-Time Fractional mZK Equation	134
	5.4.2 Influence of Fractional Order Derivative on Solitary Wave Solutions of Fractional mZK Equation	135
5.5	Application of Improved Fractional Sub-Equation Method to the Time-Fractional KdV–ZK Equation	135
	5.5.1 Numerical Simulations of Time-Fractional KdV–ZK Equation	138
5.6	Implementation of Improved Fractional Sub-Equation Method to the Space-Time Fractional Modified mKdV–ZK Equation	140
	5.6.1 Numerical Simulations of Space-Time Fractional mKdV–ZK Equation	142
5.7	Conclusion	143

6. New Exact Solutions of Fractional Differential Equations by (G'/G)-Expansion Method and Improved (G'/G)-Expansion Method 145

6.1	Introduction	145
6.2	Outline of Present Study	145
	6.2.1 Time-Fractional Fifth Order Modified Kawahara Equation	145
	6.2.2 Time-Fractional Coupled JM Equations	146
	6.2.3 Time-Fractional KK Equation	147
	6.2.4 Time-Fractional Third Order Time-Fractional Modified KdV Equation	147
	6.2.4.1 Formulation of Fractional mKdV Equation	147

6.3 Application of Proposed (G'/G)-Expansion Method to the Time-Fractional Fifth Order Modified Kawahara Equation 151

6.3.1 The Numerical Simulations for Solutions of Modified Kawahara Equation 158

6.4 Implementation of (G'/G)-Expansion Method to the Time-Fractional Coupled JM Equations 160

6.5 Implementation of the Proposed Improved (G'/G)-Expansion Method for the Time-Fractional Kaup–Kupershmidt Equation 164

6.6 Application of (G'/G)-Expansion Method and Improved (G'/G)-Expansion Method to the Time-Fractional Modified KdV Equation 167

6.6.1 Application of (G'/G)-Expansion Method to the Time-Fractional Modified KdV Equation 167

6.6.1.1 The Numerical Simulation for Solution of Time-Fractional Third Order Modified KdV Equation Using Proposed (G'/G)-Expansion Method 170

6.6.2 Implementation of Improved (G'/G)-Expansion Method for the Time-Fractional Modified KdV Equation 173

6.6.2.1 The Numerical Simulation for Solution of Time-Fractional Third Order Modified KdV Equation Using Proposed Improved (G'/G)-Expansion Method 176

6.7 Conclusion 181

7. New Exact Solutions of Fractional Differential Equations by Proposed Tanh and Modified Kudryashov Methods 183

7.1 Introduction 183

7.2 Outline of Present Study 183

7.2.1 Time-Fractional Fifth-Order S–K Equation 184

7.2.2 Time-Fractional Fifth-Order Modified S–K Equation 184

7.2.3 Time-Fractional K–S Equation 185

7.3 Implementation of Proposed Tanh Method for the Exact Solutions of Time-Fractional Fifth-Order S–K Equation 185

7.3.1 The Numerical Simulations for Solutions of Time-Fractional Fifth-Order S–K Equation Using Proposed Tanh Method 187

7.3.1.1 Numerical Results and Discussion 187

7.4 Application of Proposed Tanh and Modified Kudryashov Methods for the Exact Solitary Wave Solutions of Time-Fractional Fifth-Order mS–K Equation 190

7.4.1 Application of Proposed Tanh Method for the Exact Solitary Wave Solutions of Time-Fractional Fifth-Order mS–K Equation 190

7.4.1.1 The Numerical Simulations for Presenting the Nature of the Solutions of Fractional Fifth-Order mS–K Equations by Utilizing Proposed Tanh Method 192

7.4.2 Application of Proposed Modified Kudryashov Method for the Exact Solitary Wave Solutions of Time-Fractional Fifth-Order mS–K Equation 194

7.4.2.1 The Numerical Simulations for Solutions of Time-Fractional Fifth-Order mS–K Equations by Using Modified Kudryashov Method 196

xii | Contents

7.5   Implementation of Proposed Tanh Method for the Exact Solutions of Time-Fractional K–S Equation 198
      7.5.1   The Numerical Simulations for Time-Fractional K–S Equation Obtained by New Proposed Tanh Method 202
7.6   Implementation of Proposed Tanh Method for the Exact Solutions of Time-Fractional Coupled JM Equations 203
      7.6.1   The Numerical Simulations for Solutions of Time-Fractional Coupled JM Equations Using Tanh Method 206
7.7   Conclusion 207

8. New Exact Solutions of Fractional Differential Equations by Proposed Novel Method 209

8.1   Introduction 209
8.2   Outline of Present Study 209
      8.2.1   Time-Fractional KdV–Burgers Equation 209
      8.2.2   Time-Fractional Combined KdV–mKdV Equation 210
      8.2.3   Time-Fractional Coupled Schrödinger–KdV Equations 210
      8.2.4   Time-Fractional Coupled SB Equations 211
8.3   Exact Solutions for Time-Fractional KdV–Burgers Equation 211
      8.3.1   Numerical Simulations for Time-Fractional KdV–Burgers Equation 213
8.4   Exact Solutions for Time-Fractional KdV–mKdV Equation 214
      8.4.1   Numerical Simulations for Time-Fractional KdV–mKdV Equation 216
8.5   Exact Solutions for Time-Fractional Coupled Schrödinger–KdV Equations 217
      8.5.1   Numerical Simulations for Time-Fractional Coupled SK Equations 221
8.6   Exact Solutions for Time-Fractional Coupled SB Equations 223
      8.6.1   Numerical Simulations for Time-Fractional Coupled SB Equations 227
8.7   Conclusion 229

9. New Exact Solutions of Fractional Coupled Differential Equations by Jacobi Elliptic Function Method 231

9.1   Introduction 231
9.2   Outline of Present Study 231
9.3   Implementation of Proposed Method for Exact Solutions for Time-Fractional Coupled Drinfeld–Sokolov–Wilson Equations 232
9.4   Numerical Simulations for Time-Fractional Coupled Drinfeld–Sokolov–Wilson Equations 239
      9.4.1   Numerical Simulation for Time-Fractional Coupled DSW Equations Based on the Solutions Obtained by Case I of Set I 239
      9.4.2   Numerical Simulation for Time-Fractional Coupled DSW Equations Based on the Solutions Obtained by Case II of Set I 242
      9.4.3   Numerical Simulation for Time-Fractional Coupled DSW Equations Based on the Solutions Obtained by Case V of Set I 244
9.5   Conclusion 246

Contents xiii

10. Formulation and Solutions of Fractional Continuously Variable-Order Mass-Spring Damper Systems 247

10.1 Introduction 247

10.2 Outline of Present Study 248

10.3 Theory of Variable-Order Fractional Calculus 249

    10.3.1 Variable-Order Differential Operator 249

    10.3.2 Analysis of Variable-Order Differential Operator 253

    10.3.3 Laplace Transform of Variable-Order Operator 254

10.4 Analysis of Viscoelastic Oscillator of a Spring Damper System 255

10.5 Definition of Continuous Variable Fractional Order Frictional Damping Force 258

10.6 Formulation for Mass-Spring Damper System 259

    10.6.1 Free Oscillation with Viscoelastic Damping in Case I 260

    10.6.2 Free Oscillation with Viscous-Viscoelastic or Viscoelastic-Viscous Damping System in Case II 262

    10.6.3 Forced Oscillation with Viscous-Viscoelastic or Viscoelastic-Viscous Damping in Case III 265

    10.6.4 Forced Oscillation with Viscoelastic-Viscous Damping in Case IV 266

10.7 Application of Proposed Successive Recursion Method for Solution of Fractional Continuously Variable-Order Mass-Spring Damper System 267

    10.7.1 Implementation of Successive Recursion Method for Free Oscillation of Mass-Spring Viscoelastic Damping System in Case I 268

        10.7.1.1 First Recursion 269

        10.7.1.2 Second Recursion 269

    10.7.2 Implementation of Successive Recursion Method for Free Oscillation of Mass-Spring Viscous-Viscoelastic Damping System in Case II 272

        10.7.2.1 First Recursion 273

        10.7.2.2 Second Recursion 274

    10.7.3 Application of Successive Recursion Method for Forced Oscillation of Spring-Mass Viscous-Viscoelastic Damping System in Case III 276

        10.7.3.1 First Recursion 277

        10.7.3.2 Second Recursion 278

    10.7.4 Application of Successive Recursion Method for Forced Oscillation of Spring-Mass Viscoelastic-Viscous Damping System in Case IV 280

        10.7.4.1 First Recursion 281

        10.7.4.2 Second Recursion 282

10.8 Numerical Simulations and Discussion 284

    10.8.1 Numerical Simulation of Fractional Continuously Variable-Order Mass-Spring Damper System for Free Oscillation with Viscoelastic Damping-Case I 285

    10.8.2 Numerical Simulation of Fractional Continuously Variable-Order Mass-Spring Damper System for Free Oscillation with Viscous-Viscoelastic Damping-Case II 286

| | 10.8.3 | Numerical Simulation of Fractional Continuously Variable-Order Mass-Spring Damper System for Forced Oscillation with Viscous-Viscoelastic Damping-Case III | 286 |

10.8.3 Numerical Simulation of Fractional Continuously Variable-Order Mass-Spring Damper System for Forced Oscillation with Viscous-Viscoelastic Damping-Case III 286

10.8.4 Numerical Simulation of Fractional Continuously Variable-Order Mass-Spring-Damper System for Forced Oscillation with Viscoelastic-Viscous Damping-Case IV 288

10.8.5 Physical Interpretations 288

10.9 Conclusion 289

References 291

Index 309

List of Figures

Figure 1.1 The graph function in real axis, when $f = \Gamma(z)$ 2

Figure 3.1 Plots of the fifth-order approximations for u and v for **Case I** 54

Figure 3.2 Plots of the fifth-order approximations for u and v for **Case II** 57

Figure 3.3 Plots of the fifth-order approximations for u and v for **Case III** 63

Figure 3.4 **Case I:** For $\alpha = 2$, $\beta = 1$ (Classical order). (a) The HPTM solution for $u(x,t)$, (b) corresponding solution for $u(x,t)$ when $t = 0$ 73

Figure 3.5 **Case I:** For $\alpha = 2$, $\beta = 1$. (a) The HPTM solution for $\operatorname{Re}(v(x,t))$, (b) corresponding solution for $\operatorname{Re}(v(x,t))$ when $t = 0$ 73

Figure 3.6 **Case I:** For $\alpha = 2$, $\beta = 1$. (a) The HPTM solution for $\operatorname{Im}(v(x,t))$, (b) corresponding solution for $\operatorname{Im}(v(x,t))$ when $t = 0$ 73

Figure 3.7 **Case I:** For $\alpha = 2$, $\beta = 1$. (a) The HPTM solution for $|v(x,t)|$, (b) corresponding solution for $|v(x,t)|$ when $t = 0$ 74

Figure 3.8 **Case II:** For $\alpha = 1.75$, $\beta = 0.75$ (Fractional order). (a) The HPTM solution for $u(x,t)$, (b) corresponding solution for $u(x,t)$ when $t = 0$ 74

Figure 3.9 **Case II:** For $\alpha = 1.75$, $\beta = 0.75$. (a) The HPTM solution for $\operatorname{Re}(v(x,t))$, (b) corresponding solution for $\operatorname{Re}(v(x,t))$ when $t = 0$ 74

Figure 3.10 **Case II:** For $\alpha = 175$, $\beta = 0.75$. (a) The HPTM solution for $\operatorname{Im}(v(x,t))$, (b) corresponding solution for $\operatorname{Im}(v(x,t))$ when $t = 0$ 75

Figure 3.11 **Case II:** For $\alpha = 175$, $\beta = 0.75$. (a) The HPTM solution for $|v(x,t)|$, (b) corresponding solution for $|v(x,t)|$ when $t = 0$ 75

Figure 3.12 **Case III:** For $\alpha = 1.5$, $\beta = 0.5$ (Fractional order). (a) The HPTM solution for $u(x,t)$, (b) corresponding solution for $u(x,t)$ when $t = 0$ 75

Figure 3.13 **Case III:** For $\alpha = 1.5$, $\beta = 0.5$. (a) The HPTM solution for $\operatorname{Re}(v(x,t))$, (b) corresponding solution for $\operatorname{Re}(v(x,t))$ when $t = 0$ 76

Figure 3.14 **Case III:** For $\alpha = 1.5$, $\beta = 0.5$. (a) The HPTM solution for $\operatorname{Im}(v(x,t))$, (b) corresponding solution for $\operatorname{Im}(v(x,t))$ when $t = 0$ 76

Figure 3.15 **Case III:** For $\alpha = 1.5$, $\beta = 0.5$. (a) The HPTM solution for $|v(x,t)|$, (b) corresponding solution for $|v(x,t)|$ when $t = 0$ 76

Figure 3.16 Graphical comparison of absolute errors in the solution of $u(x,t)$ for HPM and HPTM, when $\alpha = 2$, $\beta = 1$ 77

Figure 3.17 Graphical comparison of absolute errors in the solution of $\operatorname{Re}(v(x,t))$ for HPM and HPTM, when $\alpha = 2$, $\beta = 1$ 77

Figure 3.18 Graphical comparison of absolute errors in the solution of $\operatorname{Im}(v(x,t))$ for HPM and HPTM, when $\alpha = 2$, $\beta = 1$ 78

xv

xvi   *List of Figures*

Figure 3.19 Graphical comparison of absolute errors in the solution of $|v(x,t)|$ for HPM and HPTM, when $\alpha = 2$, $\beta = 1$...78

Figure 3.20 **Case I:** For $\alpha = 2$, $\beta = 2$. (a) The MHAM solution for $\mathrm{Re}\left[u(x,t)\right]$, (b) corresponding solution for $\mathrm{Re}\left[u(x,t)\right]$ when $t = 0.1$...................................88

Figure 3.21 **Case I:** For $\alpha = 2$, $\beta = 2$. (a) The MHAM solution for $\mathrm{Im}\left[u(x,t)\right]$, (b) corresponding solution for $\mathrm{Im}\left[u(x,t)\right]$ when $t = 0.1$...................................88

Figure 3.22 **Case I:** For $\alpha = 2$, $\beta = 2$. (a) The MHAM solution for $|u(x,t)|$, (b) corresponding solution for $|u(x,t)|$ when $t = 0.1$...................................88

Figure 3.23 **Case I:** For $\alpha = 2$, $\beta = 2$. (a) The MHAM solution for $v(x,t)$, (b) corresponding solution for $v(x,t)$ when $t = 0.1$...................................89

Figure 3.24 **Case II:** For $\alpha = 1.75$, $\beta = 1.5$. (a) The MHAM solution for $\mathrm{Re}\left[u(x,t)\right]$, (b) corresponding solution for $\mathrm{Re}\left[u(x,t)\right]$ when $t = 0.1$...................................89

Figure 3.25 **Case II:** For $\alpha = 1.75$, $\beta = 1.5$. (a) The MHAM solution for $\mathrm{Im}\left[u(x,t)\right]$, (b) corresponding solution for $\mathrm{Im}\left[u(x,t)\right]$ when $t = 0.1$...................................89

Figure 3.26 **Case II:** For $\alpha = 1.75$, $\beta = 1.5$. (a) The MHAM solution for $|u(x,t)|$, (b) corresponding solution for $|u(x,t)|$ when $t = 0.1$...................................90

Figure 3.27 **Case II:** For $\alpha = 1.75$, $\beta = 1.5$. (a) The MHAM solution for $v(x,t)$, (b) corresponding solution for $v(x,t)$ when $t = 0.1$...................................90

Figure 3.28 The $\hbar$-curve for partial derivatives of $\mathrm{Re}\left[u(x,t)\right]$ at (0,0) for the third-order MHAM solution when $\alpha = 2$, $\beta = 2$...................................91

Figure 3.29 The $\hbar$-curve for partial derivatives of $|u(x,t)|$ at (0,0) for the third-order MHAM solution when $\alpha = 2$, $\beta = 2$...................................91

Figure 3.30 The $\hbar$-curve for partial derivatives of $v(x,t)$ at (0,0) for the third-order MHAM solution when $\alpha = 2$, $\beta = 2$...................................91

Figure 3.31 The result obtained by the MHAM for various $\hbar$ by third-order MHAM approximate solution for $\alpha = 2$, $\beta = 2$ in comparison with the exact solution when $-1 < x < 1$ and $t = 0.6$...................................92

Figure 3.32 The result obtained by the MHAM for various $\hbar$ by third-order MHAM approximate solution for $\alpha = 2$, $\beta = 2$ in comparison with the exact solution when $-1 < x < 1$ and $t = 0.6$...................................92

Figure 3.33 The result obtained by the MHAM for various $\hbar$ by third-order MHAM approximate solution for $\alpha = 2$, $\beta = 2$ in comparison with the exact solution when $-1 < x < 1$ and $t = 0.6$...................................92

Figure 3.34 The result obtained by the MHAM for various $\hbar$ by third-order MHAM approximate solution for $\alpha = 2$, $\beta = 2$ in comparison with the exact solution when $-1 < x < 1$ and $t = 0.6$...................................93

Figure 3.35 The absolute errors for the K–G–Z equation by the third-order MHAM approximation of $\mathrm{Re}\left[u(x,t)\right]$...................................93

Figure 3.36 The absolute errors for the K–G–Z equation by the third-order MHAM approximation of $\mathrm{Im}\left[u(x,t)\right]$...................................94

List of Figures   xvii

Figure 3.37 The absolute errors for the K–G–Z equation by the third-order
MHAM approximation of $\left|u(x,t)\right|$ 94

Figure 3.38 The absolute errors for the K–G–Z equation by the third-order
MHAM approximation of $v(x,t)$ 94

Figure 3.39 Graphical comparison of absolute errors in the solution of $\mathrm{Re}(u(x,t))$
for (a) MHAM and (b) HPTM, when $\alpha = 2$, $\beta = 2$ 95

Figure 3.40 Graphical comparison of absolute errors in the solution of $\mathrm{Im}(u(x,t))$
for (a) MHAM and (b) HPTM, when $\alpha = 2$, $\beta = 2$ 95

Figure 3.41 Graphical comparison of absolute errors in the solution of $\left|u(x,t)\right|$ for
(a) MHAM and (b) HPTM, when $\alpha = 2$, $\beta = 2$ 95

Figure 3.42 Graphical comparison of absolute errors in the solution of $v(x,t)$ for
(a) MHAM and (b) HPTM, when $\alpha = 2$, $\beta = 2$ 96

Figure 3.43 The $\hbar$-curve for partial derivatives of $u(x,t)$ at $(0,0)$ for the
fourth-order MHAM solution when $\alpha = 2$, $\beta = 2$ 103

Figure 3.44 The $\hbar$-curve for partial derivatives of $v(x,t)$ at $(0,0)$ for the fourth-order
MHAM solution when $\alpha = 2$, $\beta = 2$ 103

Figure 3.45 **Case I:** For $\alpha = 2$, $\beta = 2$. (a) The MHAM solution for $u(x,t)$,
(b) corresponding solution for $u(x,t)$ when $t = 0.1$ 104

Figure 3.46 **Case I:** For $\alpha = 2$, $\beta = 2$. (a) The MHAM solution for $v(x,t)$,
(b) corresponding solution for $v(x,t)$ when $t = 0.1$ 104

Figure 3.47 **Case II:** For $\alpha = 1.75$, $\beta = 1.5$. (a) The MHAM solution for $u(x,t)$,
(b) corresponding solution for $u(x,t)$ when $t = 0.1$ 104

Figure 3.48 **Case II:** For $\alpha = 1.75$, $\beta = 1.5$. (a) The MHAM solution for $v(x,t)$,
(b) corresponding solution for $v(x,t)$ when $t = 0.1$ 105

Figure 3.49 Graphical comparison of absolute errors in the solution of $u(x,t)$ for
MHAM and HPTM, when $\alpha = 2$, $\beta = 2$ 105

Figure 3.50 Graphical comparison of absolute errors in the solution of $v(x,t)$ for
MHAM and HPTM, when $\alpha = 2$, $\beta = 2$ 106

Figure 4.1 The $\hbar$-curve for partial derivatives of $u(x,t)$ at $(1\times10^{-5},0)$ for the
11th-order MHAM-FT solution when $\alpha = 2$ 114

Figure 4.2 **Case I:** For $a = 2$. (a) The MHAM-FT solution for $u(x,t)$,
(b) corresponding solution for $u(x,t)$ when $t = 0.4$ 114

Figure 4.3 **Case II:** For $a = 1.5$. (a) The MHAM-FT solution for $u(x,t)$,
(b) corresponding solution for $u(x,t)$ when $t = 0.4$ 115

Figure 4.4 The $\hbar$-curve for partial derivatives of $u(x,t)$ at $(1\times10^{-5},0)$ for the
11th-order MHAM-FT solution when $\alpha = 2$, $\beta = 1$ 117

Figure 4.5 **Case I:** For $\alpha = 2$, $\beta = 1$. (a) The MHAM-FT solution for $u(x,t)$,
(b) corresponding solution for $u(x,t)$ when $t = 0.5$ 118

xviii *List of Figures*

Figure 4.6 Case II: For $\alpha = 1.5$, $\beta = 0.7$. (a) The MHAM-FT solution for $u(x,t)$, (b) corresponding solution for $u(x,t)$ when $t = 0.5$......... 118

Figure 4.7 The $\hbar$-curve for partial derivatives of $u(x, t)$, at $(0, 0)$ for the MHAM solution 121

Figure 4.8 Case I: For $\alpha = 0.5$. (a) The MHAM travelling wave solution for $u(x,t)$, (b) corresponding 2-D solution for $u(x,t)$ when $t = 1$......... 122

Figure 4.9 Case II: For $\alpha = 0.75$. (a) The MHAM travelling wave solution for $u(x,t)$, (b) corresponding 2-D solution for $u(x,t)$ when $t = 1$ 122

Figure 5.1 The bell shaped solitary wave solution for $\Phi_{11}(\xi)$ obtained from eq. (5.9), when $\sigma = -1$......... 130

Figure 5.2 The solitary wave solution for $\Phi_{12}(\xi)$ obtained from eq. (5.9), when $\sigma = -1$......... 130

Figure 5.3 The solitary wave solution for $\Phi_{13}(\xi)$ obtained from eq. (5.9), when $\sigma = 1$ 131

Figure 5.4 The solitary wave solution for $\Phi_{14}(\xi)$ obtained from eq. (5.9), when $\sigma = 1$ 131

Figure 5.5 The solitary wave solution for $|\Phi_{14}(\xi)|$ obtained from eq. (5.9) for the ZK equation for different values of α 132

Figure 5.6 The solitary wave solution for $|\Phi_{13}(\xi)|$ obtained from eq. (5.18)......... 134

Figure 5.7 The solitary wave solution for $|\Phi_{14}(\xi)|$ obtained from eq. (5.18)......... 134

Figure 5.8 The solitary wave solution for $|\Phi_{14}(\xi)|$ obtained from eq. (5.18) for the mZK equation for different values of α 135

Figure 5.9 The bell shaped solitary wave solution for $\Phi_{11}(\xi)$ obtained from eq. (5.25), when $\sigma = -1$, $y = 1$, and $z = 1$......... 138

Figure 5.10 The solitary wave solution for $\Phi_{12}(\xi)$ obtained from eq. (5.25), when $\sigma = -1$, $y = 1$, and $z = 1$......... 139

Figure 5.11 The solitary wave solution for $\Phi_{13}(\xi)$ obtained from eq. (5.25), when $\sigma = 1$, $y = 1$, and $z = 1$......... 139

Figure 5.12 The solitary wave solution for $\Phi_{14}(\xi)$ obtained from eq. (5.25), when $\sigma = 1$, $y = 1$, and $z = 1$......... 140

Figure 5.13 The solitary wave solution for $|\Phi_{11}(\xi)|$ obtained from eq. (5.32), when $y = 1$ and $z = 1$......... 143

Figure 5.14 The solitary wave solution for $|\Phi_{12}(\xi)|$ obtained from eq. (5.32), when $y = 1$ and $z = 1$......... 143

Figure 6.1 I: For $\alpha = 1$ (classical order). (a) Three-dimensional solitary wave solution graph for $|u(x,t)|$ appears in Φ_{11} of eq. (6.26) of **Case I**, (b) corresponding two-dimensional solution graph for $|u(x,t)|$ when $t = 0.1$, $a = 1$, $b = 1$, $c = 1$, $k = 1$, $\lambda = 2$, $v = 1$, $C_1 = 1$, $C_2 = 0$, and $\alpha = 1$ 158

List of Figures

Figure 6.2 **II:** For $\alpha = 0.5$ (fractional order). (a) Three-dimensional solitary wave solution graph for $|u(x,t)|$ appears in Φ_{11} of eq. (6.26) of **Case I**, (b) corresponding two-dimensional solution graph for $|u(x,t)|$ when $t = 0.1$, $a = 1$, $b = 1$, $c = 1$, $k = 1$, $\lambda = 2$, $v = 1$, $C_1 = 1$, $C_2 = 0$, and $\alpha = 0.5$ 158

Figure 6.3 **III:** For $\alpha = 1$ (classical order). (a) Three-dimensional solitary wave solution graph for $|u(x,t)|$ appears in Φ_{12} of eq. (6.26) of **Case I**, (b) corresponding two-dimensional solution graph for $|u(x,t)|$ when $t = 0.1$, $a = 1$, $b = 1$, $c = 1$, $k = 1$, $\lambda = 2$, $v = 1$, $C_1 = 1$, $C_2 = 0$, and $\alpha = 1$ 159

Figure 6.4 **IV:** For $\alpha = 0.5$ (fractional order). (a) Three-dimensional solitary wave solution graph for $|u(x,t)|$ appears in Φ_{12} of eq. (6.26) of **Case I**, (b) corresponding two-dimensional solution graph for $|u(x,t)|$ when $t = 0.1$, $a = 1$, $b = 1$, $c = 1$, $k = 1$, $\lambda = 2$, $v = 1$, $C_1 = 1$, $C_2 = 0$, and $\alpha = 0.5$ 159

Figure 6.5 **Case Ia:** For $\alpha = 1$ (classical order). (a) The (G'/G)-expansion method solitary wave solution for $|u(x,t)|$ appears in eq. (6.69) of **Case I**, (b) corresponding solution for $|u(x,t)|$ when $t = 0$, $C_1 = 1$, $C_2 = 2$, $p = 6$, $q = 1$, $\mu = 1$, $\lambda = 3$, and $\alpha = 1$ 170

Figure 6.6 **Case Ib:** For $\alpha = 0.5$ (fractional order). (a) The (G'/G)-expansion method solitary wave solution for $|u(x,t)|$ appears in eq. (6.69) of **Case I**, (b) corresponding solution for $|u(x,t)|$ when $t = 0$, $C_1 = 1$, $C_2 = 2$, $p = 6$, $q = 1$, $\mu = 1$, $\lambda = 3$, and $\alpha = 0.5$ 170

Figure 6.7 **Case Ic:** For $\alpha = 0.75$ (fractional order). (a) The (G'/G)-expansion method solitary wave solution for $|u(x,t)|$ appears in eq. (6.69) of **Case I**, (b) corresponding solution for $|u(x,t)|$ when $t = 0$, $C_1 = 1$, $C_2 = 2$, $p = 6$, $q = 1$, $\mu = 1$, $\lambda = 3$, and $\alpha = 0.75$ 171

Figure 6.8 **Case IIa:** For $\alpha = 1$ (classical order). (a) The (G'/G)-expansion method solitary wave solution for $|u(x,t)|$ appears in eq. (6.69) of **Case I**, (b) corresponding solution for $|u(x,t)|$ when $t = 0$, $C_1 = 1$, $C_2 = 2$, $p = 6$, $q = 1$, $\mu = 1$, and $\alpha = 1$ 171

Figure 6.9 **Case IIb:** For $\alpha = 0.5$ (fractional order). (a) The (G'/G)-expansion method solitary wave solution for $|u(x,t)|$ appears in eq. (6.70) of **Case I**, (b) corresponding solution for $|u(x,t)|$ when $t = 0$, $C_1 = 1$, $C_2 = 2$, $p = 6$, $q = 1$, $\mu = 1$, $\lambda = 1$, and $\alpha = 0.5$ 171

Figure 6.10 **Case IIc:** For $\alpha = 0.75$ (fractional order). (a) The (G'/G)-expansion method solitary wave solution for $|u(x,t)|$ appears in eq. (6.70) of **Case I**, (b) corresponding solution for $|u(x,t)|$ when $t = 0$, $C_1 = 1$, $C_2 = 2$, $p = 6$, $q = 1$, $\mu = 1$, $\lambda = 1$, and $\alpha = 0.75$ 172

Figure 6.11 **Case IIIa:** For $\alpha = 1$ (classical order). (a) The (G'/G)-expansion method solitary wave solution for $|u(x,t)|$ appears in eq. (6.71) of **Case I**, (b) corresponding solution for $|u(x,t)|$ when $t = 0$, $C_1 = 1$, $C_2 = 2$, $p = 6$, $q = 1$, $\mu = 1$, and $\alpha = 1$ 172

Figure 6.12 **Case IIIb:** For $\alpha = 0.5$ (fractional order). (a) The (G'/G)-expansion method solitary wave solution for $|u(x,t)|$ appears in eq. (6.71) of **Case I**, (b) corresponding solution for $|u(x,t)|$ when $t = 0$, $C_1 = 1$, $C_2 = 2$, $p = 6$, $q = 1$, $\mu = 1$, and $\alpha = 0.5$ 172

Figure 6.13 **Case IIIc:** For $\alpha = 0.75$ (fractional order). (a) The (G'/G)-expansion method solitary wave solution for $|u(x,t)|$ appears in eq. (6.71) of **Case I**, (b) corresponding solution for $|u(x,t)|$ when $t = 0$, $C_1 = 1$, $C_2 = 2$, $p = 6$, $q = 1$, $\mu = 1$, and $\alpha = 0.75$ 173

Figure 6.14 **Case Ia:** For $\alpha = 1$ (classical order). (a) The improved (G'/G)-expansion method solitary wave solution for $|u(x,t)|$ appears in eq. (6.78) of **Case I**, (b) corresponding solution for $|u(x,t)|$ when $t = 0$, $C_1 = 1$, $C_2 = 2$, $p = 6$, $q = 1$, $A = 0$, $B = 1$, $C = -1$, and $\alpha = 1$ 176

Figure 6.15 **Case Ib:** For $\alpha = 0.5$ (fractional order). (a) The improved (G'/G)-expansion method solitary wave solution for $|u(x,t)|$ appears in eq. (6.78) of **Case I**, (b) corresponding solution for $|u(x,t)|$ when $t = 0$, $C_1 = 1$, $C_2 = 2$, $p = 6$, $q = 1$, $A = 0$, $B = 1$, $C = -1$, and $\alpha = 0.5$ 177

Figure 6.16 **Case Ic:** For $\alpha = 0.75$ (fractional order). (a) The improved (G'/G)-expansion method solitary wave solution for $|u(x,t)|$ appears in eq. (6.78) of **Case I**, (b) corresponding solution for $|u(x,t)|$ when $t = 0$, $C_1 = 1$, $C_2 = 2$, $p = 6$, $q = 1$, $A = 0$, $B = 1$, $C = -1$, and $\alpha = 0.75$ 177

Figure 6.17 **Case IIa:** For $\alpha = 1$ (classical order). (a) The improved (G'/G)-expansion method solitary wave solution for $|u(x,t)|$ appears in eq. (6.79) of **Case I**, (b) corresponding solution for $|u(x,t)|$ when $t = 0$, $C_1 = 1$, $C_2 = 2$, $p = 6$, $q = 1$, $A = 0$, $B = -1$, $C = -1$, and $\alpha = 1$ 177

Figure 6.18 **Case IIb:** For $\alpha = 0.5$ (fractional order). (a) The improved (G'/G)-expansion method solitary wave solution for $|u(x,t)|$ appears in eq. (6.79) of **Case I**, (b) corresponding solution for $|u(x,t)|$ when $t = 0$, $C_1 = 1$, $C_2 = 2$, $p = 6$, $q = 1$, $A = 0$, $B = -1$, $C = -1$, and $\alpha = 0.5$ 178

Figure 6.19 **Case IIc:** For $\alpha = 0.75$ (fractional order). (a) The improved (G'/G)-expansion method solitary wave solution for $|u(x,t)|$ appears in eq. (6.79) of **Case I**, (b) corresponding solution for $|u(x,t)|$ when $t = 0$, $C_1 = 1$, $C_2 = 2$, $p = 6$, $q = 1$, $A = 0$, $B = -1$, $C = -1$, and $\alpha = 0.75$ 178

Figure 6.20 **Case IIIa:** For $\alpha = 1$ (classical order). (a) The improved (G'/G)-expansion method solitary wave solution for $|u(x,t)|$ appears in eq. (6.80) of **Case I**, (b) corresponding solution for $|u(x,t)|$ when $t = 0$, $C_1 = 1$, $C_2 = 2$, $p = 6$, $q = 1$, $A = 2$, $B = 0$, $C = 4$, and $\alpha = 1$ 178

Figure 6.21 **Case IIIb:** For $\alpha = 0.5$ (fractional order). (a) The improved (G'/G)-expansion method solitary wave solution for $|u(x,t)|$ appears in eq. (6.80) of **Case I**, (b) corresponding solution for $|u(x,t)|$ when $t = 0$, $C_1 = 1$, $C_2 = 2$, $p = 6$, $q = 1$, $A = 2$, $B = 0$, $C = 4$, and $\alpha = 0.5$ 179

Figure 6.22 **Case IIIc:** For $\alpha = 0.75$ (fractional order). (a) The improved (G'/G)-expansion method solitary wave solution for $|u(x,t)|$ appears in eq. (6.80) of **Case I**, (b) corresponding solution for $|u(x,t)|$ when $t = 0$, $C_1 = 1$, $C_2 = 2$, $p = 6$, $q = 1$, $A = 2$, $B = 0$, $C = 4$, and $\alpha = 0.75$ 179

Figure 6.23 **Case IVa:** For $\alpha = 1$ (classical order). (a) The improved (G'/G)-expansion method solitary wave solution for $|u(x,t)|$ appears in eq. (6.81) of **Case I**, (b) corresponding solution for $|u(x,t)|$ when $t = 0$, $C_1 = 1$, $C_2 = 2$, $p = 6$, $q = 1$, $A = -1$, $B = 0$, $C = -2$, and $\alpha = 1$ 179

List of Figures   xxi

Figure 6.24 **Case IVb:** For $\alpha = 0.5$ (fractional order). (a) The improved (G'/G)-expansion method solitary wave solution for $|u(x,t)|$ appears in eq. (6.81) of **Case I**, (b) corresponding solution for $|u(x,t)|$ when $t = 0$, $C_1 = 1$, $C_2 = 2$, $p = 6$, $q = 1$, $A = -1$, $B = 0$, $C = -2$, and $\alpha = 0.5$ 180

Figure 6.25 **Case IVc:** For $\alpha = 0.75$ (fractional order). (a) The improved (G'/G)-expansion method solitary wave solution for $|u(x,t)|$ appears in eq. (6.81) of **Case I**, (b) corresponding solution for $|u(x,t)|$ when $t = 0$, $C_1 = 1$, $C_2 = 2$, $p = 6$, $q = 1$, $A = -1$, $B = 0$, $C = -2$, and $\alpha = 0.75$ 180

Figure 7.1 **Case 1a:** For $\alpha = 0.75$ (fractional order). (a) The tanh method solitary wave solution for $u(x,t)$ appears in eq. (7.12) of **Case I**, (b) corresponding solution for $u(x,t)$ when $t = 0$, $c = 0.5$, and $\alpha = 0.75$ 188

Figure 7.2 **Case 2a:** For $\alpha = 0.75$ (fractional order). (a) The tanh method solitary wave solution for $u(x,t)$ appears in eq. (7.12) of **Case I**, (b) corresponding solution for $u(x,t)$ when $t = 0$, $c = 0.5$, and $\alpha = 0.25$ 188

Figure 7.3 **Case 1b:** For $\alpha = 0.25$ (fractional order). (a) The tanh method solitary wave solution for $u(x,t)$ appears in eq. (7.13) of **Case II**, (b) corresponding solution for $u(x,t)$ when $t = 0$, $c = 0.5$, and $\alpha = 0.75$ 188

Figure 7.4 **Case 2b:** For $\alpha = 0.25$ (fractional order). (a) The tanh method solitary wave solution for $u(x,t)$ appears in eq. (7.13) of **Case II**, (b) corresponding solution for $u(x,t)$ when $t = 0$, $c = 0.5$, and $\alpha = 0.25$ 189

Figure 7.5 **Case 3a:** For $\alpha = 0.75$ (fractional order). (a) The tanh method solitary wave solution for $u(x,t)$ appears in eq. (7.14) of **Case III**, (b) corresponding solution for $u(x,t)$ when $t = 0$, $c = 0.5$, and $\alpha = 0.75$ 189

Figure 7.6 **Case 3b:** For $\alpha = 0.25$ (fractional order). (a) The tanh method solitary wave solution for $u(x,t)$ appears in eq. (7.14) of **Case III**, (b) corresponding solution for $u(x,t)$ when $t = 0$, $c = 0.5$, and $\alpha = 0.25$ 189

Figure 7.7 **Case 1a:** For $\alpha = 0.25$ (fractional order). The three-dimensional solitary wave solution graph obtained by using the tanh method for $u(x,t)$ appears in eq. (7.20) of **Case I**, when $c = 0.75$ and $\alpha = 0.25$............................. 192

Figure 7.8 **Case 1b:** For $\alpha = 0.75$ (fractional order). The three-dimensional solitary wave solution graph obtained by using the tanh method for $u(x,t)$ appears in eq. (7.20) of **Case I**, when $c = 0.75$ and $\alpha = 0.75$ 193

Figure 7.9 **Case 2a:** For $\alpha = 0.25$ (fractional order). The three-dimensional solitary wave solution graph obtained by using the tanh method for $u(x,t)$ appears in eq. (7.21) of **Case II**, when $c = 0.5$ and $\alpha = 0.25$ 193

Figure 7.10 **Case 2b:** For $\alpha = 0.75$ (fractional order). The three-dimensional solitary wave solution graph obtained by using the tanh method for $u(x,t)$ appears in eq. (7.21) of **Case II**, when $c = 0.5$ and $\alpha = 0.75$ 194

Figure 7.11 **Case 1a:** For $\alpha = 0.75$ (fractional order). The three-dimensional solitary wave solution graph obtained by using the modified Kudryashov method for $u(x,t)$ appears in eq. (7.26) of **Case I**, when $l = 0.5$, $a = 10$, and $\alpha = 0.75$ 196

Figure 7.12 **Case 1b:** For $\alpha = 1$ (classical order). The three-dimensional solitary wave solution graph obtained by using the modified Kudryashov method for $u(x,t)$ appears in eq. (7.26) of **Case I**, when $l = 0.5$, $a = 10$, and $\alpha = 1$ 197

Figure 7.13 **Case 2a:** For $\alpha = 0.75$ (fractional order). The three-dimensional solitary wave solution graph obtained by using the modified Kudryashov method for $u(x,t)$ appears in eq. (7.28) of **Case II**, when $l = 0.5$, $a = 10$, and $\alpha = 0.75$ 197

Figure 7.14 **Case 2b:** For $\alpha = 1$ (classical order). The three-dimensional solitary wave solution graph obtained by using the modified Kudryashov method for $u(x,t)$ appears in eq. (7.28) of **Case II**, when $l = 0.5$, $a = 10$, and $\alpha = 1$ 198

Figure 7.15 **Case 1:** For $\alpha = 0.5$ (fractional order). (a) The traveling wave solution for $u(x,t)$ appears in eq. (7.36) of **Case I**, (b) corresponding solution for $u(x,t)$, when $t = 0$, $a = 1$, $b = 1$, $k = 1$, and $\alpha = 0.5$ 202

Figure 7.16 **Case 2:** For $\alpha = 0.5$ (fractional order). (a) The traveling wave solution for $u(x,t)$ appears in eq. (7.37) of **Case II**, (b) corresponding solution for $u(x,t)$, when $t = 0$, $a = 1$, $b = 1$, $k = 1$, and $\alpha = 0.5$ 203

Figure 7.17 (a) The tanh method three-dimensional solitary wave solution for $u(x,t)$ appears in eq. (7.51) as Φ_{11} of **Case I**, (b) corresponding two-dimensional solution graph for $u(x,t)$ when $t = 0$, $v = 0.5$, and $\alpha = 1$...... 207

Figure 7.18 (a) The tanh method three dimensional solitary wave solution for $v(x,t)$ appears in eq. (7.51) as Ψ_{11} of **Case I**, (b) corresponding two dimensional solution graph for $v(x,t)$ when $t = 0$, $v = 0.5$, and $\alpha = 1$ 207

Figure 8.1 **Case I:** For $\alpha = 1$ (classical order). (a) The three-dimensional antikink-solitary wave solution graph for $u(x,t)$ appears in eq. (8.9) as Φ_{11} in **Case I**, when $\varepsilon = -6$, $v = 1$, $\eta = 1$, $k = 10$, and $\alpha = 1$, (b) the corresponding two-dimensional solution graph for $u(x,t)$ when $t = 0$........ 213

Figure 8.2 **Case II:** For $\alpha = 0.5$ (fractional order). (a) The three-dimensional antikink-solitary wave solution graph for $u(x,t)$ appears in eq. (8.9) as Φ_{11} in **Case I**, when $\varepsilon = -6$, $v = 1$, $\eta = 1$, $k = 10$, and $\alpha = 0.5$, (b) the corresponding two-dimensional solution graph for $u(x,t)$ when $t = 0$........ 214

Figure 8.3 **Case III:** For $\alpha = 0.75$ (fractional order). (a) The three-dimensional antikink-solitary wave solution graph for $u(x,t)$ appears in eq. (8.9) as Φ_{11} in **Case I**, when $\varepsilon = -6$, $v = 1$, $\eta = 1$, $k = 10$, and $\alpha = 0.75$, (b) the corresponding two-dimensional solution graph for $u(x,t)$ when $t = 0$ 214

Figure 8.4 **Case I:** For $\alpha = 1$ (classical order). (a) The three-dimensional kink-solitary wave solution graph for $|u(x,t)|$ appears in eq. (8.15) as Φ_{11} in **Case I**, when $a = 3$, $\gamma = -1/6$, $b = 0.6$, $c = 1$, and $\alpha = 1$, (b) the corresponding two-dimensional solution graph for $|u(x,t)|$ when $t = 0$ 216

Figure 8.5 **Case II:** For $\alpha = 0.5$ (fractional order). (a) The three-dimensional kink-solitary wave solution graph for $|u(x,t)|$ appears in eq. (8.15) as Φ_{11} in **Case I**, when $a = 3$, $\gamma = -1/6$, $b = 0.6$, $c = 1$, and $\alpha = 0.5$, (b) the corresponding two-dimensional solution graph for $|u(x,t)|$ when $t = 0$ 217

List of Figures xxiii

Figure 8.6 **Case III:** For $\alpha = 0.75$ (fractional order). (a) The three-dimensional kink-solitary wave solution graph for $|u(x,t)|$ appears in eq. (8.15) as Φ_{11} in **Case I**, when $a = 3$, $\gamma = -1/6$, $b = 0.6$, $c = 1$, and $\alpha = 0.75$, (b) the corresponding two-dimensional solution graph for $|u(x,t)|$ when $t = 0$ 217

Figure 8.7 **Case I:** For $\alpha = 1$ (classical order). (a) The three-dimensional solitary wave graph for $u(x,t)$ appears in eq. (8.24) as Φ_{11} in **Case 1**, when $c = 0.3$ and $\alpha = 1$, (b) the corresponding two-dimensional graph for $u(x,t)$ when $t = 0$ 222

Figure 8.8 **Case I:** For $\alpha = 1$ (classical order). (a) The three-dimensional solitary wave graph for $v(x,t)$ appears in eq. (8.24) as Ψ_{11} in **Case I**, when and $\alpha = 1$, (b) the corresponding two-dimensional graph for $v(x,t)$ when $t = 0$ 222

Figure 8.9 **Case II:** For $\alpha = 0.5$ (fractional order). (a) The three-dimensional solitary wave graph for $u(x,t)$ appears in eq. (8.24) as Φ_{11} in **Case I**, when $c = 0.3$ and $\alpha = 0.5$, (b) the corresponding two-dimensional graph for $u(x,t)$ when $t = 0$ 222

Figure 8.10 **Case II:** For $\alpha = 0.5$ (fractional order). (a) The three-dimensional solitary wave graph for $v(x,t)$ appears in eq. (8.24) as Ψ_{11} in **Case I**, when and $\alpha = 0.5$, (b) the corresponding two-dimensional graph for $v(x,t)$ when $t = 0$... 223

Figure 8.11 **Case I:** For $\alpha = 1$ (classical order). (a) The three-dimensional solitary wave graph for $|u(x,t)|$ appears in eq. (8.39) as Φ_{11} in **Case I**, when $\varepsilon = 0.5$, $c = 0.3$, and $\alpha = 1$, (b) the corresponding two-dimensional graph for $|u(x,t)|$ when $t = 0$ 228

Figure 8.12 **Case I:** For $\alpha = 1$ (classical order). (a) The three-dimensional solitary wave graph for $v(x,t)$ appears in eq. (8.39) as Ψ_{11} in **Case I**, when $c = 0.3$ and $\alpha = 1$, (b) the corresponding two-dimensional solution graph for $v(x,t)$ when $t = 0$.. 228

Figure 8.13 **Case II:** For $\alpha = 0.5$ (fractional order). (a) The three-dimensional solitary wave graph for $|u(x,t)|$ appears in eq. (8.39) as Φ_{11} in **Case I**, when $\varepsilon = 0.5$, $c = 0.3$, and $\alpha = 0.5$, (b) the corresponding two-dimensional solution graph for $|u(x,t)|$ when $t = 0$................................... 228

Figure 8.14 **Case II:** For $\alpha = 0.5$ (fractional order). (a) The three-dimensional solitary wave solution graph for $v(x,t)$ appears in eq. (8.39) as Ψ_{11} in **Case I**, when $c = 0.3$ and $\alpha = 0.5$, (b) the corresponding two-dimensional solution graph for $v(x,t)$ when $t = 0$................................... 229

Figure 9.1 **Case I:** For $\alpha = 1$ (classical order). (a) The 3-D double periodic solution graph for $v(x,t)$ appears in eq. (9.19) as Ψ_{11} in **Case I** of **Set I**, when $a = 3$, $b = 2$, $k = 1$, $\gamma = 2$, $\varepsilon = 1$, $m = 0.3$, and $\alpha = 1$, (b) the corresponding 2-D graph for $v(x,t)$, when $t = 1$.................................. 239

Figure 9.2 **Case I:** For $\alpha = 1$ (classical order). (a) The 3-D double periodic solution graph for $u(x,t)$ appears in eq. (9.19) as Φ_{11} in **Case I** of **Set I**, when $a = 3$, $b = 2$, $k = 1$, $\gamma = 2$, $\varepsilon = 1$, $m = 0.3$, and $\alpha = 1$, (b) the corresponding 2-D graph for $u(x,t)$, when $t = 1$ 240

Figure 9.3 **Case II:** For $\alpha = 0.5$ (fractional order). (a) The 3-D double periodic solution graph for $v(x,t)$ appears in eq. (9.19) as Ψ_{11} in **Case I** of **Set I**, when $a = 3$, $b = 2$, $k = 1$, $\gamma = 2$, $\varepsilon = 1$, $m = 0.3$, and $\alpha = 0.5$, (b) the corresponding 2-D graph for $v(x,t)$, when $t = 1$..240

Figure 9.4 **Case II:** For $\alpha = 0.5$ (fractional order). (a) The 3-D double periodic solution graph for $u(x,t)$ appears in eq. (9.19) as Φ_{11} in **Case I** of **Set I**, when $a = 3$, $b = 2$, $k = 1$, $\gamma = 2$, $\varepsilon = 1$, $m = 0.3$, and $\alpha = 0.5$, (b) the corresponding 2-D graph for $u(x,t)$, when $t = 1$..240

Figure 9.5 **Case III:** For $\alpha = 0.75$ (fractional order). (a) The 3-D double periodic solution graph for $v(x,t)$ appears in eq. (9.19) as Ψ_{11} in **Case I** of **Set I**, when $a = 3$, $b = 2$, $k = 1$, $\gamma = 2$, $\varepsilon = 1$, $m = 0.3$, and $\alpha = 0.75$, (b) the corresponding 2-D graph for $v(x,t)$, when $t = 1$..241

Figure 9.6 **Case I:** For $\alpha = 1$ (classical order). (a) The 3-D double periodic solution graph for $u(x,t)$ appears in eq. (9.19) as Φ_{11} in **Case I** of **Set I**, when $a = 3$, $b = 2$, $k = 1$, $\gamma = 2$, $\varepsilon = 1$, $m = 0.3$, and $\alpha = 0.75$, (b) the corresponding 2-D graph for $u(x,t)$, when $t = 1$..241

Figure 9.7 **Case I:** For $\alpha = 1$ (classical order). (a) The 3-D double periodic solution graph for $v(x,t)$ appears in eq. (9.20) as Ψ_{12} in **Case II** of **Set I**, when $a = 3$, $b = 2$, $k = 1$, $\gamma = 2$, $\varepsilon = 1$, $m = 0.3$, and $\alpha = 1$, (b) the corresponding 2-D graph for $v(x,t)$, when $t = 1$..242

Figure 9.8 **Case I:** For $\alpha = 1$ (classical order). (a) The 3-D double periodic solution graph for $u(x,t)$ appears in eq. (9.20) as Φ_{12} in **Case II** of **Set I**, when $a = 3$, $b = 2$, $k = 1$, $\gamma = 2$, $\varepsilon = 1$, $m = 0.3$, and $\alpha = 1$, (b) the corresponding 2-D graph for $u(x,t)$, when $t = 1$..242

Figure 9.9 **Case II:** For $\alpha = 0.75$ (fractional order). (a) The 3-D double periodic solution graph for $v(x,t)$ appears in eq. (9.20) as Ψ_{12} in **Case II** of **Set I**, when $a = 3$, $b = 2$, $k = 1$, $\gamma = 2$, $\varepsilon = 1$, $m = 0.3$, and $\alpha = 0.75$, (b) the corresponding 2-D graph for $v(x,t)$, when $t = 1$..243

Figure 9.10 **Case II:** For $\alpha = 0.75$ (fractional order). (a) The 3-D double periodic solution graph for $u(x,t)$ appears in eq. (9.20) as Φ_{12} in **Case II** of **Set I**, when $a = 3$, $b = 2$, $k = 1$, $\gamma = 2$, $\varepsilon = 1$, $m = 0.3$, and $\alpha = 0.75$, (b) the corresponding 2-D graph for $u(x,t)$, when $t = 1$..243

Figure 9.11 **Case I:** For $\alpha = 1$ (classical order). (a) The 3-D double periodic solution graph for $v(x,t)$ appears in eq. (9.23) as Ψ_{15} in **Case V** of **Set I**, when $a = 3$, $b = 2$, $k = 1$, $\gamma = 2$, $\varepsilon = 1$, $m = 0.3$, and $\alpha = 1$, (b) the corresponding 2-D graph for $v(x,t)$, when $t = 1$..244

Figure 9.12 **Case I:** For $\alpha = 1$ (classical order). (a) The 3-D double periodic solution graph for $u(x,t)$ appears in eq. (9.23) as Φ_{15} in **Case V** of **Set I**, when $a = 3$, $b = 2$, $k = 1$, $\gamma = 2$, $\varepsilon = 1$, $m = 0.3$, and $\alpha = 1$, (b) the corresponding 2-D graph for $u(x,t)$, when $t = 1$..244

Figure 9.13 **Case II:** For $\alpha = 0.75$ (fractional order). (a) The 3-D double periodic solution graph for $v(x,t)$ appears in eq. (9.23) as Ψ_{15} in **Case V** of **Set I**, when $a = 3$, $b = 2$, $k = 1$, $\gamma = 2$, $\varepsilon = 1$, $m = 0.3$, and $\alpha = 0.75$, (b) the corresponding 2-D graph for $v(x,t)$, when $t = 1$..245

List of Figures

xxv

Figure 9.14 **Case II:** For $\alpha = 0.75$ (fractional order). (a) The 3-D double periodic solution graph for $u(x,t)$ appears in eq. (9.23) as Φ_{15} in **Case V** of **Set I**, when $a = 3$, $b = 2$, $k = 1$, $\gamma = 2$, $\varepsilon = 1$, $m = 0.3$, and $\alpha = 0.75$, (b) the corresponding 2-D graph for $u(x,t)$, when $t = 1$ 245

Figure 10.1 Block diagram for variable-order integral operator 255

Figure 10.2 A mass-spring oscillator sliding on a variable-order guide 257

Figure 10.3 A mass-spring oscillator under viscoelastic damping when no external force is applied (**Case I**) 260

Figure 10.4 Figure representing continuously variable viscoelastic oscillator (**Case I**) 261

Figure 10.5 Viscous-viscoelastic and viscoelastic-viscous oscillators 263

Figure 10.6 A mass-spring oscillator under "viscous-viscoelastic" guide when no external force is applied (**Case II**) 263

Figure 10.7 A mass-spring oscillator sliding on a continuous order guide when external force is applied (**Case III**) 266

Figure 10.8 Viscous-viscoelastic and viscoelastic-viscous oscillators 266

Figure 10.9 A mass-spring oscillator sliding on a continuous order guide with viscoelastic damping when external force is applied (**Case IV**) 267

Figure 10.10 The displacement-time graph for fractional continuous order spring-mass damper model for free oscillation with viscoelastic damping plots for **Case I** 285

Figure 10.11 The velocity-time graph for fractional continuous order spring-mass damper model for free oscillation with viscoelastic damping plots for **Case I** 285

Figure 10.12 The displacement-time graph for fractional continuous order spring-mass damper model for free oscillation with viscous-viscoelastic damping plots for **Case II** 286

Figure 10.13 The velocity-time graph for fractional continuous order spring-mass damper model for free oscillation with viscous-viscoelastic damping plots for **Case II** 287

Figure 10.14 The displacement-time graph for fractional continuous order mass-spring damper model for forced oscillation with viscous-viscoelastic damping plots for **Case III** 287

Figure 10.15 The displacement-time graph for fractional continuous order mass-spring damper model for forced oscillation with viscoelastic-viscous damping plots for **Case IV** 288

List of Tables

Table 2.1 Families of Jacobi elliptic function solutions of elliptic equation (2.86)............42

Table 3.1 Comparison of absolute errors between four term HPM and HPTM solutions with regard to exact solutions for different values of t respectively when $x = 1.5$ in case of $\alpha = 2$ and $\beta = 1$...79

Table 3.2 Comparison of relative errors between four term HPM and HPTM solutions with regard to exact solutions for different values of t respectively when $x = 1.5$ in case of $\alpha = 2$ and $\beta = 1$...79

Table 4.1 Comparison of the solutions between third order MHAM and VIM solutions for different values of x and t when $\alpha = 0.5$ 123

Table 4.2 The L_2 and L_∞ errors for third order MHAM solutions with regard to VIM solutions for different values of x when $\alpha = 0.5$ 123

xxvii

Preface

During the past decades, fractional calculus, which deals with derivatives and integrals of any arbitrary real or complex order, has gained considerably significant prevalence and importance, mainly due to its substantiated applications in numerous extensive and diverse fields in engineering and science. The partial differential equations involving derivatives with non-integer or fractional order have shown adequate models for various physical phenomena in areas such as viscoelasticity models, fluid mechanics, advection–diffusion models, biological population models, optics, signals processing, nuclear science, etc.

Most of the time these new fractional-order models are more suitable than the integer-order models, due to their capability to describe the memory and hereditary characteristics in various materials. For these reasons, numerical methods are frequently used to find the approximate solutions of non-linear differential equations. Nevertheless, numerical solutions do not explicitly depict the nature of the physical systems and are inadequate in determining the general properties of certain systems of equations. Due to these major reasons, semi-analytical and analytical methods have been developed for solving the non-linear partial differential equations. The development of the semi-analytical and analytical methods has been seen in the end of the twentieth century. With the aid of faster processing calculation techniques, the solutions of analytical solutions of semi-analytical and analytical methods have been presented more precisely, which have a certain implementation of solutions over numerical methods. Therefore, our main focus in the present work is to analyze the various semi-analytical and analytical methods for finding approximate and exact solutions of fractional order partial differential equations.

This book has 10 chapters. To begin with, the introductory concepts of fractional calculus have been discussed in Chapter 1. In this chapter, some definitions of fractional calculus have been presented. Mainly, the definitions of the Riemann–Liouville fractional integral and derivative, Caputo fractional derivative, Grünwald–Letnikov fractional derivative, Riesz fractional derivative, modified Riemann–Liouville derivative, and local fractional derivatives have been discussed. Also, we have discussed some useful functions, viz., the gamma function, beta function, Mittag–Leffler function, and Wright function, which are useful to describe the definitions for fractional calculus.

In Chapter 2, the fundamentals and algorithms of the semi-analytical methods and analytical methods have been discussed with the brief literature review. The details and algorithms of such semi-analytical and analytical methods, viz., homotopy perturbation method (HPM), homotopy perturbation transform method (HPTM), modified homotopy analysis method (MHAM), fractional sub-equation method, improved fractional sub-equation method, (G'/G)-expansion method, improved (G'/G)-expansion method, modified Kudryashov method, Jacobi elliptical function method, and successive recursion method have been discussed. Also, some new methods like the modified homotopy analysis method with Fourier transform method (MHAM-FT), new tanh-sech method, and other new methods proposed by the authors have been reported.

In Chapter 3, we have presented schemes in order to obtain the approximate solutions of a class of time-fractional Lotka–Volterra equations of predator–prey model by using the HPM. Moreover, the HPM and HPTM have been applied for finding the solutions for fractional coupled Klein–Gordon–Schrödinger (K–G–S) equations with initial conditions.

xxix

Also, the HPTM and MHAM are applied to obtain the approximate solution of the fractional coupled Klein–Gordon–Zakharov (K–G–Z) equations and fractional coupled sine-Gordon equations.

The new analytical technique, viz., the MHAM-FT has been used to obtain the approximate solution of the Riesz fractional diffusion equation (RFDE), Riesz fractional advection-dispersion equation (RFADE), and Riesz time-fractional Camassa–Holm equation in Chapter 4. The RFDE, RFADE, and Riesz time-fractional Camassa–Holm equation have been first time solved by the MHAM-FT method in order to justify applicability of the above methods. The comparison of the above methods also has been done with the exact solutions.

In Chapter 5, we have successfully obtained exact solutions for the space-time fractional Zakharov–Kuznetsev equation, space-time fractional modified Zakharov–Kuznetsev (mZK) equation, time-fractional (3 + 1)-dimensional Korteweg–de Vries (KdV) –Zakharov–Kuznetsev and space-time fractional (3 + 1)-dimensional modified KdV–Zakharov–Kuznetsev equations. By using the fractional sub-equation method, three types of exact analytical solutions, including the generalized hyperbolic function solutions, generalized trigonometric function solutions, and rational solution for the space-time fractional Zakharov–Kuznetsev equation, and two types of exact analytical solutions, including the generalized hyperbolic function solutions and generalized trigonometric function solutions for the space-time fractional mZK equation, are addressed. We have also applied the improved fractional sub-equation method for getting the hyperbolic, trigonometric, and rational function form involving some parameters for time-fractional (3 + 1)-dimensional KdV–Zakharov–Kuznetsev and space-time fractional (3 + 1)-dimensional modified KdV–Zakharov–Kuznetsev equations.

The exact solutions of non-linear evolution equations, namely, the time-fractional modified Kawahara equation, fractional coupled Jaulent–Miodek (JM) equation, time-fractional modified KdV equation, and the time-fractional Kaup–Kupershmidt equation have been analyzed by (G'/G)-expansion method and improved (G'/G)-expansion method in Chapter 6. The fractional complex transform can easily convert a fractional differential equation into its equivalent ordinary differential equation form, so fractional complex transform is extremely effective for solving fractional differential equations. The focused methods have many advantages: they are straightforward and concise. Furthermore, this study shows that the proposed method is quite efficient. The performance of this method is reliable and effective and gives the exact solitary wave solutions which have been presented by three-dimensional and two-dimensional solutions graphs.

In Chapter 7, we determine the solitary wave exact solutions of non-linear evolution equations, namely, time-fractional fifth-order Sawada–Kotera (S–K), time-fractional fifth-order modified S–K, time-fractional fifth-order Kuramoto–Sivashinsky (K–S) equations, and fractional coupled JM equations by newly proposed tanh-sech method via fractional complex transform. We have also applied the Kudryashov method to solve time-fractional fifth-order mS–K equations. By applying the tanh-sech method, we obtain three solitary wave solutions. Both methods provide new and more general type solitary wave solutions which are significant to reveal the pertinent features of the physical phenomenon. The fractional complex transform can easily convert a fractional differential equation into its equivalent ordinary differential equation form. By using the obtained solutions, we have drawn the three-dimensional graphs, which give the nature of the solutions as a solitary wave.

The new method proposed by authors has been utilized in Chapter 8 for getting exact solutions of time-fractional KdV–Burgers, time-fractional KdV-modified (mKdV), and

Preface

time-fractional coupled Schrödinger–KdV and time-fractional coupled Schrödinger–Boussinesq equations. The most essential interest of the proposed new method is that it takes less computation for obtaining the exact solutions. The obtained exact solutions have also been used here for performing the numerical simulations. From these numerical simulations, we have analyzed the nature of solution as kink and anti-kink solitary waves.

In Chapter 9, we have implemented the Jacobi elliptical method for getting exact solutions of time-fractional coupled Drinfeld–Sokolov–Wilson equations. We have used here fractional complex transform for transformation of the non-linear fractional differential equations to non-linear ordinary differential equations. The exact solutions obtained from the proposed method have also been used for presenting the numerical simulations.

In Chapter 10, we have analyzed several definitions of variable-order operators and concluded that the variable operator must be able to return the intermediate values between 0 to 1, which correspond to the continuous order argument of the operator in any case. The variable-order operators defined in different forms by the independent researchers satisfy the criteria for modeling the dynamic systems. As an instance, we have also analyzed the dynamic nature of variable-order operators in modeling the viscoelastic oscillator, viz., mass-spring damping system. Also, we modeled the fractional continuously variable-order mass-spring damper systems for free oscillation with viscoelastic damping and viscous-viscoelastic damping and forced oscillation with viscous-viscoelastic damping and viscoelastic-viscous damping. The approach is new in the sense of changing of behavior of guide continuously with the small change of time Δt with respect to both viscoelastic and viscous-viscoelastic oscillators of order q. The analytical solutions of the fractional continuously variable-order mass-spring damper systems have been successfully obtained by the successive recursive method. The graphical plots also have been presented for different values of damping ratio.

Acknowledgments

Second author would like to sincerely express deepest sense of gratitude to Professor S. Saha Ray, Professor, National Institute of Technology Rourkela, Odisha, India, for his encouragement in the preparation of this book. Without his keen guidance, lively discussion, proficient advice, persistent encouragement, and support, this book could not have been done.

I also express my sincere gratitude to Dr. Achyuta Samanta, founder of the Kalinga Institute of Industrial Technology (KIIT) and Kalinaga Institute of Social Sciences (KISS), Bhubaneswar, India, for his kind cooperation and support. The moral support received from my colleagues at the KIIT is also acknowledged.

I would like to keep in record the unforgettable moments I spent with some special people in Dr. Ashrita Patra, Dr. Arun Kumar Gupta, Dr. Prakash Kumar Sahu, and Soumyendra Singh. I am also thankful to my friends, Siddharth, Janmenjaya, Nikash, and Srusti, for their selfless help and encouragement. There are too many of you to name individually, but you have my sincerest thanks for your stimulating scholastic conversation and fruitful help.

Moreover, I am especially grateful to CRC Press/Taylor & Francis Group for their cooperation in all aspects of the production of this book.

Finally, and most importantly, I express my heartfelt gratitude to my grandparents, late Krushna Chandra Sahoo and late Menka Sahoo; my parents, Pratap Kishore Sahoo and Saraswati Sahoo; and my elder sister, Priyadarsini Sahoo: and I am greatly indebted to them for bearing the inconvenience during the work. I owe a deep sense of indebtedness to my family members, Manmath Nath Sahoo, Sanjay Kumar Sahoo, Arpita Sahoo, Bijay Sahoo, Suresh Chandra Sahoo, Nirmala Sahoo, Mamata Sahoo, and Pitamber Satapathy, for their continuous inspiration, motivated encouragement, and enthusiastic support.

Subhadarshan Sahoo

Authors

Dr. Santanu Saha Ray is currently a professor at the Department of Mathematics, National Institute of Technology, Rourkela, India. Dr. Saha Ray completed his PhD in 2008 from Jadavpur University, Kolkata, India. He received his MCA degree in 2001 from Indian Institute of Engineering Science and Technology (IIEST), erstwhile Bengal Engineering College, Shibpur, India. He completed a master's degree in applied mathematics at the Calcutta University, Kolkata, India, in 1998 and a bachelor's (honors) degree in mathematics at St. Xavier's College (currently known as St. Xavier's University, Kolkata), Kolkata, India, in 1996. Dr. Saha Ray has about 17 years of teaching experience at the undergraduate and postgraduate levels in glorious Institutes like National Institute of Technology, Rourkela, and two renowned private engineering institutes in Kolkata, West Bengal. He has also about 16 years of research experience in various fields of applied mathematics. He has published many peer reviewed research papers in numerous fields and various international SCI journals of repute, like *Applied Mathematics and Computation, Communication in Nonlinear Science and Numerical Simulation, Nonlinear Dynamics, Transaction ASME Journal of Applied Mechanics, Journal of Computational and Nonlinear Dynamics, Computers and Mathematics with Applications, Journal of Computational and Applied Mathematics, Mathematical Methods in the Applied Sciences, Computers & Fluids, Physica Scripta, Communications in Theoretical Physics, Nuclear Engineering and Design, International Journal of Nonlinear Science and Numerical Simulation, Annals of Nuclear Energy,* and *Journal of Mathematical Chemistry,* etc. For a detailed citation overview, the reader may be referred to Scopus. To date, he has more than 149 research papers published in journals of international repute, including more than 119 SCI journal papers.

He has solely authored a book entitled *Graph Theory with Algorithms and Its Applications: In Applied Science and Technology* published by Springer. A solely authored book entitled *Fractional Calculus with Applications for Nuclear Reactor Dynamics* has been published by the CRC Press/Taylor & Francis Group. Another solely authored book entitled *Numerical Analysis with Algorithms and Programming* has been also published in by CRC Press/Taylor & Francis Group. A book entitled *Wavelet Methods for Solving Partial Differential Equations and Fractional Differential Equations* has been recently launched by CRC Press/Taylor & Francis Group.

Currently, he is acting as editor-in-chief for the Springer international journal entitled *International Journal of Applied and Computational Mathematics.* He is also an associate editor of a Springer international journal, *Mathematical Sciences* and *Nonlinear Science Letters A,* journal and reviewer of several journals of Elsevier, Springer, and Taylor & Francis Group, etc. He had also been the lead guest editor in the International SCI journals of Hindawi Publishing Corporation, USA.

He has contributed papers on several topics, such as fractional calculus, mathematical modeling, mathematical physics, stochastic modeling, integral equations, and wavelet methods. He is a member of the Society for Industrial and Applied Mathematics (SIAM) and the American Mathematical Society (AMS).

He was the principal investigator of the two *Board of Research in Nuclear Sciences* research project, with grants from by Bhabha Atomic Research Centre, Mumbai, India. He was also the principal investigator of a research project financed by the Department of Science and Technology, Government of India. Currently, he has been acting as principal investigator of another research project financed by the National Board for Higher Mathematics, Department of Atomic Energy, Government of India.

Under his sole supervision, four research scholars have been awarded with a PhD from NIT Rourkela. Currently, he is supervising two research scholars.

It is not out of place to mention that he had attended the workshop organized by West Bengal University of Technology (WBUT) on "Review of Engineering Degree Curriculum of Mathematics" held at the National Institute of Technical Teacher's Training and Research (NITTTR) Kolkata, from 26–30 July 2004. In that, the Workshop Curriculum of Mathematics in undergraduate level as well as postgraduate level of West Bengal University of Technology was revised.

He was invited to deliver a lecture in a Short Term Training Programme on "Mathematical Modelling" organized by the Department of Science, National Institute of Technical Teacher's Training and Research (NITTTR) Kolkata, from 4–8 December 2006. He also attended a Summer School on "Mathematical Modelling and Its Engineering Applications" organized jointly by West Bengal State Council of Science & Technology and Department of Science & Technology, Government of West Bengal, in association with B.P. Poddar Institute of Management & Technology, Kolkata, during 27 April to 9 May 2007.

It is worth noting to mention that he was invited to deliver a lecture in a Workshop on "Fractional Order systems" organized by Instrumentation & Electronics Engineering Department, Jadavpur University, Salt Lake Campus, Kolkata, sponsored by IEEE Kolkata Chapter, DRDL Hydrabad, BRNS (DAE) Mumbai, from 28–29 March 2008.

Moreover, he was also invited to deliver a lecture in a Short Term Training Programme on "Mathematical Modelling" organized by the Department of Science, National Institute of Technical Teacher's Training and Research Kolkata, from 24th to 28th November 2008.

He was convener of "Symposium on Recent Trends and Emerging Applications of Mathematical Sciences (SRTEAMS-2013)" sponsored by CSIR, New Delhi; DST, Government of India and INSA, New Delhi, held on 16–17 May 2013. He had attended as invited participant in the "International Conference on Mathematical Modeling in Physical Sciences" held in Madrid, Spain, from 28–31 August 2014.

Recently, he attended as invited participant in the "Global Conference on Applied Physics and Mathematics" held in Rome, Italy, from 25–27 July 2016. Travel grant was sponsored by SERB, DST, Government of India.

Dr. Subhadarshan Sahoo is currently an assistant professor in the Department of Mathematics, Kalinga Institute of Industrial Technology (KIIT), Bhubaneswar, India. He has done a PhD under the supervision of Prof. S. Saha Ray in the Department of Mathematics of the National Institute of Technology Rourkela, Odisha, India. He has been working on the BRNS research project entitled "Application of analytical methods for the solutions of generalized fractional and continuous order differential equations with the implementation in computer simulation" funded by the BRNS, BARC, Department of Atomic Energy (DAE), Government of India, under the supervision of Professor S. Saha Ray, who was the principal investigator of the aforesaid BRNS project. Also, he has been awarded a senior research fellowship (SRF) funded by CSIR. Furthermore, he has pursued a MSc in applied mathematics from the National Institute of Technology Rourkela, Odisha, India, and a BSc from the Government Autonomous College, Rourkela, Odisha, India. His current research interest includes the analytical methods for the solution of partial and fractional differential equations.

1

Introduction and Preliminaries of Fractional Calculus

1.1 Introduction

Fractional calculus is a generalization of ordinary differentiation and integration to arbitrary (non-integer) order. This was introduced at the end of seventeenth century, when Leibniz and Newton invented differential calculus. Literally, it was introduced in the year of 1695, when Leibniz wrote a letter to L'Hôpital raising the possibility of generalizing the meaning of derivatives from integer order to non-integer order. Since then, many famous mathematicians have worked on this concept and related questions, creating the field which is known today as fractional calculus.

During the past decades, fractional calculus, which deals with the derivatives and integrals of any arbitrary real or complex order, has gained considerable significant prevalence and importance, mainly due to its substantiated applications in numerous extensive and diverse fields in engineering and science. The partial differential equations involving derivatives with non-integer or fractional order have shown adequate models for various physical phenomena in areas such as viscoelasticity models, fluid mechanics, advection–diffusion models, biological population models, optics, signals processing, nuclear science [1–5], etc.

Fractional calculus is also considered as the branch of mathematical analysis that studies the possibility of taking real number, or even complex number, powers of the differential operator. Fractional calculus has resulted in a great interest for researchers due to its dynamic behavior and exact description of complex non-linear phenomena in various fields of science and engineering during the past few decades. In recent years, fractional calculus has become a very attractive subject to mathematicians, and many different forms of fractional differential and integral operators, viz., the Riemann–Liouville fractional integral and derivative, Caputo fractional derivative, Grünwald–Letnikov fractional derivative, Riesz fractional derivative, modified Riemann–Liouville derivative, and local fractional derivatives, have been introduced, which have the application for describing the non-linear fractional differential equations.

1.2 Preliminaries of Fractional Calculus

In this section, some definitions and properties of fractional derivatives and integrations have been discussed.

1

1.2.1 Useful Mathematical Functions

Before discussing the definitions of various fractional differential and integral operators, some useful functions have been presented here, those are inherently tied to fractional calculus and will commonly be encountered [1–5]. These functions include the gamma function, the Euler psi function, incomplete gamma function, beta function, incomplete beta function, Mittag–Leffler functions, Mellin–Ross function, Wright function, error function, hypergeometric functions, viz., Gauss, Kummer, generalized hypergeometric functions, and the *H*-function etc.

1.2.1.1 Gamma Function

In this section, the definitions and some properties of the gamma function have been discussed. The most basic interpretation of the gamma function is simply the generalization of the factorial for all real numbers. It can be defined also for the complex number.

Definition 1.2.1.1.1: Gamma function is defined as:

$$\Gamma(z) = \int_0^\infty e^{-t} t^{z-1} dt, \; \Re(z) > 0, \tag{1.1}$$

where $\Re(z)$ is the real part of all complex number $z \in \mathbb{C}$. The integral of eq. (1.1) is converges for all complex $z \in \mathbb{C}(\Re(z) > 0)$. Gamma function defined everywhere on real axis, except its singular points, viz., 0, –1, –2,…. So, the function domain is …∪(–2, –1)∪(–1, 0)∪(0, +∞). The graph of gamma function is depicted in Figure 1.1.

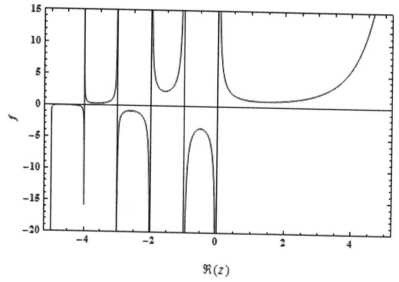

FIGURE 1.1
The graph function in real axis, when $f = \Gamma(z)$.

Introduction and Preliminaries of Fractional Calculus

Remark 1.2.1.1.1: Gamma functions have the following properties, those are given as:

1. $\Gamma(z+1) = z\Gamma(z)$ and $\Gamma\left(\frac{1}{2}\right) = \sqrt{\pi}$, $\Re(z) > 0$,
2. $\Gamma(z) = (z-1)!$, $\Re(z) \in \mathbb{N}$, and
3. $\Gamma(z)\Gamma(1-z) = \frac{\pi}{\sin(\pi z)}$, $z \notin \mathbb{N}$, $\Re(z) < 1$.

Definition 1.2.1.1.2: (Euler psi function)

Euler psi function is defined as the logarithmic derivative of gamma function, which is defined as:

$$\psi(z) = \frac{d}{dz}\log\Gamma(z) = \frac{\Gamma'(z)}{\Gamma(z)}, \ z \in \mathbb{C}. \tag{1.2}$$

Remark 1.2.1.1.2: The Euler psi function has the following property:

1. $\psi(z+m) = \psi(z) + \sum_{k=0}^{m-1}\frac{1}{z+k}, z \in \mathbb{C}, m \in \mathbb{N}.$

Definition 1.2.1.1.3: (Incomplete gamma function)

Incomplete gamma function [4,6] arises from Euler's integral for the gamma function presented in eq. (1.1), by decomposing into an integral from 0 to ω and another from ω to ∞ as:

$$\gamma(z, \omega) = \int_0^{\omega} e^{-t} t^{z-1} dt, z, \omega \in \mathbb{C}, \Re(z) > 0.$$

$$\Gamma(z, \omega) = \int_{\omega}^{\infty} e^{-t} t^{z-1} dt, \Re(z) > 0, \left|\arg(\omega)\right| < \pi. \tag{1.3}$$

Remark 1.2.1.1.3: Incomplete gamma functions have the following properties, those are given as:

1. $\gamma(z, \infty) = \Gamma(z, 0) = \Gamma(z)$ and
2. $\gamma(z, \omega) + \Gamma(z, \omega) = \Gamma(z), \Re(z) > 0.$

1.2.1.2 Beta Function

Definition 1.2.1.2.1: Beta function is defined by the following definite integral [1–4], which is given as:

$$B(z,\omega) = \int_0^1 t^{z-1}(1-t)^{\omega-1}dt, \Re(z), \Re(\omega) > 0. \tag{1.4}$$

Remark 1.2.1.2.1: Beta functions have the following properties, those are given as:

1. $B(z,\omega) = B(\omega,z)$,

2. $B(z,\omega) = 2\int_0^{\pi/2}(\sin\theta)^{2z-1}(\cos\theta)^{2\omega-1}d\theta, \Re(z), \Re(\omega) > 0$,

3. $B(z,\omega) = \int_0^\infty \dfrac{t^{z-1}}{(1+t)^{z+\omega}}dt, \Re(z), \Re(\omega) > 0$,

4. $B(z,\omega) = B(z,\omega+1) + B(z+1,\omega)$,

5. $B(z,\omega+1) = B(z,\omega)\dfrac{\omega}{z+\omega}$,

6. $B(z+1,\omega) = B(z,\omega)\dfrac{z}{z+\omega}$, and

7. $B(z,\omega)B(z+\omega,1-\omega) = \dfrac{\pi}{z\sin(\pi\omega)}$.

Remark 1.2.1.2.2: Beta functions have the following relation with gamma function, which is presented as:

$$B(z,\omega) = \frac{\Gamma(z)\Gamma(\omega)}{\Gamma(z+\omega)}.$$

Definition 1.2.1.2.2: (Incomplete beta function)

The generalized form of beta function is known as incomplete beta function, which is given as:

$$B(z;a,b) = \int_0^z t^{a-1}(1-t)^{b-1}dt, \Re(z) > 0. \tag{1.5}$$

It is noted that, when $z = 1$, the incomplete beta function becomes beta function, which has many implementations in physics, functional analysis, and integral calculus.

1.2.1.3 Mittag–Leffler Function

This section deals with the Mittag–Leffler function and its generalizations. Its importance is realized during the last few decades due to its direct involvement in the problems of

Introduction and Preliminaries of Fractional Calculus

physics, biology, engineering, and applied sciences. The Mittag–Leffler function naturally occurs as the solution of fractional order differential equations or fractional order integral equations. It is the direct generalization of the exponential functions and can be defined in terms of a power series.

Definition 1.2.1.3.1: The one-parameter representation of the Mittag–Leffler function [1–4] can be defined in terms of a power series as:

$$E_\alpha(z) = \sum_{k=0}^{\infty} \frac{z^k}{\Gamma(\alpha k + 1)}, \alpha > 0, z \in \mathbb{C}. \tag{1.6}$$

Definition 1.2.1.3.2: The two-parameter representation of the Mittag–Leffler function can be defined in terms of a power series as:

$$E_{\alpha,\beta}(z) = \sum_{k=0}^{\infty} \frac{z^k}{\Gamma(\alpha k + \beta)}, \alpha, \beta > 0, z \in \mathbb{C}. \tag{1.7}$$

Definition 1.2.1.3.3: The generalized Mittag–Leffler function [4,7,8] can be defined as:

$$E_{\alpha,\beta}^\gamma(z) = \sum_{k=0}^{\infty} \frac{(\gamma)_n}{\Gamma(\alpha k + \beta)} \frac{z^k}{k!}, \alpha, \beta, \gamma \in \mathbb{C}, \Re(\alpha), \Re(\beta), \Re(\gamma) > 0, z \in \mathbb{C}, \tag{1.8}$$

where $(\gamma)_n$ is the Pochhammer symbol, defined as:

$$(\gamma)_n = \frac{\Gamma(\gamma + n)}{\Gamma(\gamma)} = \begin{cases} 1, & n = 0, \gamma \neq 0, \\ \gamma(\gamma+1)(\gamma+2)...(\gamma+n-1), & n \in \mathbb{N}, \gamma \in \mathbb{C}. \end{cases} \tag{1.9}$$

Remark 1.2.1.3.1: The usual derivative of the Mittag–Leffler function $E_{\alpha,\beta}(z)$ can be expressed in the form of the generalized Mittag–Leffler function eq. (1.8) by [4]:

$$\frac{d^n}{dz^n}\left[E_{\alpha,\beta}(z)\right] = n! E_{\alpha,\beta+\alpha n}^{n+1}(z), n \in \mathbb{N}, z \in \mathbb{C}.$$

Remark 1.2.1.3.2: The following properties hold for the Mittag–Leffler function, which are given as [9]:

1. $E_{\alpha,\beta}(z) = \dfrac{1}{\Gamma(\beta)} + z E_{\alpha, \alpha+\beta}(z),$

2. $E_{\alpha,\beta}(z) = \beta E_{\alpha, \beta+1}(z) + \alpha z \dfrac{d}{dz} E_{\alpha, \beta+1}(z),$ and

3. $\left(\dfrac{d}{dz}\right)^m \left[z^{\beta-1} E_{\alpha,\beta}(z^\alpha)\right] = z^{\beta-m-1} E_{\alpha, \beta-m}(z^\alpha), \Re(\beta - m) > 0, m = 0, 1, 2,...$

1.2.1.4 Mellin–Ross Function

The Mellin–Ross function $E_t(v,a)$ arises when finding the fractional derivative of an exponential e^{at}. The function is closely related to both the incomplete gamma and Mittag–Leffler functions.

Definition 1.2.1.4.1: The Mellin–Ross function is defined as [9]:

$$E_t(v,a) = t^v \sum_{k=0}^{\infty} \frac{(at)^k}{\Gamma(k+v+1)} = t^v E_{1,v+1}(at). \tag{1.10}$$

1.2.1.5 Wright Function

The Wright function plays an important role in the solution of a fractional differential equation.

Definition 1.2.1.5.1: The Wright function proposed by Wright [10] in 1933, denoted by $W(z;\alpha,\beta)$, is defined as [1–4]:

$$W(z;\alpha,\beta) = \sum_{k=0}^{\infty} \frac{z^k}{k!\Gamma(\alpha k+\beta)}, \alpha > -1 \text{ and } z, \beta \in \mathbb{C}. \tag{1.11}$$

Remark 1.2.1.5.1: The Wright function can be expressed with the help of the Mittag–Leffler function as [8]:

$$\frac{d^n}{dz^n}\left[z^{\beta-1}E^{\gamma}_{\alpha,\beta+r\alpha}(z)\right] = \sum_{k=0}^{\infty} z^{\beta-1-n}(\gamma)_k W(z;\alpha,\beta+r\alpha-n).$$

1.2.1.6 Error Function

Definition 1.2.1.6.1: Error function $\mathrm{erf}(z)$ is given by:

$$\mathrm{erf}(z) = \frac{2}{\sqrt{\pi}} \int_0^z e^{-t^2} dt, z \in \mathbb{C}. \tag{1.12}$$

Remark 1.2.1.6.1: Some properties of the error function are given as:

1. (Complementary error function)

$$\mathrm{erfc}(z) = \frac{2}{\sqrt{\pi}} \int_z^{\infty} e^{-t^2} dt = 1 - \mathrm{erf}(z),$$

2. $\mathrm{erf}(-z) = -\mathrm{erf}(z)$, and

3. $E_{1/2}(z^{1/2}) = e^z(1 + \mathrm{erf}(z^{1/2}))$.

Introduction and Preliminaries of Fractional Calculus

1.2.1.7 Hypergeometric Functions

In this section, definition and some properties of Gauss, Kummer, and generalized hypergeometric functions have been presented.

When $\alpha > 0$, the following lemma yields the integral representation of $E_\alpha(z)$ as a Mellin–Barnes countour integral.

Lemma 1.2.1.7.1: For $\alpha > 0$ and $z \in \mathbb{C}\left(\left(|\arg(z)| < \pi\right)\right)$, the following relation holds:

$$E_\alpha(z) = \frac{1}{2\pi i} \int\limits_{\gamma - i\infty}^{\gamma + i\infty} \frac{\Gamma(s)\Gamma(1-s)}{\Gamma(1-\alpha s)} (-z)^{-s} ds, \tag{1.13}$$

where the path of integration separates all the poles $s = -k (k \in \mathbb{N}_0)$ to the left and all poles at $s = 1 + n$ $(n \in \mathbb{N}_0)$ to the right.

Definition 1.2.1.7.1: (Gauss hypergeometric function)

The Gauss hypergeometric function ${}_2F_1(a, b; c; z)$ is defined in the unit disk as the sum of hypergeometric series given by [4]:

$$_2F_1(a, b; c; z) = \sum_{k=0}^{\infty} \frac{(a)_k (b)_k}{(c)_k} \frac{z^k}{k!}, \tag{1.14}$$

where $|z| < 1$, $a, b \in \mathbb{C}$, $c \in \mathbb{C} \setminus \mathbb{Z}_0^- := \{0, -1, -2, \ldots\}$ and $(a)_k$ is the Pochhammer symbol defined in eq. (1.9). The series presented in eq. (1.14) is absolutely convergent for $|z| \le 1$, when $\Re(c - a - b) > 0$ and conditionally convergent for $|z| = 1 (z \neq 1)$ if $-1 < \Re(c - a - b) \le 0$.

Remark 1.2.1.7.1: The Euler integral representation of the Gauss hypergeometric function is given as:

$$_2F_1(a, b; c; z) = \frac{\Gamma(c)}{\Gamma(b)\Gamma(c-b)} \int\limits_0^1 t^{b-1} (1-t)^{c-b-1} (1-zt)^{-a} dt, 0 < \Re(b) < \Re(c), |\arg(1-z)| < \pi. \tag{1.15}$$

If $c \notin \mathbb{Z}_0^-$, then ${}_2F_1(a, b; c; z)$ has another integral representation in terms of the Mellin–Barnes countour integral given as:

$$_2F_1(a, b; c; z) = \frac{1}{2\pi i} \frac{\Gamma(c)}{\Gamma(a)\Gamma(b)} \int\limits_{\gamma - i\infty}^{\gamma + i\infty} \frac{\Gamma(a-s)\Gamma(b-s)}{\Gamma(c-s)} (-z)^{-s} \Gamma(s) ds, \tag{1.16}$$

where $|\arg(-z)| < \pi$ and the path integration starts at the point $\gamma - i\infty (\gamma \in \mathbb{R})$ and terminates at the point $\gamma + i\infty$, separating all the poles $s = -k$ $(k = 0, 1, 2, \ldots)$ to the left and all poles $s = a + n(n \in \mathbb{N}_0)$ and $s = b + m$ $(m \in \mathbb{N}_0)$ to the right.

Remark 1.2.1.7.2: Some properties of the Gauss hypergeometric function are given as [4]:

1. $_2F_1(a,b;c;z) = {}_2F_1(b,a;c;z)$,

2. $_2F_1(a,b;c;0) = {}_2F_1(0,b;c;z) = 1$,

3. $_2F_1(a,b;b;z) = (1-z)^{-a}$,

4. $_2F_1(a,b;c;1) = \dfrac{\Gamma(c)\Gamma(c-a-b)}{\Gamma(c-a)\Gamma(c-b)}, \ (\Re(c-a-b) > 0)$,

5. (Euler transformation formula)
   $_2F_1(a,b;c;z) = (1-z)^{c-a-b} {}_2F_1(c-a,c-b;c;z)$,

6. $\left(\dfrac{d}{dz}\right)^n {}_2F_1(a,b;c;z) = \dfrac{(a)_n(b)_n}{(c)_n} {}_2F_1(a+n,b+n;c+n;z), n \in \mathbb{N}$, and

7. $\left(\dfrac{d}{dz}\right)^n \left[z^{a+n-1} {}_2F_1(a,b;c;z)\right] = (a)_n z^{a-1} {}_2F_1(a+n,b;c;z), n \in \mathbb{N}$.

Definition 1.2.1.7.2: (Kummer hypergeometric function)

The confluent hypergeometric Kummer function is defined as [4]:

$$\Phi(a;c;z) = {}_1F_1(a;c;z) = \sum_{k=0}^{\infty} \frac{(a)_k}{(c)_k} \frac{z^k}{k!}, \tag{1.17}$$

where $z,a \in \mathbb{C}$, $c \in \mathbb{C} \backslash \mathbb{Z}_0^- := \{0,-1,-2,...\}$. But unlike hypergeometric series as presented in eq. (1.14), the series eq. (1.17) is convergent for any $z \in \mathbb{C}$.

Remark 1.2.1.7.3: The Euler integral representation of the Gauss hypergeometric function is given as:

$$\Phi(a;c;z) = \frac{\Gamma(c)}{\Gamma(a)\Gamma(c-a)} \int_0^1 t^{a-1}(1-t)^{c-a-1} e^{zt} dt, 0 < \Re(a) < \Re(c). \tag{1.18}$$

The function $\Phi(a;c;z)$ has another integral representation in terms of the Mellin–Barnes countour integral, given as:

$$\Phi(a;c;z) = \frac{1}{2\pi i} \frac{\Gamma(c)}{\Gamma(a)} \int_{\gamma-i\infty}^{\gamma+i\infty} \frac{\Gamma(a-s)}{\Gamma(c-s)} (-z)^{-s} \Gamma(s) ds, \tag{1.19}$$

where $|\arg(-z)| < \pi$ and the path integration separates all the poles $s = -k$ ($k = 0,1,2,...$) to the left and all poles $s = a+n$ ($n \in \mathbb{N}_0$) to the right.

Introduction and Preliminaries of Fractional Calculus

Definition 1.2.1.7.3: (Generalized hypergeometric function)

The Gauss hypergeometric series eq. (1.14) and the Kummer hypergeometric series eq. (1.17) are extended to the generalized hypergeometric function, which is defined as [4]:

$$_pF_q(a_1,...,a_p;b_1,...,b_q;z) = \sum_{k=0}^{\infty} \frac{(a_1)_k...(a_p)_k}{(b_1)_k...(b_q)_k} \frac{z^k}{k!},$$

(1.20)

where $a_i, b_j \in \mathbb{C}$, $b_j \neq 0, -1, -2,...\{i = 1,...,p; j = 1,...,q\}$. The series presented in eq. (1.20) is absolutely convergent for all values of $z \in \mathbb{C}$ if $p \leq q$. When $p = q+1$, the series in eq. (1.20) is absolutely convergent for $|z| < 1$ and for $|z| = 1$ when $\Re(\sum_{j=1}^{q} b_j - \sum_{i=1}^{p} a_i) > 0$, while it is conditionally convergent for $|z| = 1$ $(z \neq 1)$ if $-1 < \Re(\sum_{j=1}^{q} b_j - \sum_{i=1}^{p} a_i) \leq 0$.

Remark 1.2.1.7.4: If $b_j \notin \mathbb{Z}_0\{j = 1,...,q\}$, then the generalized hypergeometric function has an integral representation in terms of the Mellin–Barnes contour integral as presented in eqs. (1.16) and (1.19), which is given as:

$$_pF_q(a_1,...,a_p;b_1,...,b_q;z) = \frac{1}{2\pi i} \frac{\prod_{j=1}^{q} \Gamma(b_j)}{\prod_{i=1}^{p} \Gamma(a_i)} \int_{\gamma-i\infty}^{\gamma+i\infty} \frac{\prod_{i=1}^{p} \Gamma(a_i - s)}{\prod_{j=1}^{q} \Gamma(b_j - s)} (-z)^{-s}\Gamma(s)ds,$$

(1.21)

where $|\arg(-z)| < \pi$ and the path integration separates all the poles $s = -k$ $(k = 0,1,2,...)$ to the left and all poles $s = a_j + n$ $(n \in \mathbb{N}_0; j = 1,2,...,p)$ to the right.

Remark 1.2.1.7.5: Property of generalized hypergeometric function is given as:

1. $\left(\frac{d}{dz}\right)^n {}_pF_q(a_1,...,a_p;b_1,...,b_q;z) = \frac{(a_1)_n...(a_p)_n}{(b_1)_n...(b_q)_n} {}_pF_q(a_1 + n,...,a_p + n;b_1 + n,...,b_q + n;z), n \in \mathbb{N}.$

1.2.1.8 H-*Function*

In this section, the definition and properties of the H-function [11] have been presented.

Definition 1.2.1.8.1: For integers m,n,p,q, such that $0 \leq m \leq q$ and $0 \leq n \leq p$, for $a_i, b_j \in \mathbb{C}$ and for $\alpha_i, \beta_j \in \mathbb{R}^+(i = 1,2,...,p, j = 1,2,...,q)$, the H-function is defined as [4]:

$$H_{p,q}^{m,n}(z) = H_{p,q}^{m,n}\left[z \middle| \begin{matrix} (a_i,\alpha_i)_{1,p} \\ (b_j,\beta_j)_{1,q} \end{matrix}\right]$$

$$= H_{p,q}^{m,n}\left[z \middle| \begin{matrix} (a_1,\alpha_1),...,(a_p,\alpha_p) \\ (b_1,\beta_1),...,(b_q,\beta_q) \end{matrix}\right],$$

(1.22)

$$= \frac{1}{2\pi i} \int_C H_{p,q}^{m,n}(s)z^{-s}ds$$

where

$$H_{p,q}^{m,n} = \frac{\prod\limits_{j=1}^{m} \Gamma(b_j + \beta_j s) \prod\limits_{i=1}^{n} \Gamma(1 - a_i - \alpha_i s)}{\prod\limits_{i=n+1}^{r} \Gamma(a_i + \alpha_i s) \prod\limits_{j=m+1}^{m} \Gamma(1 - b_j - \beta_j s)}. \tag{1.23}$$

Remark 1.2.1.8.1 The relation between the H-function and the generalized Mittag–Leffler function is expressed with the help of the following expression, which is given as [9]:

$$E_{\alpha,\beta}^{\gamma}(z) = \frac{1}{\Gamma(\gamma)} H_{1,2}^{1,1} \left[-z \left| \begin{array}{l} (1-\gamma,1) \\ (0,1),(1-\beta,\alpha) \end{array} \right. \right].$$

1.2.2 Riemann–Liouville Integral and Derivative

Over the years, many mathematicians, using their own notation and approach, have found various definitions that fit the idea of a fractional (non-integer) order integral or derivative. One version that has been popularized in the world of fractional calculus is the Riemann–Liouville definition. The Riemann–Liouville definition was firstly introduced by Riemann, which is derived from Abel's integral.

Definition 1.2.2.1: A real function $f(t)$, $t > 0$ is said to be in the space C_γ, $\gamma \in \mathbb{R}$ if there exists a real number $p(> \gamma)$, such that $f(t) = t^p f_1(t)$, where $f_1(t) \in C[0, \infty]$, and it is said to be in the space C_γ^m if $f^{(m)} \in C_m$, $m \in \mathbb{N}$.

Definition 1.2.2.2: The fractional order of the Riemann–Liouville integral of order $\alpha (>0)$, of a function $f \in C_\gamma$, $\gamma \geq -1$ is defined as [1,3,5,12]:

$$J_t^\alpha f(t) = \frac{1}{\Gamma(\alpha)} \int_0^t (t - \tau)^{\alpha - 1} f(\tau) d\tau, t > 0, \alpha \in \mathbb{R}^+, \tag{1.24}$$

where $\mathbb{R}^+$ is the set of positive real numbers.

Definition 1.2.2.3: The left-hand side and right-hand side of the Riemann–Liouville fractional integral of a function $f \in C_\gamma$, $(\gamma \geq -1)$ are defined as:

$$_{-\infty}J_t^\alpha f(t) = \frac{1}{\Gamma(m-\alpha)} \int_{-\infty}^t (t - \tau)^{m-\alpha-1} f(\tau) d\tau, m-1 < \alpha < m, \, m \in \mathbb{N}$$

$$_t J_\infty^\alpha f(t) = \frac{(-1)^m}{\Gamma(m-\alpha)} \int_t^\infty (\tau - t)^{m-\alpha-1} f(\tau) d\tau, \text{ for } m-1 < \alpha < m, \, m \in \mathbb{N}. \tag{1.25}$$

Introduction and Preliminaries of Fractional Calculus

Remark 1.2.2.1: For $f \in C_\gamma, \gamma \geq -1$, we have following property:

$$J^\alpha t^\beta = \frac{\Gamma(\beta+1) t^{\alpha+\beta}}{\Gamma(\beta+\alpha+1)}, \beta > -1, \alpha > -1-\beta.$$

Definition 1.2.2.4: The fractional order of the Riemann–Liouville derivative of order $\alpha(>0)$ is defined as [1,3,5]:

$$D_t^\alpha f(t) = \begin{cases} \dfrac{1}{\Gamma(m-\alpha)} \dfrac{d^m}{dt^m} \displaystyle\int_0^t (t-\tau)^{m-\alpha-1} f(\tau)d\tau & \text{if } m-1 < \alpha < m, \, m \in \mathbb{N}, \\[4mm] \dfrac{d^m f(t)}{dt^m} & \text{if } \alpha = m, \, m \in \mathbb{N}, \end{cases} \tag{1.26}$$

Definition 1.2.2.5 The left Riemann–Liouville fractional derivative of order $\alpha(m-1 < \alpha < m, m \in \mathbb{N})$ can be defined as:

$$_{-\infty}D_t^\alpha f(t) = \frac{1}{\Gamma(m-\alpha)} \frac{d^m}{dt^m} \int_{-\infty}^t (t-\tau)^{m-\alpha-1} f(\tau)d\tau, m-1 < \alpha < m, \, m \in \mathbb{N}. \tag{1.27}$$

Definition 1.2.2.6: The right Riemann–Liouville fractional derivative of order $\alpha(m-1 < \alpha < m, m \in \mathbb{N})$ can be defined as:

$$_{t}D_\infty^\alpha f(t) = \frac{(-1)^m}{\Gamma(m-\alpha)} \frac{d^m}{dt^m} \int_t^\infty (\tau-t)^{m-\alpha-1} f(\tau)d\tau, \, m-1 < \alpha < m, \, m \in \mathbb{N}. \tag{1.28}$$

Remark 1.2.2.2: One of the interesting properties of the Riemann–Liouville fractional derivative is the derivative of a constant is not zero. So mathematically it can be written as:

$$D_t^\alpha C = \frac{Ct^{-\alpha}}{\Gamma(1-\alpha)}, \tag{1.29}$$

where C is a constant.

1.2.3 Caputo Fractional Derivative

In the case of the Riemann–Liouville derivative, the derivative of a constant is not zero. So there are less physical representations of those types of conditions. Due to that reason, in some cases, the alternate definition of fractional derivative, viz., the Caputo derivative has been used.

Definition 1.2.3.1: The general definition of the Caputo derivative [1,3] is defined as:

$$D_t^\alpha f(t) = J_t^{m-\alpha} D^m f(t) = \begin{cases} \dfrac{1}{\Gamma(m-\alpha)} \int_0^t (t-\tau)^{m-\alpha-1} \dfrac{d^m f(\tau)}{d\tau^m} d\tau & \text{if } m-1 < \alpha < m, \ m \in \mathbb{N}, \\[3mm] \dfrac{d^m f(t)}{dt^m} & \text{if } \alpha = m, \ m \in \mathbb{N}, \end{cases} \tag{1.30}$$

where the parameter α is the order of the derivative and is allowed to be real or even complex, also. In this book, only real and positive α will be considered.

Remark 1.2.3.1: For the Caputo derivative, we have the following properties:

$$D^\alpha C = 0, \ (C \text{ is a constant}). \tag{1.31}$$

$$D^\alpha t^\beta = \frac{\Gamma(\beta+1) \, t^{\beta-\alpha}}{\Gamma(\beta-\alpha+1)}, \ \beta > \alpha - 1 \text{ and } \beta > -1. \tag{1.32}$$

Lemma 1.2.3.1: If $m-1 < \alpha \le m, \ m \in \mathbb{N}$ and $f \in C_\gamma^m, \gamma \ge -1$, then:

$$D^\alpha J^\alpha f(t) = f(t),$$

and

$$J^\alpha D^\alpha f(t) = f(t) - \sum_{k=0}^{m-1} f^k(0^+) \frac{t^k}{k!}, \ t > 0. \tag{1.33}$$

1.2.4 Grünwald–Letnikov Fractional Derivative

The Grünwald–Letnikov fractional derivative was initially proposed by Anton Karl Grünwald (1838–1920) in 1867 and by Aleksey Vasilievich Letnikov (1837–1888) in 1868. The Grünwald–Letnikov fractional derivative is based on finite differences, which is equivalent to the Riemann–Liouville definition.

Definition 1.2.4.1: The Grünwald–Letnikov fractional derivative of order $\alpha(>0)$ [1,3] is given as:

$$_a D_t^\alpha f(t) = \lim_{\substack{h \to 0 \\ mh = t-a}} h^{-q} \sum_{r=0}^m \omega_r^q f(t - rh), \tag{1.34}$$

where $\omega_r^q = (-1)^r \begin{pmatrix} q \\ r \end{pmatrix}$.

Introduction and Preliminaries of Fractional Calculus

13

Remark 1.2.4.1: Some properties of Grünwald–Letnikov fractional derivative are given as:

$$\omega_0^q = 1 \text{ and } \omega_r^q = \left(1 - \frac{q+1}{r}\right)\omega_{r-1}^q, r = 1, 2, \ldots \tag{1.35}$$

1.2.5 Riesz Fractional Integral and Derivative

In this section, the definition of one special type of fraction integral and derivative, viz., Riesz fractional integral and derivative have been reported.

Definition 1.2.5.1: The Riesz fractional integral [1,3,6,13] of the order α, $n-1 \leq \alpha < n$ of a function $f \in C_\gamma$, $(\gamma \geq -1)$ is defined as:

$${}_0^R J_t^\alpha f(t) = c_\alpha \left({}_{-\infty} J_t^\alpha + {}_t J_\infty^\alpha\right) f(t)$$

$$= \frac{c_\alpha}{\Gamma(\alpha)} \int_{-\infty}^{\infty} |t - \tau|^{\alpha-1} f(\tau) d\tau,$$

where $c_\alpha = \dfrac{1}{2\cos\left(\dfrac{\pi\alpha}{2}\right)}$, $\alpha \neq 1$.

Here, ${}_{-\infty} J_t^\alpha$, ${}_t J_\infty^\alpha$ are the left-hand and right-hand sides of the Riemann–Liouville fractional integral operators defined in definition 1.3.3.

Definition 1.2.5.2: The Riesz fractional derivative of order $\alpha(n-1 < \alpha \leq n, n \in \mathbb{N})$ on the infinite domain $-\infty < t < \infty$ of a function $f \in C_\gamma$, $(\gamma \geq -1)$ is defined as [1,3,6,13,14]:

$$\frac{d^\alpha f(t)}{d|t|^\alpha} = -c_\alpha \left({}_{-\infty} D_t^\alpha f(t) + {}_t D_\infty^\alpha f(t)\right), \tag{1.36}$$

where $c_\alpha = \dfrac{1}{2\cos\left(\dfrac{\pi\alpha}{2}\right)}$, $\alpha \neq 1$.

Here, ${}_{-\infty} D_t^\alpha$ and ${}_t D_\infty^\alpha$ are the left-hand and right-hand sides of the Riemann–Liouville fractional differential operator defined in definitions 1.3.5 and 1.3.6, respectively.

In case of $a \leq t \leq b$ (i.e., t defined in a finite interval), the Riesz fractional derivative of order $\alpha(n-1 < \alpha \leq n, n \in \mathbb{N})$ can be written as:

$$\frac{d^\alpha f(t)}{d|t|^\alpha} = -\frac{1}{2\cos\left(\dfrac{\alpha\pi}{2}\right)}\left({}_a D_t^\alpha f(t) + {}_t D_b^\alpha f(t)\right),$$

where:

$$_aD_t^\alpha f(t) = \frac{1}{\Gamma(n-\alpha)} \frac{d^n}{dt^n} \int_a^t \frac{f(\tau)d\tau}{(t-\tau)^{1-n+\alpha}},$$

$$_tD_b^\alpha f(t) = \frac{(-1)^n}{\Gamma(n-\alpha)} \frac{d^n}{dt^n} \int_t^b \frac{f(\tau)d\tau}{(\tau-t)^{1-n+\alpha}}$$

Remark 1.2.5.1: Let $\alpha > 0$ and $\beta > 0$ be such that $n-1 < \alpha$, $\beta \le n$, and $\alpha + \beta \le n$, then we have following index rule:

$$_0^RD_t^\alpha \left(_0^RD_t^\beta f(t) \right) = {_0^RD_t^{\alpha+\beta}} f(t).$$

Remark 1.2.5.2: The Riesz fractional operator $_0^RD_t^{\alpha-1}f(t)$ of the order $0 < \alpha < 1$ can be expressed as Riesz fractional integral operator $_0^RJ_t^{1-\alpha}f(t)$ by the following identity, defined in [1,3,6,13]:

$$_0^RD_t^{\alpha-1}f(t) = {_0^RJ_t^{1-\alpha}} f(t), t \in T$$

Lemma 1.2.5.1: For a function $f(t)$ defined on the infinite domain $(-\infty < t < \infty)$, the following equality holds:

$$-(-\Delta)^{\frac{\alpha}{2}} f(t) = -c_\alpha (_{-\infty}D_t^\alpha + {_tD_\infty^\alpha})f(t) = \frac{d^\alpha}{d|t|^\alpha} f(t), \text{ for } n-1 < \alpha \le n, n \in \mathbb{N}.$$

Proof: According to Samko et al. [3], a fractional power of the Laplace operator is defined as follows:

$$-(-\Delta)^{\frac{\alpha}{2}} f(t) = -\mathcal{F}^{-1}|\xi|^\alpha \mathcal{F}(f(t)), \tag{1.37}$$

where $\mathcal{F}$ and $\mathcal{F}^{-1}$ denote the Fourier transform and inverse Fourier transform of $f(t)$, respectively. Hence, we have:

$$-(-\Delta)^{\frac{\alpha}{2}} f(t) = -\frac{1}{2\pi} \int_{-\infty}^\infty e^{-it\xi} |\xi|^\alpha \int_{-\infty}^\infty e^{i\xi\eta} f(\eta)d\eta d\xi$$

Supposing that $f(t)$ vanishes at $t = \pm\infty$, the integration by parts, yields:

$$\int_{-\infty}^\infty e^{i\xi\eta} f(\eta)d\eta = -\frac{1}{i\xi} \int_{-\infty}^\infty e^{i\xi\eta} f'(\eta)d\eta.$$

Introduction and Preliminaries of Fractional Calculus

Thus, we obtain:

$$-(-\Delta)^{\frac{\alpha}{2}} f(t) = -\frac{1}{2\pi} \int\limits_{-\infty}^{\infty} f'(\eta) \left[i \int\limits_{-\infty}^{\infty} e^{i\xi(\eta-t)} \frac{|\xi|^{\alpha}}{\xi} d\xi \right] d\eta.$$

Let $I = i \int\limits_{-\infty}^{\infty} e^{i\xi(\eta-t)} \frac{|\xi|^{\alpha}}{\xi} d\xi$, then:

$$I = i \left[-\int\limits_{0}^{\infty} e^{i\xi(t-\eta)} \xi^{\alpha-1} d\xi + \int\limits_{0}^{\infty} e^{i\xi(\eta-t)} \xi^{\alpha-1} d\xi \right],$$

for $0 < \alpha < 1$, we have:

$$I = i \left[\frac{-\Gamma(\alpha)}{[i(\eta-t)]^{\alpha}} + \frac{\Gamma(\alpha)}{[i(t-\eta)]^{\alpha}} \right] = \frac{sign(t-\eta)\Gamma(\alpha)\Gamma(1-\alpha)}{|t-\eta|^{\alpha} \Gamma(1-\alpha)} \left[i^{1-\alpha} + (-i)^{1-\alpha} \right].$$

Using $\Gamma(\alpha)\Gamma(1-\alpha) = \dfrac{\pi}{\sin(\pi\alpha)}$ and $i^{1-\alpha} + (-i)^{1-\alpha} = 2\sin\left(\dfrac{\alpha\pi}{2} \right)$, we obtain:

$$I = \frac{sign(t-\eta)\pi}{\cos\left(\dfrac{\alpha\pi}{2} \right) |t-\eta|^{\alpha} \Gamma(1-\alpha)}$$

Hence, for $0 < \alpha < 1$:

$$-(-\Delta)^{\frac{\alpha}{2}} f(t) = -\frac{1}{2\pi} \int\limits_{-\infty}^{\infty} f'(\eta) \frac{sign(t-\eta)\pi}{\cos\left(\dfrac{\alpha\pi}{2} \right) |t-\eta|^{\alpha} \Gamma(1-\alpha)} d\eta$$

$$= -\frac{1}{2\cos\left(\dfrac{\alpha\pi}{2} \right)} \left[\frac{1}{\Gamma(1-\alpha)} \int\limits_{-\infty}^{t} \frac{f'(\eta)}{(t-\eta)^{\alpha}} d\eta - \frac{1}{\Gamma(1-\alpha)} \int\limits_{t}^{\infty} \frac{f'(\eta)}{(\eta-t)^{\alpha}} d\eta \right].$$

Following [1,3], for $0 < \alpha < 1$, the Grünwald–Letnikov fractional derivative in $[a,t]$ is given by:

$$_a D_t^{\alpha} f(t) = \frac{f(a)(t-a)^{-\alpha}}{\Gamma(1-\alpha)} + \frac{1}{\Gamma(1-\alpha)} \int\limits_{a}^{t} \frac{f'(\eta)}{(t-\eta)^{\alpha}} d\eta.$$

Therefore, if $f(t)$ tends to zero for $a \to -\infty$, then we have:

$$_{-\infty} D_t^{\alpha} f(t) = \frac{1}{\Gamma(1-\alpha)} \int\limits_{-\infty}^{t} \frac{f'(\eta)}{(t-\eta)^{\alpha}} d\eta$$

Similarly, if $f(t)$ tends to zero for $b \to +\infty$, then we have:

$$_x D_\infty^\alpha f(t) = \frac{-1}{\Gamma(1-\alpha)} \int_t^\infty \frac{f'(\eta)}{(\eta-t)^\alpha} \, d\eta.$$

Hence, if $f(t)$ is continuous and $f'(t)$ is integrable for $t \geq a$, then for every $\alpha\,(0 < \alpha < 1)$, the Riemann–Liouville derivative exists and coincides with the Grünwald–Letnikov derivative. Finally, for $0 < \alpha < 1$, we have:

$$-(-\Delta)^{\frac{\alpha}{2}} f(t) = -\frac{1}{2\cos\left(\dfrac{\alpha\pi}{2}\right)} [_{-\infty}D_t^\alpha f(t) + {}_t D_\infty^\alpha f(t)] = \frac{d^\alpha}{d|t|^\alpha} f(t),$$

where $_{-\infty}D_t^\alpha f(t) = \dfrac{1}{\Gamma(1-\alpha)} \dfrac{d}{dt} \displaystyle\int_{-\infty}^t \dfrac{f(\eta)d\eta}{(t-\eta)^\alpha}$ and $_t D_\infty^\alpha f(t) = \dfrac{-1}{\Gamma(1-\alpha)} \dfrac{d}{dt} \displaystyle\int_t^\infty \dfrac{f(\eta)d\eta}{(\eta-t)^\alpha}$.

Following the similar argument, for $1 < \alpha < 2$, we can obtain:

$$-(-\Delta)^{\frac{\alpha}{2}} f(t) = -\frac{1}{2\cos\left(\dfrac{\alpha\pi}{2}\right)} [_{-\infty}D_t^\alpha f(t) + {}_t D_\infty^\alpha f(t)] = \frac{d^\alpha}{d|t|^\alpha} f(t),$$

where $_{-\infty}D_t^\alpha f(t) = \dfrac{1}{\Gamma(2-\alpha)} \dfrac{d^2}{dt^2} \displaystyle\int_{-\infty}^t \dfrac{f(\eta)d\eta}{(t-\eta)^{\alpha-1}}$ and $_t D_\infty^\alpha f(t) = \dfrac{1}{\Gamma(2-\alpha)} \dfrac{d^2}{dt^2} \displaystyle\int_t^\infty \dfrac{f(\eta)d\eta}{(\eta-t)^{\alpha-1}}$.

Finally, for $n-1 < \alpha < n$, we have:

$$-(-\Delta)^{\frac{\alpha}{2}} f(t) = -\frac{1}{2\cos\left(\dfrac{\alpha\pi}{2}\right)} [_{-\infty}D_t^\alpha u(x) + {}_t D_\infty^\alpha f(t)] = \frac{d^\alpha}{d|t|^\alpha} f(t),$$

where $_{-\infty}D_t^\alpha f(t) = \dfrac{1}{\Gamma(n-\alpha)} \dfrac{d^n}{dt^n} \displaystyle\int_{-\infty}^t \dfrac{f(\xi)d\xi}{(t-\xi)^{\alpha+1-n}}$ and $_t D_\infty^\alpha f(t) = \dfrac{(-1)^n}{\Gamma(n-\alpha)} \dfrac{d^n}{dt^n} \displaystyle\int_t^\infty \dfrac{f(\xi)d\xi}{(\xi-t)^{\alpha+1-n}}$.

Remark 1.2.5.3: For a function $f(t)$ defined on the finite interval $[0,L]$, the result in eq. (1.37) holds by setting:

$$f^*(t) = \begin{cases} f(t) & t \in (0,L), \\ 0 & t \notin (0,L). \end{cases}$$

That is $f^*(t) = 0$ on the boundary points and beyond the boundary points.

Introduction and Preliminaries of Fractional Calculus

Definition 1.2.5.3: The Riesz–Feller fractional derivative proposed by Feller [6] is a generalization of the Riesz fractional derivative. For $0 < \alpha \leq 2$, $\alpha \neq 1$ and free parameter θ, $|\theta| \leq \min\{\alpha, 2-\alpha\}$, the Riesz–Feller fractional derivative is defined as [15]:

$$^F D_\theta^\alpha f(x) = (C_+(\alpha,\theta)D_+^\alpha + C_-(\alpha,\theta)D_-^\alpha)f(x), \tag{1.38}$$

where the coefficients $C_\pm(\alpha,\theta)$ are given by:

$$C_+(\alpha,\theta) = \frac{\sin\left((\alpha-\theta)\dfrac{\pi}{2}\right)}{\sin(\alpha\pi)}, C_-(\alpha,\theta) = \frac{\sin\left((\alpha+\theta)\dfrac{\pi}{2}\right)}{\sin(\alpha\pi)},$$

and D_+^α and D_-^α are the left- and right-sided Weyl fractional derivatives of order α, defined for $x \in \mathbb{R}$ and $\alpha > 0$, $n-1 < \alpha \leq n$, $n \in \mathbb{N}$ as:

$$(D_+^\alpha)f(x) := \left(\frac{d}{dx}\right)^n \left(I_+^{n-\alpha}f\right)(x),$$

$$(D_-^\alpha)f(x) := \left(\frac{d}{dx}\right)^n \left(I_-^{n-\alpha}f\right)(x).$$

In the above formulae, $I_\pm^{n-\alpha}$ are the left-and right-sided Weyl fractional integrals given by:

$$I_+^\alpha f(x) = \frac{1}{\Gamma(\alpha)} \int_{-\infty}^{x} (x-\xi)^{\alpha-1} f(\xi)d\xi,$$

and

$$I_-^\alpha f(x) = \frac{1}{\Gamma(\alpha)} \int_{x}^{\infty} (\xi-x)^{\alpha-1} f(\xi)d\xi.$$

Lemma 1.2.5.2: The integration formula of Riemann–Liouville fractional derivatives of the order $0 < \alpha < 1$ [16]:

$$\int_\Gamma f(x,t)\,_0D_t^\alpha g(x,t)dt = \int_\Gamma g(x,t)\,_tD_\tau^\alpha f(x,t)dt,$$

is valid under the assumption that $f, g \in C(\Omega \times \Gamma)$, and that for arbitrary $x \in \Omega$, $_0D_t^\alpha f(x,t)$, $_tD_\tau^\alpha g(x,t)$ exist at every point $t \in \Gamma$ and are continuous in t.

1.2.6 Modified Riemann–Liouville Derivative

In this section, another form of fractional derivative, viz., modified Riemann–Liouville derivative [17,18] of order α has been proposed, which is defined by the expression:

$$f^{(\alpha)}(x) = \lim_{h \downarrow 0} \left(\frac{\sum_{k=0}^{\infty} (-1)^k \binom{\alpha}{k} f[x+(\alpha-k)h]}{h^{\alpha}} \right), \alpha \in \mathbb{R}, 0 < \alpha \le 1, \tag{1.39}$$

which can be written as

$$D_x^{\alpha} f(x) = \begin{cases} \dfrac{1}{\Gamma(1-\alpha)} \dfrac{d}{dx} \displaystyle\int_0^x (x-\xi)^{-\alpha} \left[f(\xi) - f(0) \right] d\xi, & \text{if } 0 < \alpha < 1, \\[4mm] \left(f^{(\alpha-n)}(x) \right)^{(n)}, & \text{if } n \le \alpha < n+1, n \ge 1. \end{cases} \tag{1.40}$$

1.2.7 Local Fractional Derivative

In this section, the theory and definitions of local fractional derivative has been discussed.

1.2.7.1 Local Fractional Continuity of a Function

Definition 1.2.7.1.1: Suppose that $f(x)$ is defined throughout some interval containing x_0 and all points near x_0, then $f(x)$ is said to be local fractional continuous at $x = x_0$, denoted by $\lim_{x \to x_0} f(x) = f(x_0)$, if to each positive ε and some positive constant k corresponds some positive δ, such that [19–21]:

$$\left| f(x) - f(x_0) \right| < k\varepsilon^{\alpha}, 0 < \alpha \le 1, \tag{1.41}$$

whenever $|x - x_0| < \delta$, $\varepsilon, \delta > 0$ and $\varepsilon, \delta \in \mathbb{R}$. Consequently, the function $f(x)$ is called local fractional continuous on the interval (a, b), denoted by:

$$f(x) \in C_{\alpha}(a, b), \tag{1.42}$$

where α is fractal dimension with $0 < \alpha \le 1$.

Definition 1.2.7.1.2: A function $f(x): \mathbb{R} \to -\mathbb{R}$, $X \mapsto f(X)$ is called a non-differentiable function of exponent α, $0 < \alpha \le 1$, which satisfies the Hölder function of exponent α, then for $x, y \in X$, we have [19–21]:

$$\left| f(x) - f(y) \right| \le C|x - y|^{\alpha}. \tag{1.43}$$

Introduction and Preliminaries of Fractional Calculus

Definition 1.2.7.1.3: A function $f(x): \mathbb{R} \to \text{-}\mathbb{R}$, $X \mapsto f(X)$ is called to be local fractional continuous of order α, $0 < \alpha \leq 1$, or shortly α-local fractional continuous, when we have [19–21]:

$$f(x) - f(x_0) = O\left((x - x_0)^\alpha\right). \tag{1.44}$$

Remark 1.2.7.1.1: A function $f(x)$ is said to be in the space $C_\alpha[a, b]$ if and only if it can be written as [19–21]:

$$f(x) - f(x_0) = O\left((x - x_0)^\alpha\right),$$

with any $x_0 \in [a, b]$ and $0 < \alpha \leq 1$.

Theorem 1.2.7.1.1: *(Generalized Hadamard's Theorem)* [22]

Any function $f(x) \in C^\alpha(I)$ in a neighborhood of a point x_0 can be decomposed in the form

$$f(x) = f(x_0) + \frac{(x - x_0)^\alpha}{\Gamma(1 + \alpha)} r(x),$$

where $r(x) \in C^{m\alpha}(I)$ (where m times α th differentiable on $I \subset \mathbb{R}$).

1.2.7.2 Local Fractional Derivative

If a function is not differentiable at $x = x_0$, but has a fractional derivative of order α at this point, then it is locally equivalent to the function:

$$f(x) = f(x_0) + \frac{(x - x_0)^\alpha}{\Gamma(\alpha + 1)} f^{(\alpha)}(x_0) + O\left((x - x_0)^{2\alpha}\right), \tag{1.45}$$

Definition 1.2.7.2.1: Following to eq. (1.45), the local fractional derivative of $f(x) \in C_\alpha(a, b)$ of order α at $x = x_0$ is defined as [19–21]:

$$f^{(\alpha)}(x_0) = \left.\frac{d^\alpha f(x)}{dx^\alpha}\right|_{x = x_0} = \lim_{x \to x_0} \frac{\Delta^\alpha(f(x) - f(x_0))}{(x - x_0)^\alpha}, \tag{1.46}$$

where $\Delta^\alpha(f(x) - f(x_0)) \cong \Gamma(1 + \alpha)(f(x) - f(x_0))$ and $0 < \alpha \leq 1$.

Another definition of local fractional derivative has been proposed by Kolwankar and Gangal [23] by means of theory on the Cantor space, which is given as follows.

Definition 1.2.7.2.2: Local fractional derivative of order $\alpha(0<\alpha<1)$ of a function $f \in C^0 : \mathbb{R} \to \mathbb{R}$ is defined as:

$$D^\alpha f(x) = \lim_{\zeta \to x} D^\alpha_x (f(\zeta) - f(x)),$$

(1.47)

if the limit exists in $\mathbb{R} \cup \infty$.

If $f(x)$ is differentiable at the point other than $x = x_0$, with non-zero value of the derivative, then it can be approximated locally as:

$$f(x) = f(x_0) + f'(x_0)(x - x_0) + o(x - x_0).$$

(1.48)

So the local fractional derivative of $f(x)$ at $x = x_0$ becomes:

$$D^\alpha(x_0) = \lim_{x \to x_0} \frac{d^\alpha(f(x) - f(x_0))}{d(x - x_0)^\alpha}$$
$$= f'(x_0) \lim_{x \to x_0} \frac{d^\alpha(x - x_0)}{d(x - x_0)^\alpha},$$

(1.49)

Remark 1.2.7.2.1: The following rules are hold [24]:

1. $\dfrac{d^\alpha x^{k\alpha}}{dx^\alpha} = \dfrac{\Gamma(1+k\alpha)}{\Gamma(1+(k-1)\alpha)} x^{(k-1)\alpha}$, and

2. $\dfrac{d^\alpha E_\alpha(kx^\alpha)}{dx^\alpha} = kE_\alpha(kx^\alpha)$, k is a constant.

Remark 1.2.7.2.2: [19–21,24]

1. If $y(x) = (f \circ u)(x)$, where $u(x) = g(x)$, then we have:

$$\frac{d^\alpha y(x)}{dx^\alpha} = f^{(\alpha)}\big(g(x)\big)\big(g^{(1)}(x)\big)^\alpha,$$

(1.50)

when $f^{(\alpha)}\big(g(x)\big)$ and $g^{(1)}(x)$ exist, and

2. If $y(x) = (f \circ u)(x)$, where $u(x) = g(x)$, then we have:

$$\frac{d^\alpha y(x)}{dx^\alpha} = f^{(1)}\big(g(x)\big)g^{(\alpha)}(x),$$

(1.51)

when $f^{(1)}\big(g(x)\big)$ and $g^{(\alpha)}(x)$ exist.

Introduction and Preliminaries of Fractional Calculus

1.3 Conclusion

The purpose of this chapter is to introduce and review some useful definitions and properties of fractional calculus. Mainly, the definitions and properties of the Riemann–Liouville fractional integral and derivative, Caputo fractional derivative, Grünwald–Letnikov fractional derivative, Riesz fractional derivative, modified Riemann–Liouville derivative, and local fractional derivative have been discussed. However, the definitions of the Riemann–Liouville of fractional differentiation played an important role in the development of fractional calculus. The Riemann–Liouville definition is not extensively useful due to the property, as the derivative of constant is not zero. In the field of the fractional differential equations, the Caputo derivative and Riemann–Liouville ones are mostly used. It seems that the former is more welcome since the initial value of fractional differential equation with the Caputo derivative is the same as that of integer differential equation. Also, the Riemann–Liouville fractional derivative is mostly used by mathematicians, but this is not suitable for real world physical problems, as it requires the definition of fractional order initial conditions, which have no physically significant explanation yet. An alternative definition introduced by Caputo has the advantage of defining integer order initial conditions for fractional order differential equations. As a result, on this work, we will use the Caputo fractional derivative by means of Caputo in his work on the theory of viscoelasticity.

Also, the above definitions are non-local and the promising tools for describing memory phenomena as the kernel function of fractional derivative, which define the memory function of the system. It is worth to mention that, the fractional order representations possess long memory characteristics that show the system behavior in more realistic manner. By using the properties of various fractional derivatives and integrals, the fractional differential equations have been solved, which have been presented in the later part of the book.

There are certain special functions that have been also used for defining and analyzing the definitions of fractional calculus. In this chapter, we have also discussed some useful functions, which are useful to describe the mathematical framework for fractional calculus. We have presented the gamma and Mittag–Leffler functions, which turn out to be well established extensions of the factorial and the exponential function. These functions play an important role for practical applications of the fractional calculus. In addition to that, the functions like the Euler psi function, incomplete gamma function, beta function, incomplete beta function, Wright function, Mellin–Ross function, the error function, the hypergeometric functions, viz., Gauss, Kummer and generalized hypergeometric functions, and the H-function have been described in this chapter. Moreover, the motivation, objective, and the organization of book have been also discussed in this chapter.

2

Overview of Numerous Analytical Methods

2.1 Introduction

Numerous physical and engineering problems are modeled by differential equations. In many cases, the solutions of these problems are difficult to obtain due to their non-linear arrangement. Thus, the numerical methods are introduced to find the approximate solutions of non-linear differential equations. Nevertheless, numerical solutions do not explicitly depict the nature of the physical systems and are inadequate in determining the general properties of certain systems of equations.

Due to these major reasons, the semi-analytical and analytical methods have been developed for solving the non-linear partial differential equations (NPDEs). The developments of the semi-analytical and analytical methods have been seen in the end of the twentieth century. With the aid of faster processing calculation techniques, the analytical solutions of semi-analytical and analytical methods have been presented more precisely.

In this chapter, the theory and algorithms of various semi-analytical and analytical methods have been discussed, which tend to give the analytical approximate and exact solutions of non-linear fractional partial differential equations.

2.2 Homotopy Perturbation Method with Adomian Polynomial

The homotopy perturbation method (HPM) is a series expansion method which is used in the solution of NPDEs. HPM was first introduced by He [25] in the year of 1999 and modification of HPM done by Yan [26]. The method has been shown to solve effectively, easily, and accurately a large class of non-linear problems [27–32]. The main idea of the method is based on to construct the homotopy of the governing NPDEs by using the embedding parameter p. By using the proposed theory, the general operator has been split into a linear and a non-linear component. By the using the definition of topology, as p changes from zero to one, the approximate solution approaches the exact solution. The embedding parameter $p \in [0,1]$ and the approximate solution can be expressed as a series solution of the power of p. This series is then substituted into the homotopy equation and solved recursively to obtain the approximate solution for governing equations.

In this section, the basic idea of the homotopy perturbation method has been discussed. For discussing the basic idea, consider the following non-linear differential equation:

$$A(u) - f(r) = 0, \qquad r \in \Omega, \tag{2.1}$$

with the boundary condition:

$$B\left(u, \frac{\partial u}{\partial n}\right) = 0, \qquad r \in \Gamma, \tag{2.2}$$

where A is a general differential operator, B is a boundary operator, $f(r)$ is known as analytical function, Γ is the boundary of domain Ω, and $\partial/\partial n$ denotes differentiation along the normal drawn outwards from Ω.

A can be divided into two parts, which are linear L and non-linear N. Therefore eq. (2.1) can be rewritten as follows:

$$L(u) + N(u) - f(r) = 0. \tag{2.3}$$

We construct a homotopy of eq. (2.1) $v(r,p) : \Omega \times [0,1] \to \mathbb{R}$ which satisfies:

$$H(v,p) = (1-p)[L(v) - L(u_0)] + p\left[A(v) - f(r)\right] = 0, p \in [0,1], r \in \Omega, \tag{2.4}$$

which is equivalent to:

$$H(v,p) = L(v) - L(u_0) + pL(u_0) + p\left[N(v) - f(r)\right] = 0, \tag{2.5}$$

where $p \in [0,1]$ is an embedding parameter and u_0 is an initial approximation of eq. (2.1), which satisfies the boundary conditions. It follows from eqs. (2.4) and (2.5) that:

$$H(v,0) = L(v) - L(u_0) = 0 \text{ and } H(v,1) = A(v) - f(r) = 0. \tag{2.6}$$

In topology, $L(v) - L(u_0)$, $A(v) - f(r)$ are called homotopy. We assume the solution of eq. (2.5) can be written as a power series in p, as following:

$$v = v_0 + pv_1 + p^2 v_2 + \dots . \tag{2.7}$$

The approximate solution of eq. (2.1) can be obtained as:

$$u = \lim_{p \to 1} v = v_0 + v_1 + v_2 + \dots . \tag{2.8}$$

The convergence of the series (2.8) has been discussed in [29,30,33,34].

2.2.1 Adomian Polynomial

It is quite difficult to handle the various non-linear terms of governing equations that appear in homotopy series. So to handle such non-linear terms appearing in the governing equations, the Adomian polynomial has been used.

Overview of Numerous Analytical Methods

An Adomian polynomial [35–37] is defined as the following mathematical expression, which is given as:

$$N(u) = \sum_{n=0}^{\infty} H_n(u_0, u_1, ..., u_n),$$

where H_n is the appropriate Adomian polynomial, which is generated according to an algorithm determined in [35–37]. For non-linear operator $N(u)$, the polynomials can be defined as:

$$H_n(u_0, u_1, ..., u_n) = \frac{1}{n!} \frac{d^n}{dp^n} N\left(\sum_{k=0}^{\infty} p^k u_k(t)\right)\Bigg|_{p=0}, n \geq 0. \tag{2.9}$$

2.3 Homotopy Perturbation Transform Method

The homotopy perturbation transform method (HPTM) [38–40] is a combined form of the Laplace transform method with the homotopy perturbation method. The method is efficient due to its elegant combination of the Laplace transformation for solving NPDEs.

To illustrate the basic idea of the HPTM, we consider a general fractional non-linear non-homogeneous partial differential equation with the initial conditions of the form:

$$_0^C D_t^\alpha u(x,t) + Ru(x,t) + Nu(x,t) = f(x,t), 0 < \alpha < 1, \tag{2.10}$$

with following initial conditions:

$$u(x,0) = h(x), u_t(x,0) = g(x), \tag{2.11}$$

where $_0^C D_t^\alpha u(x,t)$ is the Caputo fractional derivative of the function $u(x,t)$, R is the linear differential operator, N represents the general non-linear differential operator, and $f(x,t)$ is the source term. Taking the Laplace transform on both sides of eq. (2.10), we get:

$$\mathscr{L}\left[_0^C D_t^\alpha u(x,t)\right] + \mathscr{L}\left[Ru(x,t)\right] + \mathscr{L}\left[Nu(x,t)\right] = \mathscr{L}\left[f(x,t)\right]. \tag{2.12}$$

Using the property of the Laplace transform, we have:

$$\mathscr{L}\left[u(x,t)\right] = \frac{h(x)}{s} + \frac{g(x)}{s^2} + \frac{1}{s^\alpha} \mathscr{L}\left[f(x,t)\right] - \frac{1}{s^\alpha} \mathscr{L}\left[Ru(x,t)\right] - \frac{1}{s^\alpha} \mathscr{L}\left[Nu(x,t)\right]. \tag{2.13}$$

Operating with the Laplace inverse on both sides of eq. (2.13) yields:

$$u(x,t) = G(x,t) - \mathscr{L}^{-1}\left[\frac{1}{s^\alpha} \mathscr{L}\left[Ru(x,t) + Nu(x,t)\right]\right]. \tag{2.14}$$

where $G(x,t) = \frac{h(x)}{s} + \frac{g(x)}{s^2} + \frac{1}{s^\alpha}\mathscr{L}[f(x,t)]$ represents the term arising from the source term and the prescribed initial conditions. By the homotopy perturbation method, we will construct the homotopy of eq. (2.14) as:

$$u(x,t) = G(x,t) - p\left(\mathscr{L}^{-1}\left[\frac{1}{s^\alpha}\mathscr{L}\left[Ru(x,t) + Nu(x,t)\right]\right]\right). \tag{2.15}$$

By substituting, $u(x,t) = \sum_{n=0}^{\infty} p^n u_n(x,t)$ in eq. (2.15), and subsequently equating the like powers of p, we can finally obtain the analytical approximate solution $u(x,t)$ in truncated series as:

$$u(x,t) = \lim_{N \to \infty} \sum_{n=0}^{N} u_n(x,t). \tag{2.16}$$

2.4 Modified Homotopy Analysis Method

The homotopy analysis method (HAM) was first introduced by Liao [41] in 1992. The HAM provides convenient way to adjust the convergence region of the solution by convergent control parameter. The method also has the ability to use different base functions to approximate a non-linear problem. Due to the above major advantages over the other perturbation techniques, the homotopy analysis method got popular in the past decade.

In this section, the basic idea of the HAM [41–48] has been discussed. To show the basic idea, let us consider the following differential equation:

$$N\left[u(x,t)\right] = 0, \tag{2.17}$$

where N is a non-linear differential operator, x and t denote independent variables, and $u(x,t)$ is an unknown function. For simplicity, we ignore all boundary or initial conditions, which can be treated in the similar way. By means of the HAM, one first constructs the zero-order deformation equation:

$$(1-p)L[\phi(x,t;p) - u_0(x,t)] = p\hbar N\left[\phi(x,t;p)\right], \tag{2.18}$$

where L is an auxiliary linear operator, $\phi(x,t;p)$ is an unknown function, $u_0(x,t)$ is an initial guess of $u(x,t)$, $\hbar \neq 0$ is an auxiliary parameter, and $p \in [0,1]$ is the embedding parameter. For the sake of convenience, the expression in non-linear operator form has been modified in the HAM. In this modified homotopy analysis method (MHAM), the non-linear term that appeared in expression for non-linear operator form has been expanded using Adomian type of polynomials as $\sum_{n=0}^{\infty} A_n p^n$ [49].

Obviously, when $p = 0$ and $p = 1$, one has:

$$\phi(x,t;0) = u_0(x,t), \phi(x,t;1) = u(x,t). \tag{2.19}$$

Overview of Numerous Analytical Methods 27

respectively. Thus, as p increases from 0 to 1, the solution $\phi(x,t;p)$ varies from the initial guess $u_0(x,t)$ to the solution $u(x,t)$. Expanding $\phi(x,t;p)$ in a Taylor series with respect to the embedding parameter p, one has:

$$\phi(x,t;p) = u_0(x,t) + \sum_{m=1}^{+\infty} p^m u_m(x,t), \tag{2.20}$$

where

$$u_m(x,t) = \frac{1}{m!} \frac{\partial^m \phi(x,t;p)}{\partial p^m}\bigg|_{p=0}. \tag{2.21}$$

The convergence of the series (2.21) depends upon the auxiliary parameter $\hbar$. If it is convergent at $p=1$, we have:

$$u(x,t) = u_0(x,t) + \sum_{m=1}^{+\infty} u_m(x,t), \tag{2.22}$$

which must be one of the solutions of the original non-linear equation, as proved in [41,42].

Differentiating the zeroth-order deformation eq. (2.18) m-times with respect to p, and then setting $p = 0$ and finally dividing them by $m!$, we obtain the following mth-order deformation equation:

$$L\big[u_m(x,t) - \chi_m u_{m-1}(x,t)\big] = \hbar \Re_m(u_0, u_1, ..., u_{m-1}), \tag{2.23}$$

where

$$\Re_m(u_0, u_1, ..., u_{m-1}) = \frac{1}{(m-1)!} \frac{\partial^{m-1} N\big[\varphi(x,t;p)\big]}{\partial p^{m-1}}\bigg|_{p=0}$$

and

$$\chi_m = \begin{cases} 1, & m > 1, \\ 0, & m \le 1. \end{cases} \tag{2.24}$$

It should be noted that $u_m(x,t)$ for $m \ge 1$ is governed by the linear eq. (2.23), which can be solved by symbolic computational software.

2.5 Modified Homotopy Analysis Method with Fourier Transform

In this section, the Fourier transform has been used to deal with Riesz fractional operators, and then the modified homotopy analysis method has been proposed for getting analytical approximate solutions for reduced differential equation. The modified homotopy analysis method with the Fourier transform method (MHAM-FT) is specially proposed for getting analytical approximate solutions for Riesz fractional differential equations.

To show the basic idea, let us consider the following fractional differential equation:

$$N[u(x,t)] = 0, \tag{2.25}$$

where N is a non-linear differential operator containing Riesz fractional derivative, x and t denote independent variables, and $u(x,t)$ is an unknown function. For simplicity, we ignore all boundary or initial conditions, which can be treated in the similar way.

Then applying the Fourier transform and using eq. (1.37) of Lemma 1.2.5.1 in Chapter 1, we can reduce fractional differential eq. (2.25) to the following Fourier transformed differential equation:

$$N[\hat{u}(k,t)] = 0, \tag{2.26}$$

where $\hat{u}(k,t)$ is the Fourier transform of $u(x,t)$.

By means of the HAM, one first constructs the zeroth-order deformation equation of eq. (2.26) as:

$$(1-p)L[\phi(k,t;p) - \hat{u}_0(k,t)] = p\hbar N[\phi(k,t;p)], \tag{2.27}$$

where L is an auxiliary linear operator, $\phi(k,t;p)$ is an unknown function, $\hat{u}_0(k,t)$ is an initial guess of $\hat{u}(k,t)$, $\hbar \neq 0$ is an auxiliary parameter, and $p \in [0,1]$ is the embedding parameter. For the sake of convenience, the expression in non-linear operator form has been modified in the HAM. In this modified homotopy analysis method, the non-linear term that appeared in expression a for non-linear operator form has been expanded using Adomian type of polynomials as $\sum_{n=0}^{\infty} A_n p^n$ [49].

Obviously, when $p = 0$ and $p = 1$, we have:

$$\phi(k,t;0) = \hat{u}_0(k,t), \phi(k,t;1) = \hat{u}(k,t), \tag{2.28}$$

respectively. Thus, as p increases from 0 to 1, the solution $\phi(k,t;p)$ varies from the initial guess $\hat{u}_0(k,t)$ to the solution $\hat{u}(k,t)$. Expanding $\phi(k,t;p)$ in a Taylor series with respect to the embedding parameter p, we have:

$$\phi(k,t;p) = \hat{u}_0(k,t) + \sum_{m=1}^{+\infty} p^m \hat{u}_m(k,t), \tag{2.29}$$

where $\hat{u}_m(k,t) = \dfrac{1}{m!} \dfrac{\partial^m}{\partial p^m} \phi(k,t;p) \Big|_{p=0}$.

The convergence of the series (2.29) depends upon the auxiliary parameter $\hbar$. If it is convergent at $p = 1$, we have:

$$\hat{u}(k,t) = \hat{u}_0(k,t) + \sum_{m=1}^{+\infty} \hat{u}_m(k,t),$$

Overview of Numerous Analytical Methods

29

which must be one of the solutions of the original non-linear equation, as proved in [42].

Differentiating the zeroth-order deformation eq. (2.27) m-times with respect to p, and then setting $p = 0$ and finally dividing them by $m!$, we obtain the following mth-order deformation equation:

$$L\left[\hat{u}_m(k,t) - \chi_m \hat{u}_{m-1}(k,t)\right] = \hbar \Re_m\left(\hat{u}_0, \hat{u}_1, ..., \hat{u}_{m-1}\right), \tag{2.30}$$

where:

$$\Re_m\left(\hat{u}_0, \hat{u}_1, ..., \hat{u}_{m-1}\right) = \frac{1}{(m-1)!} \left.\frac{\partial^{m-1} N\left[\phi(k,t;p)\right]}{\partial p^{m-1}}\right|_{p=0}.$$

and

$$\chi_m = \begin{cases} 1, & m > 1, \\ 0, & m \leq 1. \end{cases} \tag{2.31}$$

It should be noted that $\hat{u}_m(k,t)$ for $m \geq 1$ is governed by the linear eq. (2.30), which can be solved by symbolic computational software. Then, by applying inverse the Fourier transformation, we can get $u_m(x,t)$.

2.6 Fractional Sub-Equation Method

In this section, the algorithm of fractional sub-equation method [50–54] has been discussed. This method has been used to get the exact solutions of fractional differential equations. The main steps of this method are described as follows:

Step 1: Suppose that a non-linear fractional partial differential equation (FPDE), say in three independent variables x, y, and t is given by:

$$P(u, u_x, u_y, u_t, {}_0D_x^\alpha u, {}_0D_y^\alpha u, {}_0D_t^\alpha u, ...) = 0, 0 < \alpha \leq 1, \tag{2.32}$$

where $u = u(x,y,t)$ is an unknown function. Here, P is a polynomial in u, and its various partial derivatives in which the highest order derivatives and non-linear terms are involved.

Step 2: By using the traveling wave transformation:

$$u(x,y,t) = \Phi(\xi), \xi = k_1 x + k_2 y + vt, \tag{2.33}$$

where k_1, k_2 and v are constants to be determined later, the FPDE (2.32) is reduced to the following non-linear fractional ordinary differential equation (ODE) for $u(x,y,t) = \Phi(\xi)$:

$$P(\Phi, k_1\Phi', k_2\Phi', v\Phi', k_1^\alpha D_\xi^\alpha \Phi, k_2^\alpha D_\xi^\alpha \Phi, v^\alpha D_\xi^\alpha \Phi, ...) = 0. \tag{2.34}$$

Step 3: We suppose that eq. (2.34) has the following solution:

$$\Phi(\xi) = \sum_{i=0}^{n} a_i \phi^i, \tag{2.35}$$

where $a_i (i = 0,1,2,...,n)$ are constants to be determined later, n is a positive integer determined by balancing the highest order derivative term and non-linear term in eq. (2.34), and $\phi = \phi(\xi)$ satisfies the following fractional Riccati equation:

$$D_\xi^\alpha \phi = \sigma + \phi^2, \tag{2.36}$$

where σ is a constant. By using the generalized Exp-function method via Mittag–Leffler functions, Zhang et al. [55] first obtained the following solutions of fractional Riccati eq. (2.36):

$$\phi(\xi) = \begin{cases} -\sqrt{-\sigma}\,\tanh_\alpha\left(\sqrt{-\sigma}\,\xi\right), & \sigma < 0, \\[2mm] -\sqrt{-\sigma}\,\coth_\alpha\left(\sqrt{-\sigma}\,\xi\right), & \sigma < 0, \\[2mm] \sqrt{\sigma}\,\tan_\alpha\left(\sqrt{\sigma}\,\xi\right), & \sigma > 0, \\[2mm] -\sqrt{\sigma}\,\cot_\alpha\left(\sqrt{\sigma}\,\xi\right), & \sigma > 0, \\[2mm] -\dfrac{\Gamma(1+\alpha)}{\xi^\alpha + \omega}, \quad \omega = \text{constant}, & \sigma = 0, \end{cases} \tag{2.37}$$

where the generalized hyperbolic and trigonometric functions are defined as:

$$\sinh_\alpha(x) = \frac{E_\alpha(x^\alpha) - E_\alpha(-x^\alpha)}{2}, \quad \cosh_\alpha(x) = \frac{E_\alpha(x^\alpha) + E_\alpha(-x^\alpha)}{2}, \quad \tanh_\alpha(x) = \frac{\sinh_\alpha(x)}{\cosh_\alpha(x)},$$

$$\coth_\alpha(x) = \frac{\cosh_\alpha(x)}{\sinh_\alpha(x)}, \quad \sin_\alpha(x) = \frac{E_\alpha(ix^\alpha) - E_\alpha(-ix^\alpha)}{2i}, \quad \cos_\alpha(x) = \frac{E_\alpha(ix^\alpha) + E_\alpha(-ix^\alpha)}{2},$$

$$\tan_\alpha(x) = \frac{\sin_\alpha(x)}{\cos_\alpha(x)},$$

where $i = \sqrt{-1}$ and $E_\alpha(z)$ denotes the Mittag–Leffler function, given as:

$$E_\alpha(z) = \sum_{k=0}^{\infty} \frac{z^k}{\Gamma(1+k\alpha)}. \tag{2.38}$$

Step 4: Substituting eq. (2.35) along with eq. (2.36) into eq. (2.34), and using Remark 1.2.7.2.2 of Chapter 1, we can get a polynomial in $\phi^m(\xi)$ ($m = 0,1,2,...$). Equating each coefficient of $\phi^m(\xi)$ ($m = 0,1,2,...$) to zero, yields a set of non-linear algebraic equations for $a_i (i = 0,1,2,...,n)$, σ, k_1, k_2, and v.

Overview of Numerous Analytical Methods

31

Step 5: Solving the algebraic equations system in step 4, substituting these constants a_i $(i=0,1,2,...,n)$ and σ, solutions of eq. (2.36) given in eq. (2.37) into eq. (2.35), we can obtain the explicit solutions of eq. (2.32) immediately.

2.7 Improved Fractional Sub-Equation Method

In this section, the algorithm of the improved fractional sub-equation method [56] has been proposed, which intends to solve the fractional differential equations. The main steps of this method are described as follows:

Step 1: Suppose that a non-linear FPDE, say in four independent variables x, y, z, and t is given by:

$$P(u, u_x, u_y, u_z, u_t, {}_0D_x^\alpha u, {}_0D_y^\alpha u, {}_0D_z^\alpha u, {}_0D_t^\alpha u, ...) = 0, 0 < \alpha \leq 1, \tag{2.39}$$

where $u = u(x,y,z,t)$ is an unknown function. Also, P is a polynomial in u and its various partial derivatives in which the highest order derivatives and non-linear terms are involved.

Step 2: By using the traveling wave transformation:

$$u(x,y,z,t) = \Phi(\xi), \xi = k_1 x + k_2 y + k_3 z + vt, \tag{2.40}$$

where k_1, k_2, k_3, and v are constants to be determined later, the FPDE (2.39) is reduced to the following non-linear fractional ODE for $u(x,y,z,t) = \Phi(\xi)$:

$$P(\Phi, k_1\Phi', k_2\Phi', k_3\Phi', v\Phi', k_1^\alpha D_\xi^\alpha\Phi, k_2^\alpha D_\xi^\alpha\Phi, k_3^\alpha D_\xi^\alpha\Phi, v^\alpha D_\xi^\alpha\Phi, ...) = 0. \tag{2.41}$$

Step 3: Suppose that eq. (2.41) has the following solution:

$$\Phi(\xi) = a_0 + \sum_{i=1}^{n} a_i \left(\frac{D_\xi^\alpha \phi}{\phi} \right)^i, \tag{2.42}$$

where a_i $(i=0,1,2,...,n)$ are constants to be determined later, n is a positive integer determined by balancing the highest order derivative term and non-linear term in eq. (2.41), and $\phi = \phi(\xi)$ satisfies the following fractional Riccati equation:

$$D_\xi^\alpha \phi = \sigma + \phi^2, \tag{2.43}$$

where σ is a constant.

Step 4: Substituting eq. (2.42) along with eq. (2.43) into eq. (2.41) and using Remark 1.2.7.2.2 of Chapter 1, we can get a polynomial in $\left(\frac{D_\xi^\alpha \phi}{\phi} \right)^i$ $(i=0,1,2,...)$. Equating each coefficient of $\left(\frac{D_\xi^\alpha \phi}{\phi} \right)^i$ $(i=0,1,2,...)$ to zero, yields a set of non-linear algebraic equations for a_i $(i=0,1,2,...,n)$, σ, k_1, k_2, k_3, and v.

Step 5: Solving the algebraic equations system in step 4 and substituting these constants a_i $(i = 0, 1, 2, ..., n)$, σ and solutions of eq. (2.43) given in eq. (2.37) into eq. (2.42), we can obtain the explicit solutions of eq. (2.39) immediately.

2.8 (G'/G)-Expansion Method

In this section, we proposed the algorithm of the (G'/G)-expansion method [57–59] via fractional complex transformation [60–64]. The main steps of this method are described as follows:

Step 1: Suppose that a non-linear FPDE, say in two independent variables x and t is given by:

$$P(u, u_x, u_{xx}, u_{xxx}, ..._0 D_t^\alpha u, ...) = 0, 0 < \alpha \le 1, \tag{2.44}$$

where $u = u(x, t)$ is an unknown function. Also, P is a polynomial in u and its various partial derivatives in which the highest order derivatives and non-linear terms are involved.

Step 2: By using the complex transformation [60–64], we have:

$$u(x, t) = \Phi(\xi), \xi = kx + \frac{vt^\alpha}{\Gamma(\alpha + 1)}, \tag{2.45}$$

where k and v are constants to be determined later on.

By using the chain rule of local fractional derivative [61,63], we have:

$$D_t^\alpha u = \sigma_t \Phi_\xi D_t^\alpha \xi, \tag{2.46}$$

$$D_x^\alpha u = \sigma_x \Phi_\xi D_x^\alpha \xi, \tag{2.47}$$

where σ_t and σ_x are the fractal indices [61,62]. Without loss of generality, we can take $\sigma_t = \sigma_x = \kappa$, where κ is a constant.

The FPDE (2.44) is reduced to the following non-linear ODE for $u(x, t) = \Phi(\xi)$:

$$P(\Phi, k\Phi', k^2\Phi'', k^3\Phi''', ...v\Phi', ...) = 0. \tag{2.48}$$

Step 3: Suppose that the solution of eq. (2.48) can be expressed by a polynomial in (G'/G) as follows:

$$\Phi(\xi) = a_0 + \sum_{i=1}^{n} a_i \left(\frac{G'}{G}\right)^i, \tag{2.49}$$

where $G = G(\xi)$ satisfies the second order ODE in the form:

$$G'' + \lambda G' + \mu G = 0, \tag{2.50}$$

Overview of Numerous Analytical Methods

the integer n can be determined by balancing the highest order derivative term and non-linear term appearing in eq. (2.48). Further, the eq. (2.50) can be changed into:

$$\frac{d}{d\xi}\left(\frac{G'}{G}\right) = -\left(\frac{G'}{G}\right)^2 - \lambda\left(\frac{G'}{G}\right) - \mu.$$

(2.51)

By the generalized solutions of eq. (2.50) we have:

$$\left(\frac{G'}{G}\right) = \begin{cases} \dfrac{\sqrt{\lambda^2 - 4\mu}}{2}\left(\dfrac{C_1 \sinh\left(\dfrac{\sqrt{\lambda^2 - 4\mu}}{2}\xi\right) + C_2 \cosh\left(\dfrac{\sqrt{\lambda^2 - 4\mu}}{2}\xi\right)}{C_1 \cosh\left(\dfrac{\sqrt{\lambda^2 - 4\mu}}{2}\xi\right) + C_2 \sinh\left(\dfrac{\sqrt{\lambda^2 - 4\mu}}{2}\xi\right)}\right) - \dfrac{\lambda}{2}, & \lambda^2 - 4\mu > 0 \\[3em] \dfrac{\sqrt{-\lambda^2 + 4\mu}}{2}\left(\dfrac{-C_1 \sin\left(\dfrac{\sqrt{-\lambda^2 + 4\mu}}{2}\xi\right) + C_2 \cos\left(\dfrac{\sqrt{-\lambda^2 + 4\mu}}{2}\xi\right)}{C_1 \cos\left(\dfrac{\sqrt{-\lambda^2 + 4\mu}}{2}\xi\right) + C_2 \sin\left(\dfrac{\sqrt{-\lambda^2 + 4\mu}}{2}\xi\right)}\right) - \dfrac{\lambda}{2}, & \lambda^2 - 4\mu < 0 \\[3em] \left(\dfrac{C_2}{C_1 + C_2\xi}\right) - \dfrac{\lambda}{2}, & \lambda^2 - 4\mu = 0 \end{cases}$$

(2.52)

where C_1 and C_2 are arbitrary constants.

Step 4: By substituting eq. (2.49) into eq. (2.48) and using eqs. (2.50) and (2.51), collecting all terms with the same degree of (G'/G) together, the eq. (2.48) is converted into another polynomial in (G'/G). Equating each coefficient of this polynomial to zero yields a set of algebraic equations for a_i $(i = 0,1,2,...,n)$, λ, k, v, and μ.

Step 5: Solving the algebraic equations system in step 4, substituting these constants a_i $(i = 0, 1, 2,...,n)$, λ, k, v, and μ, solutions of eq. (2.51) into eq. (2.49), we can obtain the explicit solutions of eq. (2.44) immediately.

2.9 Improved (G'/G)-Expansion Method

In this part, we deal with the improved (G'/G)-expansion method [57,65]. The main steps of this method are described as follows:

Step 1: Suppose that a non-linear FPDE, say in two independent variables x and t is given by

$$P(u, u_x, u_{xx}, u_{xxx}, ..._0 D_t^\alpha u, ...) = 0, 0 < \alpha \le 1,$$

(2.53)

where $u = u(x,t)$ is an unknown function. Also, P is a polynomial in u and its various partial derivatives in which the highest order derivatives and non-linear terms are involved.

Step 2: By using the fractional complex transformation [60–64], we have:

$$u(x,t) = \Phi(\xi), \xi = kx + \frac{vt^\alpha}{\Gamma(\alpha+1)}, \tag{2.54}$$

where k and v are constants to be determined later.

By using the chain rule [61,63], we have:

$$D_t^\alpha u = \sigma_t \Phi_\xi D_t^\alpha \xi,$$

$$D_x^\alpha u = \sigma_x \Phi_\xi D_x^\alpha \xi,$$

where σ_t and σ_x are the fractal indices [61,62]. Without loss of generality, we can take $\sigma_t = \sigma_x = \kappa$, where κ is a constant.

The FPDE (2.53) is reduced to the following non-linear ODE for $u(x,t) = \Phi(\xi)$:

$$P(\Phi, k\Phi', k^2\Phi'', k^3\Phi''', \dots v\Phi', \dots) = 0. \tag{2.55}$$

Step 3: Suppose that the solution of eq. (2.55) can be expressed by a polynomial in $F(\xi)$ as follows:

$$\Phi(\xi) = \sum_{i=1}^{n} a_i F^i(\xi). \tag{2.56}$$

The integer n can be determined by balancing the highest order derivative term and non-linear term appearing in eq. (2.55).

Step 4: Suppose that:

$$F(\xi) = \frac{G'(\xi)}{G(\xi)}, \tag{2.57}$$

where $G = G(\xi)$ satisfies the second order ODE in the form:

$$GG'' = AG^2 + BGG' + C(G')^2, \tag{2.58}$$

where the prime denotes derivative with respect to ξ. Here, A, B, and C are real parameters.

Further, the eq. (2.58) reduces into the following Riccati equation as:

$$\frac{d}{d\xi}\left(\frac{G'}{G}\right) = A + B\left(\frac{G'}{G}\right) + (C-1)\left(\frac{G'}{G}\right)^2. \tag{2.59}$$

Overview of Numerous Analytical Methods

Step 5: By the generalized solutions of eq. (2.59), we have:

Case I: if $B \neq 0$ and $\Delta = B^2 + 4A - 4AC \geq 0$, then:

$$F(\xi) = \frac{B}{2(1-C)} + \frac{B\sqrt{\Delta}}{2(1-C)}\left(\frac{C_1\exp\left(\frac{\sqrt{\Delta}}{2}\xi\right) + C_2\exp\left(\frac{-\sqrt{\Delta}}{2}\xi\right)}{C_1\exp\left(\frac{\sqrt{\Delta}}{2}\xi\right) - C_2\exp\left(\frac{-\sqrt{\Delta}}{2}\xi\right)}\right). \tag{2.60}$$

Case II: if $B \neq 0$ and $\Delta = B^2 + 4A - 4AC < 0$, then:

$$F(\xi) = \frac{B}{2(1-C)} + \frac{B\sqrt{-\Delta}}{2(1-C)}\left(\frac{iC_1\cos\left(\frac{\sqrt{-\Delta}}{2}\xi\right) - C_2\sin\left(\frac{\sqrt{-\Delta}}{2}\xi\right)}{iC_1\sin\left(\frac{\sqrt{-\Delta}}{2}\xi\right) + C_2\cos\left(\frac{\sqrt{-\Delta}}{2}\xi\right)}\right). \tag{2.61}$$

Case III: if $B = 0$ and $\Delta = A(1-C) \geq 0$, then:

$$F(\xi) = \frac{\sqrt{\Delta}}{(1-C)}\left(\frac{C_1\cos\left(\sqrt{\Delta}\xi\right) + C_2\sin\left(\sqrt{\Delta}\xi\right)}{C_1\sin\left(\sqrt{\Delta}\xi\right) - C_2\cos\left(\sqrt{\Delta}\xi\right)}\right). \tag{2.62}$$

Case IV: if $B = 0$ and $\Delta = A(1-C) < 0$, then:

$$F(\xi) = \frac{\sqrt{-\Delta}}{(1-C)}\left(\frac{iC_1\cosh\left(\sqrt{-\Delta}\xi\right) - C_2\sinh\left(\sqrt{-\Delta}\xi\right)}{iC_1\sinh\left(\sqrt{-\Delta}\xi\right) - C_2\cosh\left(\sqrt{-\Delta}\xi\right)}\right), \tag{2.63}$$

where $\xi = kx + \dfrac{vt^\alpha}{\Gamma(\alpha+1)}$ and A, B, C, C_1, C_2 are real parameters.

Step 6: By substituting eq. (2.56) into eq. (2.55) and using eq. (2.59), then after collecting all terms with the same order of (G'/G) together, the eq. (2.55) is converted into another polynomial in (G'/G). Equating each coefficient of this polynomial to zero yields a set of algebraic equations for a_i $(i = 0, 1, 2, \ldots, n)$, A, B, C, k, and v.

Step 7: Solving the algebraic equations system in step 6 and subsequently substituting these constants a_i $(i = 0, 1, 2, \ldots, n)$, A, B, C, k, and v, and also solutions of eq. (2.59) into eq. (2.56) along with eqs. (2.60) through (2.63), we can obtain the explicit solutions of eq. (2.53) immediately.

2.10 Proposed Tanh Method for FDE

In this section, the basic idea of the proposed tanh method [66–70] for solving fractional differential equations has been discussed. The fundamental steps of the proposed method are illustrated as follows:

Step 1: We have considered the most general form of non-linear fractional partial differential equations with two independent variables x and t is given by:

$$G(u, u_x, u_{xx}, u_{xxx}, \ldots {}_0D_t^\alpha u, \ldots) = 0, 0 < \alpha \leq 1,$$ (2.64)

where G is a polynomial in $u(x,t)$. It may be noted that in eq. (2.64), some non-linear terms with higher order partial derivatives are included.

Step 2: Let us consider:

$$u(x,t) = \Phi(\xi), \xi = c\left(x - \frac{vt^\alpha}{\Gamma(\alpha + 1)}\right),$$ (2.65)

be a fractional complex transformation [60–64], which can be used for reducing eq. (2.64) into a non-linear ordinary differential equation. Here c and v are arbitrary constants.

By using the chain rule, eq. (1.51) of Chapter 1 [61,63], we have:

$$D_t^\alpha u = \sigma_t \Phi_\xi D_t^\alpha \xi,$$

where σ_t is the fractal indices [61,62]. Let us assume without loss of generality that $\sigma_t = k$, where k is an arbitrary constant.

By applying eq. (2.65), the eq. (2.64) can be written as:

$$G(\Phi, c\Phi', c^2\Phi'', c^3\Phi''', \ldots, -cv\Phi', \ldots) = 0.$$ (2.66)

Step 3: By the proposed method, the solution of eq. (2.66) can be written as follows:

$$\Phi(\xi) = a_0 + \sum_{i=1}^{n} a_i Y^i.$$ (2.67)

Using the homogenous balance principle, equating non-linear term and highest order derivative term of eq. (2.66), the value of n can be determined.

By the tanh method, let $Y = \tanh(\xi)$. So by using chain rule, we have the derivatives of $\Phi(\xi)$, which are given as follows:

$$\frac{d\Phi}{d\xi} \to (1 - Y^2)\frac{d\Phi}{dY},$$ (2.68)

$$\frac{d^2\Phi}{d\xi^2} \to (1 - Y^2)\left(-2Y\frac{d\Phi}{dY} + (1 - Y^2)\frac{d^2\Phi}{dY^2}\right).$$

$$\frac{d^3\Phi}{d\xi^3} \to \left(1-Y^2\right)^3 \frac{d^3\Phi}{dY^3} - 6Y\left(1-Y^2\right)^2 \frac{d^2\Phi}{dY^2} + 2Y\left(1-Y^2\right)\left(3Y^2-1\right)\frac{d\Phi}{dY},$$

$$\frac{d^4\Phi}{d\xi^4} \to -8Y\left(1-Y^2\right)\left(3Y^2-2\right)\frac{d\Phi}{dY} + 4\left(1-Y^2\right)^2\left(9Y^2-2\right)\frac{d^2\Phi}{dY^2},$$

$$-12Y\left(1-Y^2\right)^3 \frac{d^3\Phi}{dY^3} + \left(1-Y^2\right)^4 \frac{d^4\Phi}{dY^4}.$$

Similarly the higher order derivatives can be found.

Step 4: Then by substituting eq. (2.67) into eq. (2.66) and making use of eq. (2.68), followed by way of collecting all terms with the same degree of Y^i $(i=0,1,2,...)$ together, the eq. (2.66) is transformed into an alternate polynomial in Y^i $(i=0,1,2,...)$]. Equating every coefficient of this polynomial to zero, we will get a set of algebraic equations for a_i $(i=0,1,2,...,n)$, v, and c.

Step 5: Solving the obtained algebraic systems in step 4, and in this means substituting these constants a_i $(i=0,1,2,...,n)$, v, and c into eq. (2.67), we can get the explicit solutions of eq. (2.64) immediately.

2.11 Modified Kudryashov Method

In this section, the algorithm of the modified Kudryashov method [71,72] has been proposed, which helps to get the exact analytical solutions of some FPDEs. The fundamental steps of the proposed method are illustrated as follows:

Step 1: We have considered the most general form of non-linear fractional partial differential equations with two independent variables x and t, given by:

$$H(u, u_x, u_{xx}, u_{xxx}, ..._0 D^\alpha u, ...) = 0, 0 < \alpha \le 1, \tag{2.69}$$

where H is a polynomial in $u(x,t)$. It may be noted that in eq. (2.69) some non-linear terms with higher order partial derivatives are included.

Step 2: Let us consider:

$$u(x,t) = \Psi(\zeta), \zeta = lx + \frac{\gamma t^\alpha}{\Gamma(\alpha+1)}, \tag{2.70}$$

be a fractional complex transformation [60–64], which can be used for reducing eq. (2.69) into a non-linear ordinary differential equation. Here l and γ are arbitrary constants.

By using the chain rule, eq. (1.51) of Chapter 1 [61,63], we have:

$$D_t^\alpha u = \sigma_t \Psi_\xi D_t^\alpha \zeta,$$

where σ_t is the fractal index [61,62]. Without loss of generality, let us assume that $\sigma_t = k$, where k is an arbitrary constant.

By applying complex transformation of eq. (2.70), the eq. (2.69) can be written as:

$$H(\Psi, l\Psi', l^2\Psi'', l^3\Psi''', ..., \gamma\Psi', ...) = 0. \tag{2.71}$$

Step 3: Using the modified Kudryashov method, the solution of eq. (2.71) can be written as follows:

$$\Psi(\zeta) = \sum_{j=0}^{m} b_i Q^i(\zeta), \tag{2.72}$$

where b_i $(i = 0, 1, 2, ..., m)$ are arbitrary constants to be determined later, such that $b_m \neq 0$, while $Q(\zeta)$ has the following form:

$$Q(\zeta) = \frac{1}{1 \pm a^\zeta}, \tag{2.73}$$

where the eq. (2.73) satisfies the following first order differential equation, which is given as:

$$Q_\zeta(\zeta) = Q(\zeta)(Q(\zeta) - 1)\ln a. \tag{2.74}$$

Step 4: By the proposed modified Kudryashov method, we substitute $\Psi = \zeta^{-p}$ in all terms of eq. (2.71) for determining the highest order singularity. Then the degree of all terms of eq. (2.71) has been taken into study and consequently the two or more terms of lower degree are chosen. The maximum value of p is known as the pole, and it is denoted as m. If m is an integer, then the method only can be implemented, otherwise, if m is a non-integer, the above eq. (2.71) may be transferred and the above procedure is to be repeated.

Step 5: The required number of derivatives of $\Psi(\zeta)$ with respect to ζ can be calculated by using any mathematical software.

Step 6: Then, by substituting eq. (2.72) into eq. (2.71), in the case of the proposed modified Kudryashov method, the eq. (2.71) can be written as following form:

$$\Psi[Q(\zeta)] = 0, \tag{2.75}$$

where $\Psi[Q(\zeta)]$ is polynomial in $Q(\zeta)$. Then, by collecting all terms with the same degree of $Q(\zeta)$ together and equating every coefficient of this polynomial to zero, we will get a set of algebraic equations for b_i $(i = 0, 1, 2, ..., m)$, l and γ.

Step 7: Solving the obtained algebraic systems in step 6, and in this means substituting these constants b_i $(i = 0, 1, 2, ..., m)$, l, and γ into resulting equation system, we can get the explicit solutions of eq. (2.69) immediately. The obtained solutions may involve the symmetrical hyperbolic Fibonacci functions [73,74]. We have used following symmetrical Fibonacci cosine, sine, tangent, and cotangent, which are listed below:

$$sFs(\zeta) = \frac{a^\zeta - a^{-\zeta}}{\sqrt{5}}, \, cFs(\zeta) = \frac{a^\zeta + a^{-\zeta}}{\sqrt{5}}, \, \tan Fs(\zeta) = \frac{a^\zeta - a^{-\zeta}}{a^\zeta + a^{-\zeta}}, \, \cot Fs(\zeta) = \frac{a^\zeta + a^{-\zeta}}{a^\zeta - a^{-\zeta}}.$$

Overview of Numerous Analytical Methods 39

2.12 Proposed New Method

In the present section, the algorithm of a new method has been proposed, which is newly established by the author. The steps of the proposed new method are described as follows:

Step 1: The non-linear coupled time-fractional partial differential equation with two independent variables x and t is considered here as in the following form:

$$F(u,u_x,u_{xx},u_{xxx},...D_t^\alpha u,v,v_x,v_{xx},v_{xxx},...D_t^\alpha v...)=0,$$

$$P(u,u_x,u_{xx},u_{xxx},...D_t^\alpha u,v,v_x,v_{xx},v_{xxx},...D_t^\alpha v...)=0, 0<\alpha\le 1. \tag{2.76}$$

where $u(x,t)$ and $v(x,t)$ are unknown function. Also, F and P are the functions in $u(x,t)$ and $v(x,t)$ along with their highest order partial derivatives and non-linear terms of $u(x,t)$ and $v(x,t)$, respectively.

Step 2: The travelling wave solution of eq. (2.76) is considered here with the help of fractional complex transformation [60–64], which is given by:

$$u(x,t)=\Phi(\xi)e^{i\eta},v(x,t)=\Psi(\xi),\xi=cx+\frac{\gamma t^\alpha}{\Gamma(\alpha+1)},\eta=kx+\frac{rt^\alpha}{\Gamma(\alpha+1)}, \tag{2.77}$$

where c,γ,r,k are constants, which are determined later.

By using the chain rule [61,63], we have:

$$D_t^\alpha u=\sigma_t\Phi_\xi D_t^\alpha\xi$$

$$D_t^\alpha v=\sigma_t\Psi_\xi D_t^\alpha\xi$$

where σ_t is the fractal index [61,62]. Without loss of generality, we can take $\sigma_t=\kappa$, where κ is a constant.

Using eq. (2.77), the FPDE (2.76) is reduced to the following non-linear ODE:

$$F(\Phi e^{i\eta},ic\Phi'e^{i\eta},-c^2\Phi''e^{i\eta},ic^3\Phi'''e^{i\eta},..,\gamma i\Phi'e^{i\eta},\Psi,c\Psi',c^2\Psi'',c^3\Psi'''...,\gamma\Psi')=0,$$

$$P(\Phi e^{i\eta},ic\Phi'e^{i\eta},-c^2\Phi''e^{i\eta},ic^3\Phi'''e^{i\eta},..,\gamma i\Phi'e^{i\eta},\Psi,c\Psi',c^2\Psi'',c^3\Psi'''...,\gamma\Psi')=0. \tag{2.78}$$

Step 3: Here, the exact solution of eq. (2.76) is to be assumed in the polynomial $\phi(\xi)$ of the following form [75]:

$$\Phi(\xi)=a_0+\sum_{i=1}^n a_i\phi^i(\xi),\Psi(\xi)=b_0+\sum_{i=1}^m b_i\phi^i(\xi), \tag{2.79}$$

where $\phi(\xi)$ is the logistic function (the sigmoid function), which is determined by the following formula:

$$\phi(\xi) = \frac{e^{\xi}}{1+e^{\xi}}.$$

This function can be used for finding exact solutions of non-linear differential equations [76–78]. It is also observed that the logistic function $\phi(\xi)$ is the solution of the first order differential equation called the Riccati equation that means general solution of the Riccati equation can be expressed via the logistic function [76–78], which is given as:

$$\phi_{\xi} = \phi - \phi^2. \tag{2.80}$$

Step 4: According to the proposed method, we substitute $\Phi = \xi^{-p}$ and $\Psi = \xi^{-q}$ in all terms of eq. (2.78) for determining the highest order singularity. Then the degree of all terms of eq. (2.78) has been taken in to study and, consequently, the two or more terms of lower degree are chosen. The maximum value of p and q are known as the pole and denoted as n and m, respectively. For integer values of n and m, this proposed method only can be implemented. However, if n and m are non-integer, the above eq. (2.78) can be transferred, and then the above procedure can be repeated.

Step 5: The derivatives of the function $\Phi(\xi)$ and $\Psi(\xi)$ can be calculated by using eq. (2.80). Some derivatives of $\Phi(\xi)$ and $\Psi(\xi)$ are presented as follows:

$$\Phi'(\xi) = \sum_{i=1}^{n} a_i i (1-\phi)\phi^i;$$

$$\Psi'(\xi) = \sum_{i=1}^{m} b_i i (1-\phi)\phi^i;$$

$$\Phi''(\xi) = \sum_{i=1}^{n} a_i i [\phi + i(1-\phi)](1-\phi)\phi^i; \tag{2.81}$$

$$\Psi'(\xi) = \sum_{i=1}^{m} b_i i [\phi + i(1-\phi)](1-\phi)\phi^i;$$

$$\Phi'''(\xi) = \sum_{i=1}^{n} a_i i (i+1)(1-\phi)^2\phi^{i+1} + \sum_{i=1}^{n} a_i i [\phi + i(1-\phi)](1-\phi)\phi^{i+1};$$

$$\Psi'''(\xi) = \sum_{i=1}^{m} b_i i (i+1)(1-\phi)^2\phi^{i+1} + \sum_{i=1}^{m} b_i i [\phi + i(1-\phi)](1-\phi)\phi^{i+1}.$$

Overview of Numerous Analytical Methods 41

Step 6: Substituting eq. (2.81) into eq. (2.78) and equating the coefficients of ϕ^i ($i = 0,1,2,...$) to zero, we obtain the set of algebraic equations. By solving the obtained algebraic equations, we can get the unknowns a_i ($i = 0,1,2,...,n$), b_i ($i = 0,1,2,...,m$), and other constants. Then putting the all obtained unknowns in eq. (2.79), we get the required exact solutions for eq. (2.76) immediately.

2.13 Jacobi Elliptic Function Method

In the present section, the algorithm of the Jacobi elliptic function method [79–85] has been presented. The major steps of the proposed method are described as follows:

Step 1: The non-linear-coupled time-fractional partial differential equation with two independent variables x and t is considered here as in the following form:

$$F(u, u_x, u_{xx}, u_{xxx}, ...D_t^\alpha u, ...) = 0, 0 < \alpha \leq 1, \tag{2.82}$$

where $u(x,t)$ is an unknown function. Also, F is the function in $u(x,t)$ along with their highest order partial derivatives and non-linear terms of $u(x,t)$.

Step 2: The exact solution of eq. (2.82) is considered here with the help of fractional complex transformation [60–64], which is given by:

$$u(x,t) = \Phi(\zeta), \zeta = kx + \frac{ct^\alpha}{\Gamma(\alpha+1)}, \tag{2.83}$$

where c and k are constants, which are to be determined later.

By using the chain rule [61,63], we have:

$$D_t^\alpha u = \sigma_t \Phi_\zeta D_t^\alpha \zeta,$$

where σ_t is the fractal index [61,62], without loss of generality, we can take $\sigma_t = \kappa$, where κ is a constant.

Using eq. (2.83), the FPDE (2.82) is reduced to the following non-linear ODEs

$$F(\Phi, k\Phi', k^2\Phi'', k^3\Phi''', .., c\Phi') = 0. \tag{2.84}$$

Step 3: Here, the exact solutions of eq. (2.82) are assumed in the polynomial $\phi(\zeta)$ as follows:

$$\Psi(\zeta) = A_0 + \sum_{i=1}^{N} \left(A_i \phi^i(\zeta) + B_i \phi^{-i}(\zeta) \right), \tag{2.85}$$

where $\phi(\zeta)$ also satisfies following elliptic equation:

$$\phi_\zeta^2(\zeta) = r + p\phi^2(\zeta) + q\phi^4(\zeta), \tag{2.86}$$

where $r, p, q, A_i, (i = 1, 2, ..., N)$, and $B_i (i = 1, 2, ..., N)$ are the constants, those can be evaluated later on.

Now, the solutions of eq. (2.86) can be expressed in the following form [80].

It is noted that, for simplicity, some other Jacobi elliptic function solutions of eq. (2.86) have been omitted here.

In Table 2.1, $sn\zeta = sn(\zeta, m^2)$, $cn\zeta = cn(\zeta, m^2)$, $dn\zeta = dn(\zeta, m^2)$, $ns\zeta = ns(\zeta, m^2)$ $cs\zeta = cs(\zeta, m^2)$, $ds\zeta = ds(\zeta, m^2)$, $sc\zeta = sc(\zeta, m^2)$, $sd\zeta = sd(\zeta, m^2)$, and $nc\zeta = nc(\zeta, m^2)$ are Jacobi elliptic functions and m denotes the modulus of the Jacobi elliptic functions with $0 < m < 1$.

When $m \to 1$, the above functions degenerate into hyperbolic functions, which are defined as follows:

$$sn\zeta \to \tanh\zeta, cn\zeta \to \operatorname{sech}\zeta, dn\zeta \to \operatorname{sech}\zeta, ns\zeta \to$$

$$\coth\zeta, cs\zeta \to \operatorname{csch}\zeta, ds\zeta \to \operatorname{csch}\zeta, sc\zeta \to \sinh\zeta, sd\zeta \to$$

$$\sinh\zeta, \text{ and, } nc\zeta \to \cosh\zeta.$$

When $m \to 0$, the above functions degenerate into trigonometric functions, which are defined as follows:

$$sn\zeta \to \sin\zeta, cn\zeta \to \cos\zeta, dn\zeta \to 1, ns\zeta \to \csc\zeta \, cs\zeta \to$$

$$\cot\zeta, ds\zeta \to \csc\zeta, sc\zeta \to \tan\zeta, sd\zeta \to \sin\zeta, \text{ and }, nc\zeta \to \sec\zeta.$$

The above presented Jacobi functions also satisfy the following relations:

$$sn^2\zeta + cn^2\zeta = 1, dn^2\zeta + m^2 sn^2\zeta = 1.$$

TABLE 2.1

Families of Jacobi Elliptic Function Solutions of Elliptic Equation (2.86)

No.	r	p	q	$\phi(\zeta)$
1	1	$-(1+m^2)$	m^2	$sn\zeta$
2	$1-m^2$	$2m^2-1$	$-m^2$	$cn\zeta$
3	m^2	$-(1+m^2)$	1	$ns\zeta = (sn\zeta)^{-1}$
4	$-m^2$	$2m^2-1$	$1-m^2$	$nc\zeta = (cn\zeta)^{-1}$
5	$\dfrac{1}{4}$	$\dfrac{1-2m^2}{2}$	$\dfrac{1}{4}$	$ns\zeta \pm cs\zeta$
6	$\dfrac{1-m^2}{4}$	$\dfrac{1+m^2}{2}$	$\dfrac{1-m^2}{4}$	$nc\zeta \pm sc\zeta$

Overview of Numerous Analytical Methods   43

Step 4: According to the proposed method, we substitute $\Phi = \zeta^{-p}$ in all terms of eq. (2.84) for determining the highest order singularity. Then the degree of all terms of eq. (2.84) has been taken in to study and, consequently, the two or more terms of lower degree are chosen. The maximum value of p is known as the pole and denoted as M. For integer value of M, this proposed method only can be implemented. However, if M is non-integer, the above eq. (2.84) can be transferred and then the above procedure can be repeated.

Step 5: Substituting eq. (2.85) along with (2.86) into eq. (2.84) and equating the coefficient of $\phi^i (i = 0, 1, 2, ...)$ into zero, we obtain the set of algebraic equations. By solving the obtained algebraic equations, we can get the unknowns A_i, B_i $(i = 1, 2, ..., N)$, and other constants. Then putting the all obtained unknowns in eq. (2.85), we get the required exact solutions for eq. (2.82) instantly.

2.14 Proposed Successive Recursion Method

In this section, the successive recursion method has been introduced, which is used for getting analytical approximate solutions for some practical problems in fractional differential equations.

For describing the brief outline of proposed method, let us consider the differential equation in the following form:

$$Lu + Ru = g. \tag{2.87}$$

where L is an invertible linear operator and R denotes the operator for the remaining part of the equation under consideration. The symbolic part described here, in this section, will be clear in the solutions in subsequent sections.

The general solution of the eq. (2.87) can be written as:

$$u = \sum_{n=0}^{\infty} u_n, \tag{2.88}$$

where the complete solution of $Lu = g$ is u_0. By using the property of invertible linear operator, we can write the equivalent expression of eq. (2.87) as in the following form:

$$L^{-1}Lu = L^{-1}g - L^{-1}Ru. \tag{2.89}$$

For initial value problem, we define the inverse linear operator L^{-1} for $L = D_t^n$ that is n-fold derivative operator, and its inverse that is L^{-1} will be n-fold integration operation from 0 to t. If we take $L = D_t^2$, then we have $L^{-1}Lu = u - u(0) - tu'(0)$. So eq. (2.89) reduces to

$$u = u(0) + tu'(0) + L^{-1}g - L^{-1}Ru, \tag{2.90}$$

and for boundary value problem, we have:

$$u = A + Bt + L^{-1}g - L^{-1}Ru, \tag{2.91}$$

where the integration constants A and B are determined from the given conditions. Let:

$$u_0 = u(0) + tu'(0) + L^{-1}g.$$

So the general solution of eq. (2.87) becomes:

$$u = u_0 - L^{-1}R\sum_{n=0}^{\infty} u_n, \tag{2.92}$$

where:

$$u_0 = \phi + L^{-1}g, \phi = u(0) + tu'(0). \tag{2.93}$$

And ϕ is the solution of $Lu = 0$, so that $L\phi = 0$.

Thus, we obtain the following recursive formula from eq. (2.92):

$$u_{n+1} = -L^{-1}Ru_n, n \geq 0. \tag{2.94}$$

So by using eq. (2.94), we can find the value of $u_1, u_2,...$ etc.

2.15 Conclusion

In this chapter, the fundamentals and algorithms of the semi-analytical methods and analytical methods have been discussed. We have also presented the brief literature review for the proposed semi-analytical methods and analytical methods.

The detail algorithms of such semi-analytical and analytical methods, viz., homotopy perturbation method, homotopy perturbation transform method, modified homotopy analysis method, fractional sub-equation method, improved fractional sub-equation method, (G'/G)-expansion method, improved (G'/G)-expansion method, modified Kudryashov method, Jacobi elliptic function method, and successive recursion method have been discussed. Also, some newly proposed methods like the modified homotopy analysis method with the Fourier transform method, proposed tanh method, and other methods for solving FDEs have been reported by the author.

The homotopy perturbation method does not need any transformation techniques and linearization of the equations, whereas in the case of the homotopy perturbation transform method, the Laplace transformation techniques have been used. Additionally, the homotopy perturbation transform method does not need any discretization method to have numerical solutions. This method thus eliminates the difficulties and massive computation work.

The MHAM-FT provides us with a convenient way to control the convergence of approximate series solution and solves the problem without any need for discretization of the variables. To control the convergence of the solution, we can choose the proper values of $\hbar$. The proposed MHAM-FT method is very simple and efficient for solving Riesz fractional differential equations.

Overview of Numerous Analytical Methods

The fractional sub-equation and improved sub-equation methods have many advantages: they are straight forward and concise. From the obtained results, it manifests that the proposed methods are powerful, effective, and convenient for non-linear fractional PDEs. The improved (G'/G)-expansion method provides new and more general type traveling wave solutions than the (G'/G)-expansion method, which are significant to reveal the pertinent features of the physical phenomenon. Also, the (G'/G)-expansion method yields many parameters and arbitrary constants, so in this case, the degree of freedom of solution is high for which it is difficult to handle the solutions in practical purpose. In the proposed tanh method, the solution contains only one parameter. Hence, the degree of freedom of the solution obtained by the tanh method is much less than other solutions obtained by the modified Kudryashov method. So, the proposed tanh method is more effective than the modified Kudryashov method. The most essential interest of the proposed new method is that it takes less computation for obtaining the exact solutions. In the Jacobi elliptic method, we obtained the double periodic wave solutions, which is unique in nature.

3

New Analytical Approximate Solutions of Fractional Differential Equations

3.1 Introduction

In many practical applications regarding the field of science and engineering, the physical systems are modeled by non-linear partial differential equations (NPDEs). Because, in many of the cases exact solutions are very difficult or even impossible to obtain for NPDEs, the approximate analytical solutions are particularly important for the study of dynamic systems for analyzing their physical nature. In the case of approximate analytical solutions, the success of a certain approximation method depends on the non-linearities that occur in the studied problem, and thus a general algorithm for the construction of such approximate solutions does not exist in the general cases.

In this chapter, the approximate analytical solutions for dynamical systems which are modeled by non-linear fractional partial differential equations have been presented. Such problems are frequently encountered in science and engineering and also intensely studied in recent years. In the present chapter, we have studied prey–predator biological models, viz., continuous population models of single and interacting species, time-fractional coupled non-linear Klein–Gordon–Schrödinger (K–G–S) equations, time-fractional coupled non-linear Klein–Gordon–Zakharov (K–G–Z) equations, and the time-fractional non-linear-coupled sine-Gordon equations.

Here, the approximate analytical solutions for the above dynamical and biological systems are obtained by using the analytical methods, viz., homotopy perturbation method (HPM), homotopy perturbation method (HPTM), and modified homotopy analysis method (MHAM).

3.2 Outline of Present Study

In this chapter, we shall focus our study on the non-linear fractional partial differential equations (FPDEs) that have particular applications appearing in engineering and sciences. We have considered analytical approximation methods for solving prey–predator biological models, viz., continuous population models of single and interacting species, time-fractional coupled non-linear Klein–Gordon–Schrödinger equations, time-fractional coupled non-linear K–G–Z equations, and the time-fractional non-linear-coupled sine-Gordon equations.

3.2.1 Time-Fractional Predator–Prey Models

The dynamical relationship between predator and prey is one of the dominant themes in ecology. In recent years, prey–predator models appearing in various fields of mathematical biology have been proposed and studied extensively due to their universal existence and importance. It was observed from the population data that interaction between a pair of predator–prey influences the population growth of both species. These non-linear equations are frequently used for describing interaction between two species in the dynamics of biological systems [32,86–93]. In this section, we have considered the three non-linear time-fractional biological problems which are as follows: the first problem is a time-fractional prey–predator model: Lotka–Volterra system, the second problem is a simple time-fractional two-species Lotka–Volterra competition model, and the third one is a time-fractional prey–predator model with limit cycle periodic behavior.

3.2.1.1 Time-Fractional Lotka–Volterra Model for an Interacting Species Model (Case I)

Let us consider the time-fractional predator–prey model, viz., the Lotka–Volterra system as an interacting species model to be governed by [90,91]:

$$
{}_0^C D_t^\alpha u = u - uv,
\tag{3.1}
$$

$$
{}_0^C D_t^\beta v = k(uv - v),
\tag{3.2}
$$

with the initial conditions:

$$
u(0) = u_0, \, v(0) = v_0,
\tag{3.3}
$$

where α, β are the parameters representing the order of the fractional derivatives which satisfy $m - 1 < \alpha, \beta \leq m$. Here, in this present analysis, we have taken $m = 1$ and $t > 0$.

3.2.1.2 Time-Fractional Simple Two-Species Lotka–Volterra Competition Model (Case II)

Let us consider the simple two-species time-fractional Lotka–Volterra competition model [90,91] as:

$$
{}_0^C D_t^\alpha u = u - u^2 - kuv,
\tag{3.4}
$$

$$
{}_0^C D_t^\beta v = \rho(v - v^2 - buv),
\tag{3.5}
$$

with the initial conditions:

$$
u(0) = u_0, \, v(0) = v_0,
\tag{3.6}
$$

where α, β are the parameters representing the order of the fractional derivatives which satisfy $m - 1 < \alpha, \beta \leq m$. Here, in this present analysis, we have taken $m = 1$ and $t > 0$.

New Analytical Approximate Solutions of Fractional Differential Equations

3.2.1.3 Time-Fractional Lotka–Volterra Competition Model with Limit Cycle Periodic Behavior (Case III)

Let us consider a time-fractional prey–predator model with limit cycle periodic behavior [90,91] as:

$$
{}_0^C D_t^\alpha u = u - u^2 - \frac{auv}{u+d},
\tag{3.7}
$$

$$
{}_0^C D_t^\beta v = b\left(v - \frac{v^2}{u}\right),
\tag{3.8}
$$

with initial conditions:

$$
u(0) = u_0, v(0) = v_0,
\tag{3.9}
$$

where α, β are the parameters representing the order of the fractional derivatives which satisfy $m - 1 < \alpha, \beta \le m$. Here, also we have taken $m = 1$ and $t > 0$.

3.2.2 Time-Fractional Coupled Non-Linear K–G–S Equations

The time-fractional coupled non-linear K–G–S equations are considered in the following form:

$$
{}_0^C D_t^\alpha u - u_{xx} + u - |v|^2 = 0,
\tag{3.10}
$$

$$
i\,{}_0^C D_t^\beta v + v_{xx} + uv = 0,
\tag{3.11}
$$

where:
$v(x,t)$ represents a complex scalar nucleon field
$u(x,t)$ a real scalar meson field
$i = \sqrt{-1}$

The eqs. (3.10) and (3.11) describe a system of conserved scalar nucleons interacting with neutral scalar meson coupled with the Yukawa interaction [94]. Here α, β are the parameters standing for the order of the fractional derivatives which satisfy $m - 1 < \alpha \le m$, $n - 1 < \beta \le n$, where $m = 2$, $n = 1$, and $t > 0$. When $\alpha = 2$ and $\beta = 1$, the fractional equation reduces to the classical-coupled K–G–S equations.

Darwish and Fan [95] have proposed an algebraic method to obtain the explicit exact solutions for coupled K–G–S equations. Recently, the Jacobi elliptic function expansion method has been applied to obtain the solitary wave solutions for coupled K–G–S equations [96]. Xia et al [97] have applied the homogenous balance principle to obtain the exact solitary wave solutions of the K–G–S equations. Hioe [98] has obtained periodic solitary waves for two coupled non-linear Klein–Gordon and Schrödinger equations. Bao and Yang [99] have presented efficient, unconditionally stable, and accurate numerical methods for approximations of K–G–S equations. Recently, Naber [100] has constructed the time-fractional Schrödinger equation which is solved for a free particle and for a potential well. In reference [101], a fractional non-linear Schrödinger equation has been solved by the Adomian decomposition method. The modified decomposition method for the solution of integer order classical-coupled K–G–S equations has been applied by Saha Ray [35].

Here, the HPM [25,29] and HPTM [38–40] have been applied for solving fractional coupled K–G–S equations which play an important role in quantum physics. The HPM does not depend upon a small parameter in the equation. On the other hand, the HPTM is a combined form of the Laplace transform method with the homotopy perturbation method. The above methods find the solution without any discretization or restrictive assumptions and avoid the round-off errors.

3.2.3 Time-Fractional Coupled Non-Linear K–G–Z Equations

Here, the time-fractional coupled non-linear K–G–Z equations are considered in the following form [102,103]:

$$
{}_0^C D_t^\alpha u - u_{xx} + u + uv + |u|^2 u = 0, \tag{3.12}
$$

$$
{}_0^C D_t^\beta v - v_{xx} = \left(|u|^2 \right)_{xx}. \tag{3.13}
$$

Here, α, β are the parameters standing for the order of the fractional derivatives which satisfy $m-1 < \alpha \le m$, $n-1 < \beta \le n$, where $m = 2$, $n = 2$ and $t > 0$. When $\alpha = 2$ and $\beta = 2$, the fractional equations reduce to the classical-coupled Klein–Gordon–Zakharov equations. K–G–Z equations are a classical model that describe the interaction of the Langmuir wave and the ion acoustic wave in plasma. The complex function $u(x,t)$ denotes the fast time scale component of the electric field raised by electrons, and the real function $v(x,t)$ denotes the deviation of ion density from its equilibrium.

Here, the homotopy perturbation transform method [38–40] and homotopy analysis method [41–48] with modification have been applied for solving fractional coupled Klein–Gordon–Zakharov equations which play an important role in plasma physics. The homotopy perturbation transform method is a combined form of the Laplace transform method with the homotopy perturbation method. Recently, an analytical method, namely, homotopy analysis method (HAM) [41–48] has been developed and successfully applied for getting the approximate solutions for various non-linear problems in science and engineering. The HAM contains an auxiliary parameter $\hbar$ which provides us with a simple way to adjust the convergence region and rate of the series solution. Moreover, by means of the so-called $\hbar$-curve, it is easy to find the valid region of $\hbar$ to gain a convergent series solution. Thus, through the HAM, explicit analytic solutions of non-linear problems are possible to obtain. The above methods find the solution without any discretization or restrictive assumptions and avoid the round-off errors.

Also, the HAM has been modified for the sake of convenience and easier computation for finding components in approximate solutions for $u(x,t)$ and $v(x,t)$, the individual non-linear terms in non-linear operator expression form has been expanded in Adomian type of polynomials.

New Analytical Approximate Solutions of Fractional Differential Equations   51

3.2.4 Time-Fractional Order Non-Linear-Coupled Sine-Gordon Equations

Next here, the time-fractional order non-linear-coupled sine-Gordon equations [104] have been considered in the following form:

$$\,_0^C D_t^\alpha u(x,t) - u_{xx}(x,t) = -\delta^2 \sin\big(u(x,t) - v(x,t)\big),\ x \in \mathbb{R},\ t > 0, \tag{3.14}$$

$$\,_0^C D_t^\beta v(x,t) - c^2 v_{xx}(x,t) = \sin\big(u(x,t) - v(x,t)\big),\ x \in \mathbb{R},\ t > 0, \tag{3.15}$$

where c is the ratio of the acoustic velocities of the components u and v. The dimensionless parameter δ^2 is equal to the ratio of masses of particles in the "lower" and the "upper" parts of the crystal in a generalized Frenkel–Kontorova dislocation model [104–106]. Here, $\alpha,\ \beta$ are the parameters representing the order of fractional derivatives which satisfy $m-1 < \alpha \le m, n-1 < \beta \le n$, and $t > 0$. When $\alpha = 2$ and $\beta = 2$, the fractional equation reduces to the classical-coupled sine-Gordon equations.

Coupled sine-Gordon equations were introduced by Khusnutdinova and Pelinovsky [104]. The coupled sine-Gordon equations generalize the Frenkel–Kontorova dislocation model [105,106]. The eqs. (3.14) and (3.15) with $c = 1$ were also proposed to describe the open states in DNA [107]. Khusnutdinova and Pelinovsky [104] have analyzed the linear and non-linear wave process involving the exchange of energy between the two physical components of the system. They have obtained the linear solutions by considering $|u - v| \ll 1$, the exact non-linear solutions for the case $c = 1$ and the weakly non-linear solutions, for the general case, by means of asymptotic methods. In the recent past, notable researchers Saha Ray [108] and Kaya [109] had given the solutions of sine-Gordon equations by using the modified decomposition method (MDM).

Here, the homotopy perturbation transform method [38–40] and homotopy analysis method [41–48] with modification have been applied for solving fractional coupled sine-Gordon (S-G) equations which play an important role in non-linear physics. The homotopy perturbation transform method is a combined form of the Laplace transform method with the homotopy perturbation method.

The main objective here is to employ two reliable independent analytical methods, such as the MHAM and HPTM for solving fractional coupled sine-Gordon equations. The capability, effectiveness, and convenience of the methods have been established by obtaining the analytical solutions and comparing with that of MDM. In this regard, the solutions obtained by the MDM method have been assumed as classical solutions for coupled S-G equations.

3.3 Application of HPM Methods for Solutions of Fractional Predator–Prey Model

In this section, the homotropy perturbation method has been used for getting the analytical approximate solutions for the predator–prey models. The analytical approximate solutions, numerical results, and stability analysis have been studied for each case of the Lotka–Volterra model.

3.3.1 Application of HPM for Analytical Approximate Solutions for Time-Fractional Lotka–Volterra Model for an Interacting Species Model (Case I)

Here, we have considered the eqs. (3.1) and (3.2) for study of analytical approximate solutions by using the homotopy perturbation method.

By applying the Riemann–Liouville integral to both sides of eqs. (3.1) and (3.2), respectively, we have following equations as:

$$J_t^\alpha D_t^\alpha u = J_t^\alpha (u - uv), \tag{3.16}$$

$$J_t^\beta D_t^\beta v = k J_t^\beta (uv - v). \tag{3.17}$$

After simplification of eqs. (3.16) and (3.17), we get:

$$u(t) = u(0) + J_t^\alpha (u - uv), \tag{3.18}$$

$$v(t) = v(0) + k J_t^\beta (uv - v). \tag{3.19}$$

By using the homotopy perturbation method, we will construct the homotopy of eqs. (3.18) and (3.19) as:

$$u(t) = u(0) + p J_t^\alpha (u - uv), \tag{3.20}$$

$$v(t) = v(0) + p k J_t^\beta (uv - v). \tag{3.21}$$

By substituting $u(t) = \sum_{n=0}^{\infty} p^n u_n(t)$ and $v(t) = \sum_{n=0}^{\infty} p^n v_n(t)$, in the eqs. (3.20) and (3.21), we obtain:

$$\sum_{n=0}^{\infty} p^n u_n(t) = u(0) + p J_t^\alpha \left(\sum_{n=0}^{\infty} p^n u_n(t) - \left(\left(\sum_{n=0}^{\infty} p^n u_n(t) \right) \left(\sum_{n=0}^{\infty} p^n v_n(t) \right) \right) \right), \tag{3.22}$$

$$\sum_{n=0}^{\infty} p^n v_n(t) = v(0) + p k J_t^\beta \left(\left(\left(\sum_{n=0}^{\infty} p^n u_n(t) \right) \left(\sum_{n=0}^{\infty} p^n v_n(t) \right) \right) - \sum_{n=0}^{\infty} p^n v_n(t) \right). \tag{3.23}$$

Comparing the coefficients of different powers in p for eqs. (3.22) and (3.23), we have the following system of linear differential equations.

$$\text{Coefficients of } p^0 : u_0 = u(0), \ v_0 = v(0). \tag{3.24}$$

$$\text{Coefficients of } p : u_1 = J_t^\alpha \left(u_0(t) - u_0(t) v_0(t) \right), \ v_1 = k J_t^\beta \left(-v_0(t) + u_0(t) v_0(t) \right). \tag{3.25}$$

New Analytical Approximate Solutions of Fractional Differential Equations 53

Coefficients of $p^2 : u_2 = J_t^\alpha \left(u_1(t) - v_0(t)u_1(t) - u_0(t)v_1(t) \right)$,

$$v_2 = kJ_t^\beta \left(v_0(t)u_1(t) - v_1(t) + u_0(t)v_1(t) \right), \tag{3.26}$$

and so on.

By putting $u(0) = u_0(t) = u_0$ and $v(0) = v_0(t) = v_0$ in eqs. (3.25) and (3.26), then solving them, we obtain:

$$u_1 = \frac{t^\alpha (u_0 - v_0 u_0)}{\Gamma(1+\alpha)},$$

$$v_1 = \frac{t^\beta k(-v_0 + v_0 u_0)}{\Gamma(1+\beta)},$$

$$u_2 = t^\alpha u_0 \left(\frac{t^\alpha (-1+v_0)^2}{\Gamma(1+2\alpha)} - \frac{kt^\beta (-1+u_0)v_0}{\Gamma(1+\alpha+\beta)} \right),$$

$$v_2 = kt^\beta v_0 \left(-\frac{t^\alpha u_0(-1+v_0)}{\Gamma(1+\alpha+\beta)} + \frac{kt^\beta (-1+u_0)^2}{\Gamma(1+2\beta)} \right),$$

and so on.

Finally, the third-order approximate solutions for **Case I** are obtained as:

$$u = u_0(t) + u_1(t) + u_2(t)$$

$$= u_0 + \frac{t^\alpha (u_0 - v_0 u_0)}{\Gamma(1+\alpha)} + t^\alpha u_0 \left(\frac{t^\alpha (-1+v_0)^2}{\Gamma(1+2\alpha)} - \frac{kt^\beta (-1+u_0)v_0}{\Gamma(1+\alpha+\beta)} \right), \tag{3.27}$$

and

$$v = v_0(t) + v_1(t) + v_2(t)$$

$$= v_0 + \frac{t^\beta k(-v_0 + v_0 u_0)}{\Gamma(1+\beta)} + kt^\beta v_0 \left(-\frac{t^\alpha u_0(-1+v_0)}{\Gamma(1+\alpha+\beta)} + \frac{kt^\beta (-1+u_0)^2}{\Gamma(1+2\beta)} \right). \tag{3.28}$$

3.3.1.1 Numerical Results and Discussion for Time-Fractional Lotka–Volterra Model for an Interacting Species (Case I)

In this section, numerical results of the predator and prey populations are calculated by using the initial conditions $u(0) = u_0 = 1.3$, $v(0) = v_0 = 0.6$, $k = 1$, $\alpha = 0.75$, and $\beta = 0.75$. By using these initial conditions, we plot solution graphs for **Case I** as shown in Figure 3.1.

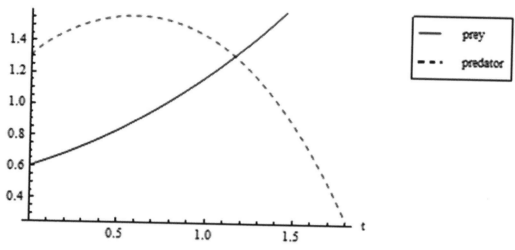

FIGURE 3.1
Plots of the fifth-order approximations for u and v for **Case I**.

Figure 3.1 shows the time evolution of population for u and v for fractional order $\alpha = 0.75$ and $\beta = 0.75$. It may be observed from Figure 3.1 when prey increases, the predator parabolically decreases with time t.

3.3.1.2 Stability Analysis for Time-Fractional Lotka–Volterra Model for an Interacting Species Model (Case I)

For analyzing the stability [110–112] of **Case I**, let us consider eqs. (3.1) and (3.2):

$$D_t^\alpha u = u - uv = F_1, \tag{3.29}$$

$$D_t^\beta v = k(uv - v) = F_2. \tag{3.30}$$

Now, the equilibrium points $E_0 = (0, 0)$ and $E_1 = (1, 1)$ have been obtained by equating eqs. (3.29) and (3.30) to zero and solving them for u and v. By taking partial derivatives of eqs. (3.29) and (3.30) with respect to u and v, we get:

$$\frac{\partial F_1}{\partial u} = 1 - v, \quad \frac{\partial F_1}{\partial v} = u,$$

$$\frac{\partial F_2}{\partial u} = kv, \quad \frac{\partial F_2}{\partial v} = k(u - 1).$$

1. At the equilibrium point $E_0 = (0, 0)$, the Jacobian matrix is given as:

$$J(E_0) = \begin{bmatrix} 1 & 0 \\ 0 & -k \end{bmatrix}.$$

New Analytical Approximate Solutions of Fractional Differential Equations   55

The eigenvalues of $J(E_0)$ are 1 and $-k$. So for all values of k, E_0 is unstable. So for given value of k, E_0 is unstable.

2. At the equilibrium point $E_1 = (1,1)$, the Jacobian matrix is given as:

$$J(E_1) = \begin{bmatrix} 0 & 1 \\ k & 0 \end{bmatrix}.$$

Now, the trace of $J(E_1) = 0$. So E_1 is unstable.

3.3.2 Application of HPM for Analytical Approximate Solutions for Time-Fractional Simple Two-Species Lotka–Volterra Competition Model (Case II)

Here, we have considered the eqs. (3.4) and (3.5) for study of analytical approximate solutions by using the homotopy perturbation method.

By applying the Riemann–Liouville integral to both sides of eqs. (3.4) and (3.5), respectively, we have the following equations as:

$$J_t^\alpha D_t^\alpha u = J_t^\alpha (u - u^2 - kuv), \tag{3.31}$$

$$J_t^\beta D_t^\beta v = \rho J_t^\beta \left(v - v^2 - buv \right). \tag{3.32}$$

After simplification of eqs. (3.31) and (3.32), we get:

$$u(t) = u(0) + J_t^\alpha (u - u^2 - kuv), \tag{3.33}$$

$$v(t) = v(0) + \rho J_t^\beta \left(v - v^2 - buv \right). \tag{3.34}$$

By using the homotopy perturbation method, we will construct the homotopy of eqs. (3.33) and (3.34) as:

$$u(t) = u(0) + p J_t^\alpha (u - u^2 - kuv), \tag{3.35}$$

$$v(t) = v(0) + p\rho J_t^\beta \left(v - v^2 - buv \right). \tag{3.36}$$

By substituting $u(t) = \sum_{n=0}^\infty p^n u_n(t)$ and $v(t) = \sum_{n=0}^\infty p^n v_n(t)$ in corresponding eqs. (3.35) and (3.36), we have:

$$\sum_{n=0}^\infty p^n u_n(t) = u(0) + p J_t^\alpha \left(\sum_{n=0}^\infty p^n u_n(t) - \left(\sum_{n=0}^\infty p^n u_n(t) \right)^2 - \alpha \left(\left(\sum_{n=0}^\infty p^n u_n(t) \right) \left(\sum_{n=0}^\infty p^n v_n(t) \right) \right) \right), \tag{3.37}$$

$$\sum_{n=0}^\infty p^n v_n(t) = v(0) + p\rho J_t^\beta \left(\sum_{n=0}^\infty p^n v_n(t) - \left(\sum_{n=0}^\infty p^n v_n(t) \right)^2 - \left(\left(\sum_{n=0}^\infty p^n u_n(t) \right) \left(\sum_{n=0}^\infty p^n v_n(t) \right) \right) \right). \tag{3.38}$$

Comparing the coefficients of different powers in p for eqs. (3.37) and (3.38), we have the following system of linear differential equations.

Coefficients of p^0: $u_0 = u(0)$, $v_0 = v(0)$. $\hspace{2cm}$ (3.39)

Coefficients of p: $u_1 = J_t^\alpha \left(u_0(t) - u_0(t)^2 - k u_0(t) v_0(t) \right)$,

$$v_1 = \rho J_t^\beta (v_0(t) - b u_0(t) v_0(t) - v_0(t)). \hspace{1cm} (3.40)$$

Coefficients of p^2: $u_2 = J_t^\alpha \left(u_1(t) - 2 u_0(t) u_1(t) - k u_1(t) v_0(t) - k u_0(t) v_1(t) \right)$, $\hspace{0.5cm}$ (3.41)

$$v_2 = \rho J_t^\beta (-b u_1(t) v_0(t) + v_1(t) - b u_0(t) v_1(t) - 2 v_0(t) v_1(t)),$$

and so on.

By putting $u(0) = u_0(t) = u_0$ and $v(0) = v_0(t) = v_0$ in eqs. (3.40) and (3.41), then solving them, we get:

$$u_1 = \frac{t^\alpha (u_0 - u_0^2 - k v_0 u_0)}{\Gamma(1+\alpha)},$$

$$v_1 = \frac{t^\beta \rho(v_0 - b v_0 u_0 - v_0^2)}{\Gamma(1+\beta)},$$

$$u_2 = t^\alpha u_0 \left(\frac{t^\alpha(-1 + u_0 + k v_0)(-1 + 2u_0 + k v_0)}{\Gamma(1+2\alpha)} - \frac{k t^\beta(-1 + b u_0 + v_0)v_0}{\Gamma(1+\alpha+\beta)} \right),$$

$$v_2 = t^\beta v_0 \rho \left(\frac{b t^\alpha u_0(-1 + u_0 + k v_0)}{\Gamma(1+\alpha+\beta)} + \frac{t^\beta \rho(-1 + b u_0 + v_0)(-1 + b u_0 + 2v_0)}{\Gamma(1+2\beta)} \right),$$

and so on.

Finally, the third-order approximate solutions for **Case II** are obtained as:

$$u = u_0(t) + u_1(t) + u_2(t)$$

$$= u_0 + \frac{t^\alpha (u_0 - u_0^2 - k v_0 u_0)}{\Gamma(1+\alpha)} \hspace{2cm} (3.42)$$

$$+ t^\alpha u_0 \left(\frac{t^\alpha(-1 + u_0 + k v_0)(-1 + 2u_0 + k v_0)}{\Gamma(1+2\alpha)} - \frac{k t^\beta(-1 + b u_0 + v_0)v_0}{\Gamma(1+\alpha+\beta)} \right),$$

and

$$v = v_0(t) + v_1(t) + v_2(t)$$

$$= v_0 + \frac{t^\beta \rho(v_0 - b v_0 u_0 - v_0^2)}{\Gamma(1+\beta)} \hspace{2cm} (3.43)$$

$$+ t^\beta v_0 \rho \left(\frac{b t^\alpha u_0(-1 + u_0 + k v_0)}{\Gamma(1+\alpha+\beta)} + \frac{t^\beta \rho(-1 + b u_0 + v_0)(-1 + b u_0 + 2v_0)}{\Gamma(1+2\beta)} \right).$$

3.3.2.1 Numerical Results and Discussion for Time-Fractional Dimple Two-Species Lotka–Volterra Competition Model (Case II)

In this section, numerical results of the predator and prey populations are calculated by using the initial conditions $u(0) = u_0(t) = 1$, $v(0) = v_0(t) = 1$, and the values of the parameters are taken as $a = 1$, $k = 1$, $\alpha = 0.75$, $\beta = 0.75$, $\rho = 1$, and $b = 0.8$.

By using these conditions, we plot solution graphs for **Case II** as shown in Figure 3.2.

Figure 3.2 shows the time evolution of population for u and v for fractional order $\alpha = 0.75$ and $\beta = 0.75$. It may be observed from Figure 3.2, both prey and predator exponentially increase with time t.

3.3.2.2 Stability Analysis for Time-Fractional Simple Two-Species Lotka–Volterra Competition Model (Case II)

For analyzing the stability [110–112] of **Case II**, let us consider eqs. (3.4) and (3.5) as follows.

$$D_t^\alpha u = u - u^2 - kuv = G_1, \qquad (3.44)$$

$$D_t^\beta v = \rho(v - v^2 - buv) = G_2. \qquad (3.45)$$

So by equating eqs. (3.44) and (3.45) to zero and solving them for u and v, we get the equilibrium points $E_0 = (0,0)$ and $E_1 = \left(\frac{1-k}{1-bk}, \frac{1-b}{1-bk}\right)$.

By taking partial derivatives of eqs. (3.58) and (3.59) with respect to u and v, we get:

$$\frac{\partial G_1}{\partial u} = 1 - 2u - kv, \quad \frac{\partial G_1}{\partial v} = -ku,$$

$$\frac{\partial G_2}{\partial u} = -\rho bv, \quad \frac{\partial G_2}{\partial v} = \rho(1 - 2v - bu).$$

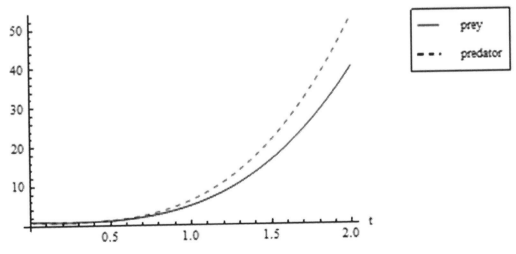

FIGURE 3.2
Plots of the fifth-order approximations for u and v for **Case II**.

1. At the equilibrium point $E_0 = (0,0)$, the Jacobian matrix is:

$$J(E_0) = \begin{bmatrix} 1 & 0 \\ 0 & \rho \end{bmatrix}.$$

The eigenvalues of $J(E_0)$ are 1 and ρ. So for all values of ρ, E_0 is unstable.

2. At the equilibrium point $E_1 = \left(\frac{1-k}{1-bk}, \frac{1-b}{1-bk} \right)$, the Jacobian matrix is given as:

$$J(E_1) = \begin{bmatrix} \dfrac{1-k}{-1+bk} & \dfrac{(-1+k)k}{-1+bk} \\[3mm] -\dfrac{(-1+b)b\rho}{-1+bk} & \dfrac{(-1+b)\rho}{-1+bk} \end{bmatrix}.$$

Characteristics equation of $J(E_1)$ is:

$$\lambda^2 + a_1\lambda - a_2 = 0, \tag{3.46}$$

where $a_1 = \frac{(-1+b)(-1+k)}{-1+bk}$ and $a_2 = \frac{\rho(-1+b)(1-k)}{-1+bk}$, this implies:

$$\lambda = -\frac{a_1}{2} \pm \frac{\sqrt{a_1^2 + 4a_2}}{2}.$$

1. For $a_1 < 0$,
   a. if $a_1^2 + 4a_2 < 0$, then the eigenvalues λ have positive real part. So the equilibrium point E_1 is unstable.
   b. if $a_1^2 + 4a_2 = 0$, then the eigenvalues λ have two positive same values. So the equilibrium point E_1 is unstable.
   c. if $a_1^2 + 4a_2 > 0$, then the eigenvalues λ have two distinct values and one of them is positive. So the equilibrium point E_1 is unstable.

2. For $a_1 > 0$,
   a. if $a_1^2 + 4a_2 < 0$, then the eigenvalues λ are imaginary and both have negative real parts. So the equilibrium point E_1 is locally asymptotically stable.
   b. if $a_1^2 + 4a_2 > 0$, then the eigenvalues λ have two distinct values say λ_1 and λ_2, where:

$$\lambda_1 = -\frac{a_1}{2} + \frac{\sqrt{a_1^2 + 4a_2}}{2} \text{ and } \lambda_2 = -\frac{a_1}{2} - \frac{\sqrt{a_1^2 + 4a_2}}{2}.$$

   Now, $\lambda_1 < 0$, if $\sqrt{a_1^2 + 4a_2} < a_1$, then $a_2 < 0$.

   So the equilibrium point E_1 is locally asymptotically stable for $a_2 < 0$.
   c. if $a_1^2 + 4a_2 = 0$, then the eigenvalues λ have same negative value of multiplicity 2. So the equilibrium point E_1 is locally asymptotically stable.

3. For $a_1 = 0$:

$$\lambda = \pm\sqrt{a_2}.$$

New Analytical Approximate Solutions of Fractional Differential Equations

a. if $a_2 > 0$, then the characteristic eq. (3.46) has at least one positive eigenvalue. So the equilibrium point E_1 is unstable.

b. if $a_2 < 0$, then the eigenvalues λ are imaginary with real part zero. So the equilibrium point E_1 is unstable.

c. if $a_2 = 0$, then the eigenvalues for λ are zero. So the equilibrium point E_1 is unstable.

4. For $a_2 < 0$,

a. if $a_1 > 0$, then all the roots of the characteristic eq. (3.46) have negative real parts. Therefore, the equilibrium point E_1 is locally asymptotically stable.

b. if $a_1 = 0$, then all the roots of the characteristic eq. (3.46) are imaginary with real part zero. So the equilibrium point E_1 is unstable.

c. if $a_1 < 0$,

  i. $a_1^2 + 4a_2 < 0$, then the eigenvalues λ are imaginary with real part positive. So the equilibrium point E_1 is unstable.

  ii. $a_1^2 + 4a_2 = 0$, then the eigenvalues λ is positive. So the equilibrium point E_1 is unstable.

  iii. $a_1^2 + 4a_2 > 0$, therefore the characteristic eq. (3.46) has at least one positive real eigenvalue. So the equilibrium point E_1 is unstable.

5. For $a_2 = 0$, the eigenvalues are $\lambda = 0$ and $-a_1$. So the equilibrium point is unstable.

6. For $a_2 > 0$,

a. If $a_1 < 0$, then always $a_1^2 + 4a_2 > 0$. Therefore the characteristic eq. (3.60) has at least one positive real eigenvalue. So the equilibrium point E_1 is unstable.

b. If $a_1 > 0$, then always $a_1^2 + 4a_2 > 0$. Then the eigenvalues λ have two distinct values say λ_1 and λ_2, where:

$$\lambda_1 = -\frac{a_1}{2} + \frac{\sqrt{a_1^2 + 4a_2}}{2} \text{ and } \lambda_2 = -\frac{a_1}{2} - \frac{\sqrt{a_1^2 + 4a_2}}{2}.$$

Now, always $\lambda_2 < 0$ and $\lambda_1 < 0$, if $\sqrt{a_1^2 + 4a_2} < a_1$, then $a_2 < 0$, which contradicts the fact that $a_2 > 0$. So the equilibrium point E_1 is unstable.

c. if $a_1 = 0$, then the characteristic eq. (3.46) has at least one positive real eigenvalue. So the equilibrium point E_1 is unstable.

3.3.3 Application of HPM for Analytical Approximate Solutions for Time-Fractional Lotka–Volterra Competition Model with Limit Cycle Periodic Behavior (Case III)

Here, we have considered the eqs. (3.7) and (3.8) for study of analytical approximate solutions by using the homotopy perturbation method.

By applying the Riemann–Liouville integral to both sides of eqs. (3.7) and (3.8), respectively, we have following equations as:

$$J_t^\alpha D_t^\alpha u = J_t^\alpha \left(u - u^2 - \frac{auv}{u+d} \right), \tag{3.47}$$

$$J_t^\beta D_t^\beta v = b J_t^\beta \left(v - \frac{v^2}{u} \right).$$

(3.48)

After simplification of eqs. (3.61) and (3.62), we get:

$$u(t) = u(0) + J_t^\alpha \left(u - u^2 - \frac{auv}{u+d} \right),$$

(3.49)

$$v(t) = v(0) + b J_t^\beta \left(v - \frac{v^2}{u} \right).$$

(3.50)

By using the homotopy perturbation method, we construct the homotopy of eqs. (3.49) and (3.50) as:

$$u(t) = u(0) + p J_t^\alpha \left(u - u^2 - \frac{auv}{u+d} \right),$$

(3.51)

$$v(t) = v(0) + p b J_t^\beta \left(v - \frac{v^2}{u} \right).$$

(3.52)

By substituting $u(t) = \sum_{n=0}^\infty p^n u_n(t)$ and $v(t) = \sum_{n=0}^\infty p^n v_n(t)$, in the eqs. (3.51) and (3.52), we have:

$$\sum_{n=0}^\infty p^n u_n(t) = u(0) + p J_t^\alpha \left(\sum_{n=0}^\infty p^n u_n(t) - \left(\sum_{n=0}^\infty p^n u_n(t) \right)^2 - a \sum_{n=0}^\infty H_n \right),$$

(3.53)

and

$$\sum_{n=0}^\infty p^n v_n(t) = v(0) + p b J_t^\beta \left(\left(\sum_{n=0}^\infty p^n v_n(t) \right) - \sum_{n=0}^\infty Q_n \right).$$

(3.54)

Here,

$$N(u,v) = \frac{uv}{u+d} = \sum_{n=0}^\infty H_n(u_0, u_1, ..., u_n; v_0, v_1, ..., v_n)$$

and

$$M(u,v) = \frac{v^2}{u} = \sum_{n=0}^\infty Q_n(u_0, u_1, ..., u_n; v_0, v_1, ..., v_n),$$

where

$$H_n(u_0, u_1, ..., u_n; v_0, v_1, ..., v_n) = \frac{1}{n!} \frac{d^n}{dp^n} N \left(\sum_{k=0}^\infty p^k u_k(t), \sum_{k=0}^\infty p^k v_k(t) \right) \Bigg|_{p=0},$$

$$Q_n(u_0, u_1, \ldots, u_n; v_0, v_1, \ldots, v_n) = \frac{1}{n!} \frac{d^n}{dp^n} M\left(\sum_{k=0}^{\infty} p^k u_k(t), \sum_{k=0}^{\infty} p^k v_k(t)\right)\Bigg|_{p=0}.$$

It is very challenging to solve these non-linear equations. So the non-linear terms of eqs. (3.53) and (3.54) have been approximated by using the Adomian polynomial.

Using the Adomian polynomial defined in eq. (3.24), we obtain:

$$H_0 = \frac{a u_0(t) v_0(t)}{u_0(t) + d},$$

$$H_1 = \frac{a(d\, u_1(t) v_0(t) + u_0(t)(d + u_0(t) v_1(t)))}{(d + u_0(t))^2},$$

$$H_2 = \frac{a(-d u_1(t)^2 v_0(t) + d(d + u_1(t) u_0(t) v_1(t) + d u_0(t) v_0(t)) + u_0(t)(d + u_0(t) v_2(t)))}{(d + u_0(t))^3},$$

and

$$Q_0 = \frac{v_0(t)^2}{u_0(t)},$$

$$Q_1 = \frac{v_0(t)(-u_1(t) v_0(t) + 2 u_0(t) v_1(t))}{u_0(t)^2},$$

$$Q_2 = \frac{u_1(t)^2 v_0(t)^2 - 2 u_0(t) u_1(t) v_0(t) v_1(t) - u_0(t) u_1(t)(-2 u_2(t) v_0(t)^2 + u_0(t)(v_1(t)^2 + 2 v_2(t) v_0(t)))}{u_0(t)^3}.$$

Substituting these values of H_n and Q_n in eqs. (3.53) and (3.54), respectively, and comparing the coefficients of different powers in p for eqs. (3.53) and (3.54), we obtain the following system of linear differential equations.

Coefficients of p^0: $u_0 = u(0)$, $v_0 = v(0)$. $\hfill (3.55)$

Coefficients of p: $u_1 = J_t^\alpha (u_0 - u_0(t)^2 - H_0)$, $v_1 = b J_t^\beta (v_0(t) - Q_0)$. $\hfill (3.56)$

Coefficients of p^2: $u_2 = J_t^\alpha (u_1(t) - 2 u_0(t) u_1(t) - H_1)$, $v_2 = b J_t^\beta (v_1(t) - Q_1)$, $\hfill (3.57)$

and so on.

By putting $u(0) = u_0(t) = u_0$ and $v(0) = v_0(t) = v_0$ in eqs. (3.56) to (3.57) and solving them, we get:

$$u_1(t) = -t^\alpha u_0 \left(\frac{d(-1 + u_0) - u_0 + u_0^2 + a v_0}{\Gamma(1 + \alpha)(u_0 + d)}\right),$$

$$v_1(t) = b v_0 t^\beta \left(\frac{(u_0 - v_0)}{\Gamma(1 + \beta) u_0}\right),$$

$$u_2(t) = \frac{t^\alpha}{(d+u_0)^3}$$

$$\left(-\frac{ab(d+u_0)^2(u_0-v_0)v_0}{\Gamma(1+\alpha+\beta)} + \frac{t^\alpha u_0((-1+u_0)(d+u_0)+av_0)((d+u_0)^2(-1+2u_0)+dav_0)}{\Gamma(1+2\alpha)} \right),$$

$$v_2(t) = \frac{t^\alpha}{(d+u_0)u_0^2} \left(\frac{bt^\beta(d+u_0)(u_0-2v_0)(u_0-v_0)}{\Gamma(1+2\beta)} - \frac{t^\alpha u_0 v_0(d(-1+u_0)-u_0+u_0^2+av_0)}{\Gamma(1+\alpha+\beta)} \right),$$

and so on.

Finally, the third-order approximate solutions for **Case III** are given as:

$$u = u_0(t) + u_1(t) + u_2(t)$$

$$= u_0 - t^\alpha u_0 \left(\frac{d(-1+u_0)-u_0+u_0^2+av_0}{\Gamma(1+\alpha)(u_0+d)} \right) + \frac{t^\alpha}{(d+u_0)^3} \left(-\frac{ab(d+u_0)^2(u_0-v_0)v_0}{\Gamma(1+\alpha+\beta)} \right. \qquad (3.58)$$

$$\left. + \frac{t^\alpha u_0((-1+u_0)(d+u_0)+av_0)((d+u_0)^2(-1+2u_0)+dav_0)}{\Gamma(1+2\alpha)} \right)$$

and

$$v = v_0(t) + v_1(t) + v_2(t)$$

$$= v_0 + bv_0 t^\beta \left(\frac{(u_0-v_0)}{\Gamma(1+\beta)u_0} \right) + \frac{t^\alpha}{(d+u_0)u_0^2} \left(\frac{bt^\beta(d+u_0)(u_0-2v_0)(u_0-v_0)}{\Gamma(1+2\beta)} \right. \qquad (3.59)$$

$$\left. - \frac{t^\alpha u_0 v_0(d(-1+u_0)-u_0+u_0^2+av_0)}{\Gamma(1+\alpha+\beta)} \right).$$

3.3.3.1 Numerical Results and Discussion for Time-Fractional Lotka–Volterra Competition Model with Limit Cycle Periodic Behavior (Case III)

In this section, numerical results of the predator and prey populations are calculated by using the specific initial conditions $u(0) = u_0(t) = 1.3$, $v(0) = v_0(t) = 1.2$ and the values of $a = 1$, $b = 5$, $d = 10$, $\alpha = 0.75$, and $\beta = 0.75$. By using these initial conditions, we obtain the fifth-order approximation solutions for **Case III** as shown in Figure 3.3.

Figure 3.3 shows the time evolution of population for u and v for fractional order $\alpha = 0.75$ and $\beta = 0.75$. From Figure 3.3, it may be observed that prey increases with time, but predator decreases up to near $t = 1$ and again from near time $t = 1$ exponentially increases with time t.

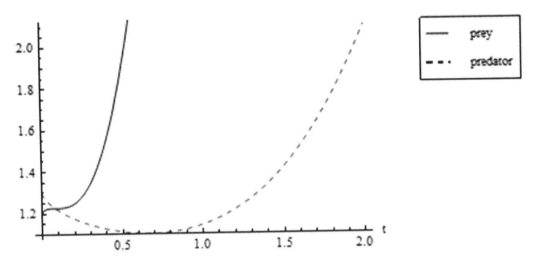

FIGURE 3.3
Plots of the fifth-order approximations for u and v for **Case III**.

3.3.3.2 Stability Analysis for Time-Fractional Lotka–Volterra Competition Model with Limit Cycle Periodic Behavior (Case III)

Let us consider the eqs. (3.7) and (3.8). For stability analysis [110–112] of this system, let:

$$D_t^\alpha u = u - u^2 - \frac{auv}{u+d} = M_1, \qquad (3.60)$$

$$D_t^\beta v = b\left(v - \frac{v^2}{u}\right) = M_2. \qquad (3.61)$$

So by equating eqs. (3.60) and (3.61) to zero and solving them for u and v, we get the equilibrium points are $E_0 = (1,0)$, $E_1 = (u^*, v^*)$, $E_2 = (\bar{u}, \bar{v})$, where:

$$u^* = \frac{1}{2}\left(1 - a - d + \sqrt{4d + (-1 + a + d)^2}\right),$$

$$v^* = \frac{1}{2}\left(1 - a - d + \sqrt{4d + (-1 + a + d)^2}\right),$$

$$\bar{u} = \frac{1}{2}\left(1 - a - d - \sqrt{4d + (-1 + a + d)^2}\right),$$

$$\bar{v} = \frac{1}{2}\left(1 - a - d - \sqrt{4d + (-1 + a + d)^2}\right).$$

By taking partial derivatives of eqs. (3.60) and (3.61) with respect to u and v, we get:

$$\frac{\partial M_1}{\partial u} = 1 - 2u + \frac{a\,u\,v}{(d+u)^2} - \frac{av}{(d+u)}, \quad \frac{\partial M_1}{\partial v} = -\frac{au}{(d+u)},$$

$$\frac{\partial M_2}{\partial u} = \frac{bv^2}{u^2}, \quad \frac{\partial M_2}{\partial v} = b\left(1 - \frac{2v}{u}\right).$$

1. At the equilibrium points $E_0 = (1,0)$, the Jacobian matrix is:

$$J(E_0) = \begin{bmatrix} -1 & -\dfrac{a}{1+d} \\ 0 & b \end{bmatrix}.$$

The eigenvalues of the matrix $J(E_0)$ are 1 and b.

a. For $b > 0$ and $b = 0$, the E_0 is unstable.

b. For $b < 0$, the E_0 is locally asymptotically stable.

2. At the equilibrium points $E_1 = (u^*, v^*)$, let us consider $u^* = v^* = k$. Then, the Jacobian matrix is:

$$J(E_1) = \begin{bmatrix} 1 - 2k + \dfrac{ak^2}{(d+k)^2} - \dfrac{ak}{d+k} & -\dfrac{ak}{d+k} \\ b & -b \end{bmatrix}.$$

Then $trace\ J(E_1) = 1 - 2k - \frac{akd}{(d+k)^2} - b$ and $det\ J(E_1) = b\left(-1 + 2k + \frac{akd}{(d+k)^2} + \frac{ak}{d+k}\right)$.
So the characteristic equation of $J(E_1)$ is given as:

$$\lambda^2 - trace\ (J(E_1))\lambda + det(J(E_1)) = 0.$$

According to Routh–Hurwitz criterion [113] for stability of a system:
$trace\ (J(E_1)) < 0.$
This implies $1 - 2k - \frac{akd}{(d+k)^2} - b < 0.$
Therefore,

$$2k + \frac{akd}{(d+k)^2} + b > 1, \text{ and } det\ (J(E_1)) > 0. \tag{3.62}$$

This implies $b\left(-1 + 2k + \frac{akd}{(d+k)^2} + \frac{ak}{d+k}\right) > 0.$
Therefore,

$$2k + \frac{akd}{(d+k)^2} + \frac{ak}{d+k} > 1. \tag{3.63}$$

Let:

$$2k + \frac{akd}{(d+k)^2} = R.$$

So eqs. (3.62) and (3.63) change to:

$$R + b > 1, \tag{3.64}$$

and

$$R + \frac{ak}{d+k} > 1, \tag{3.65}$$

respectively.

a. For $b > \frac{ak}{d+k}$

We have,

$$R > 1 - \frac{ak}{d+k} > 1 - b. \tag{3.66}$$

b. For $b < \frac{ak}{d+k}$

we have,

$$R > 1 - b > 1 - \frac{ak}{d+k}. \tag{3.67}$$

For eqs. (3.66) and (3.67), the equilibrium point E_1 is stable. A similar argument follows for stability analysis of equilibrium point E_2.

3.4 Implementation of the HPM and HPTM for the Solution of Fractional Coupled K–G–S Equations

In this section, the two reliable methods, viz., the HPM and HPTM have been implemented for solving fractional coupled K–G–S equations.

3.4.1 Implementation of the HPM for the Solution of Fractional Coupled K–G–S Equations

Let us consider the application of the HPM for the solution of the fractional coupled K–G–S eqs. (3.10) and (3.11) with given initial conditions:

$$u(x,0) = 6B^2 \operatorname{sech}^2(Bx),$$

$$u_t(x,0) = -12B^2 c \operatorname{sech}^2(Bx) \tanh(Bx),$$

$$v(x,0) = 3B \operatorname{sech}^2(Bx) e^{idx}, \tag{3.68}$$

where:

$B (\geq 1/2)$

c and d are arbitrary constants with $c = \sqrt{4B^2 - 1}/2$,

$d = -c/2B$ for the fractional coupled K–G–S equations (3.10) and (3.11).

Applying the Riemann–Liouville integral to both sides of eqs. (3.10) and (3.11), respectively, we get:

$$J_t^\alpha D_t^\alpha u = J_t^\alpha (u_{xx} - u + |v|^2),$$ (3.69)

$$J_t^\beta D_t^\beta v = i J_t^\beta (v_{xx} + uv).$$ (3.70)

After simplification of eqs. (3.69) and (3.70), we get:

$$u(x,t) = u(x,0) + t u_t(x,0) + J_t^\alpha (u_{xx} - u + |v|^2),$$ (3.71)

$$v(x,t) = v(x,0) + i J_t^\beta (v_{xx} + uv).$$ (3.72)

By using the homotopy perturbation method, we will construct the homotopy of eqs. (3.71) and (3.72) as:

$$u(x,t) = p(u(x,0) + t u_t(x,0) + J_t^\alpha (u_{xx} - u + |v|^2)),$$ (3.73)

$$v(x,t) = p(v(x,0) + i J_t^\beta (v_{xx} + uv)).$$ (3.74)

By substituting $u(x,t) = \sum_{n=0}^\infty p^n u_n(x,t)$ and $v(x,t) = \sum_{n=0}^\infty p^n v_n(x,t)$ in eqs. (3.73) and (3.74), respectively, we get:

$$\sum_{n=0}^\infty p^n u_n(x,t) = p\left(u(x,0) + t u_t(x,0) + J_t^\alpha \left(\left(\sum_{n=0}^\infty p^n u_n(x,t) \right)_{xx} \right) \right.$$
$$\left. - \sum_{n=0}^\infty p^n u_n(x,t) + \left(\left(\sum_{n=0}^\infty p^n v_n(x,t) \right) \left(\sum_{n=0}^\infty p^n \bar{v}_n(x,t) \right) \right) \right),$$ (3.75)

$$\sum_{n=0}^\infty p^n v_n(x,t) = p\left(v(x,0) + i J_t^\beta \left(\left(\sum_{n=0}^\infty p^n v_n(x,t) \right)_{xx} + \left(\left(\sum_{n=0}^\infty p^n u_n(x,t) \right) \left(\sum_{n=0}^\infty p^n v_n(x,t) \right) \right) \right) \right).$$ (3.76)

Comparing the coefficients of like powers in p for both sides of eqs. (3.75) and (3.76), we have the following system of fractional differential equations.

Coefficients of p^0: $u_0 = 0$, $v_0 = 0$. (3.77)

Coefficients of p: $u_1 = u(x,0) + t u_t(x,0) + J_t^\alpha \left(\dfrac{\partial^2 u_0}{\partial x^2} - u_0 + v_0 \bar{v}_0 \right),$

(3.78)

$$v_1 = v(x,0) + i J_t^\beta \left(\dfrac{\partial^2 v_0}{\partial x^2} + u_0 v_0 \right).$$

Coefficients of

$$p^2 : u_2 = J_t^\alpha \left(\frac{\partial^2 u_1}{\partial x^2} - u_1 + v_0 \bar{v}_1 + v_1 \bar{v}_0 \right), v_2 = i J_t^\beta \left(\frac{\partial^2 v_1}{\partial x^2} + u_0 v_1 + v_0 u_1 \right). \tag{3.79}$$

Coefficients of

$$p^3 : u_3 = J_t^\alpha \left(\frac{\partial^2 u_2}{\partial x^2} - u_2 + v_1 \bar{v}_1 + v_2 \bar{v}_0 + v_0 \bar{v}_2 \right), v_3 = i J_t^\beta \left(\frac{\partial^2 v_2}{\partial x^2} + u_1 v_1 + v_2 u_0 + v_0 u_2 \right). \tag{3.80}$$

By considering the initial conditions of equation (3.68) in eqs. (3.77) to (3.80) and solving them, we obtain:

$$u_0(x,t) = 0,$$

$$v_0(x,t) = 0,$$

$$u_1(x,t) = 6B^2 \operatorname{sech}^2(Bx) - 12B^2 c \operatorname{sech}^2(Bx) \tanh(Bx),$$

$$v_1(x,t) = 3B \operatorname{sech}^2(Bx) e^{idx},$$

$$u_2(x,t) = 6B^2 \sec^2(Bx) \frac{t^\alpha}{\Gamma(\alpha+1)} \left(-1 - 2B^2 \operatorname{sech}^2(Bx) + 4B^2 \tanh^2(Bx) \right)$$

$$+ 6B^2 \sec^2(Bx) \frac{t^{\alpha+1}}{\Gamma(\alpha+2)} \left(16B^2 c \operatorname{sech}^2(Bx) \tanh^2(Bx) - 2c \tanh(Bx)(-1 + 4B^2 \tanh^2(Bx)) \right),$$

$$v_2(x,t) = -3iB \operatorname{sech}^2(Bx) e^{idx} \frac{t^\beta}{\Gamma(\beta+1)} \left(2B^2 \operatorname{sech}^2(Bx) + (d + 2iB \tanh^2(Bx))^2 \right),$$

$$u_3(x,t) = 3B^2 \operatorname{sech}^2(Bx)(-4c \frac{t^{2\alpha+1}}{\Gamma(2\alpha+2)} \tanh(Bx) (136B^4 \operatorname{sech}^4(Bx)$$

$$+ (1 - 4B^2 \tanh^2(Bx))^2 + \operatorname{sech}^2(Bx) (16B^2 - 208B^4 \tanh^2(Bx))$$

$$+ \frac{3t^\alpha}{\Gamma(\alpha+1)} \operatorname{sech}^2(Bx) + \frac{2t^{2\alpha}}{\Gamma(2\alpha+1)} (16B^4 \operatorname{sech}^4(Bx)$$

$$+ (1 - 4B^2 \tanh^2(Bx))^2 + \operatorname{sech}^2(Bx)(4B^2 - 88B^4 \tanh^2(Bx)))),$$

$$v_3(x,t) = -3iB \operatorname{sech}^2(Bx) e^{idx} (-6B^2 \operatorname{sech}^2(Bx) \left(\frac{-t^\beta}{\Gamma(\beta+1)} + \frac{2ct^{\beta+1} \tanh(Bx)}{\Gamma(\beta+2)} \right)$$

$$+ \frac{it^{2\beta}}{\Gamma(2\beta+1)} (16B^4 \operatorname{sech}^4(Bx) + (d + 2iB \tanh^2(Bx))^4$$

$$+ 4B^2 \operatorname{sech}^2(Bx)(3d^2 + 16iBd \tanh(Bx) - 22B^2 \tanh^2(Bx)))),$$

and so on.

In this manner, the other components of the homotopy series can be easily obtained by which $u(x,t)$ and $v(x,t)$ can be evaluated in a series form as:

$$u(x,t) = u_0(x,t) + u_1(x,t) + u_2(x,t) + u_3(x,t)$$

$$= 6B^2 \operatorname{sech}^2(Bx) - 12B^2 c \operatorname{sech}^2(Bx)\tanh(Bx) + 6B^2 \sec^2(Bx)\frac{t^\alpha}{\Gamma(\alpha+1)}$$

$$\left(-1 - 2B^2 \operatorname{sech}^2(Bx) + 4B^2 \tanh^2(Bx)\right) + 6B^2 \sec^2(Bx)\frac{t^{\alpha+1}}{\Gamma(\alpha+2)}$$

$$\left(16B^2 c \operatorname{sech}^2(Bx)\tanh^2(Bx) - 2c\tanh(Bx)\left(-1 + 4B^2 \tanh^2(Bx)\right)\right)$$

$$+ 3B^2 \operatorname{sech}^2(Bx)(-4c\frac{t^{2\alpha+1}}{\Gamma(2\alpha+2)}\tanh(Bx)(136B^4 \operatorname{sech}^4(Bx)$$

$$+ \left(1 - 4B^2 \tanh^2(Bx)\right)^2 + \operatorname{sech}^2(Bx)\left(16B^2 - 208B^4 \tanh^2(Bx)\right))$$

$$+ \frac{3t^\alpha}{\Gamma(\alpha+1)}\operatorname{sech}^2(Bx) + \frac{2t^{2\alpha}}{\Gamma(2\alpha+1)}(16B^4 \operatorname{sech}^4(Bx)$$

$$+ (1 - 4B^2 \tanh^2(Bx))^2 + \operatorname{sech}^2(Bx)(4B^2 - 88B^4 \tanh^2(Bx)))),$$

(3.81)

$$v(x,t) = v_0(x,t) + v_1(x,t) + v_2(x,t) + v_3(x,t)$$

$$= 3B \operatorname{sech}^2(Bx)e^{idx} - 3iB \operatorname{sech}^2(Bx)e^{idx}\frac{t^\beta}{\Gamma(\beta+1)}$$

$$\left(2B^2 \operatorname{sech}^2(Bx) + \left(d + 2iB\tanh^2(Bx)\right)^2\right)$$

$$- 3iB \operatorname{sech}^2(Bx)e^{idx}(-6B^2 \operatorname{sech}^2(Bx)\left(\frac{-t^\beta}{\Gamma(\beta+1)} + \frac{2ct^{\beta+1}\tanh(Bx)}{\Gamma(\beta+2)}\right)$$

(3.82)

$$+ \frac{it^{2\beta}}{\Gamma(2\beta+1)}(16B^4 \operatorname{sech}^4(Bx) + \left(d + 2iB\tanh^2(Bx)\right)^4$$

$$+ 4B^2 \operatorname{sech}^2(Bx)(3d^2 + 16iBd\tanh(Bx) - 22B^2 \tanh^2(Bx)))).$$

3.4.2 Implementation of the HPTM for the Solution of Fractional Coupled K–G–S Equations

In this section, we apply the HPTM to eqs. (3.10) and (3.11) with considering the initial conditions eq. (3.68).

Applying Laplace transformation on both sides of eqs. (3.10) and (3.11), respectively, we get:

$$\mathcal{L}[u(x,t)] = \frac{u(x,0)}{s} + \frac{u_t(x,0)}{s^2} + \frac{1}{s^\alpha}\mathcal{L}[u_{xx}] - \frac{1}{s^\alpha}\mathcal{L}[u] + \frac{1}{s^\alpha}\mathcal{L}[|v|^2],$$

(3.83)

New Analytical Approximate Solutions of Fractional Differential Equations 69

$$\mathcal{L}[v(x,t)] = \frac{v(x,0)}{s} + i\frac{1}{s^\beta}\mathcal{L}[v_{xx}] + i\frac{1}{s^\beta}\mathcal{L}[uv]. \tag{3.84}$$

Then, applying inverse Laplace transformation on both sides of eqs. (3.83) and (3.84), respectively, we get:

$$u(x,t) = u(x,0) + t\,u_t(x,0) + \mathcal{L}^{-1}\left[\frac{1}{s^\alpha}\mathcal{L}\left[u_{xx} - u + |v|^2\right]\right], \tag{3.85}$$

$$v(x,t) = v(x,0) + i\mathcal{L}^{-1}\left[\frac{1}{s^\beta}\mathcal{L}[v_{xx} + uv]\right]. \tag{3.86}$$

By using the homotopy perturbation method, we will construct the homotopy of eqs. (3.85) and (3.86) as:

$$u(x,t) = u(x,0) + t u_t(x,0) + p\left(\mathcal{L}^{-1}\left[\frac{1}{s^\alpha}\mathcal{L}\left[u_{xx} - u + |v|^2\right]\right]\right), \tag{3.87}$$

$$v(x,t) = v(x,0) + ip\left(\mathcal{L}^{-1}\left[\frac{1}{s^\beta}\mathcal{L}[v_{xx} + uv]\right]\right). \tag{3.88}$$

By substituting $u(x,t) = \sum_{n=0}^{\infty} p^n u_n(x,t)$ and $v(x,t) = \sum_{n=0}^{\infty} p^n v_n(x,t)$ in eqs. (3.87) and (3.88), we get:

$$\sum_{n=0}^{\infty} p^n u_n(x,t) = u(x,0) + u_t(x,0) + p\left(\mathcal{L}^{-1}\left[\frac{1}{s^\alpha}\mathcal{L}\left(\left(\sum_{n=0}^{\infty} p^n u_n(x,t)\right)_{xx}\right.\right.\right.$$

$$\left.\left.\left. - \sum_{n=0}^{\infty} p^n u_n(x,t) + \left(\left(\sum_{n=0}^{\infty} p^n v_n(x,t)\right)\left(\sum_{n=0}^{\infty} p^n \bar{v}_n(x,t)\right)\right)\right)\right]\right), \tag{3.89}$$

$$\sum_{n=0}^{\infty} p^n v_n(x,t) = v(x,0) + ip\left(\mathcal{L}^{-1}\left[\frac{1}{s^\alpha}\mathcal{L}\left(\left(\sum_{n=0}^{\infty} p^n v_n(x,t)\right)_{xx}\right.\right.\right.$$

$$\left.\left.\left. + \left(\left(\sum_{n=0}^{\infty} p^n u_n(x,t)\right)\left(\sum_{n=0}^{\infty} p^n v_n(x,t)\right)\right)\right)\right]\right). \tag{3.90}$$

Comparing the coefficients of like powers in p for eqs. (3.89) and (3.90), we have the following system of fractional differential equations.

Coefficients of p^0: $u_0 = u(x,0) + t u_t(x,0)$, $v_0 = v(x,0)$. $\tag{3.91}$

Coefficients of p: $u_1 = \mathcal{L}^{-1}\left[\frac{1}{s^\alpha}\mathcal{L}\left[\left(\frac{\partial^2 u_0}{\partial x^2} - u_0 + v_0\bar{v}_0\right)\right]\right]$, $v_1 = i\mathcal{L}^{-1}\left[\frac{1}{s^\beta}\mathcal{L}\left[\left(\frac{\partial^2 v_0}{\partial x^2} + u_0 v_0\right)\right]\right]$. $\tag{3.92}$

Coefficients of p^2: $u_2 = \mathcal{L}^{-1}\left[\dfrac{1}{s^\alpha}\mathcal{L}\left[\left(\dfrac{\partial^2 u_1}{\partial x^2} - u_1 + v_0\bar{v}_1 + v_1\bar{v}_0\right)\right]\right]$,

$$(3.93)$$

$$v_2 = i\mathcal{L}^{-1}\left[\dfrac{1}{s^\beta}\mathcal{L}\left[\left(\dfrac{\partial^2 v_1}{\partial x^2} + u_0 v_1 + v_0 u_1\right)\right]\right].$$

Coefficients of p^3: $u_3 = \mathcal{L}^{-1}\left[\dfrac{1}{s^\alpha}\mathcal{L}\left[\left(\dfrac{\partial^2 u_2}{\partial x^2} - u_2 + v_1\bar{v}_1 + v_2\bar{v}_0 + v_0\bar{v}_2\right)\right]\right]$,

$$(3.94)$$

$$v_3 = i\mathcal{L}^{-1}\left[\dfrac{1}{s^\beta}\mathcal{L}\left[\left(\dfrac{\partial^2 v_2}{\partial x^2} + u_1 v_1 + v_2 u_0 + v_0 u_2\right)\right]\right],$$

and by putting the initial conditions in eq. (3.68) into eqs. (3.91) to (3.94) and solving them, we obtain:

$$u_0(x,t) = 6B^2\operatorname{sech}^2(Bx) - 12B^2 c\,\operatorname{sech}^2(Bx)\tanh(Bx),$$

$$v_0(x,t) = 3B\operatorname{sech}^2(Bx)e^{idx},$$

$$u_1(x,t) = 3B^2\sec^2(Bx)t^\alpha\left(\frac{(-1+4B^2)(-2+\cosh(2Bx)\operatorname{sech}^2(Bx))}{\Gamma(\alpha+1)}\right.$$
$$\left. + \frac{4ct\tanh(Bx)}{\Gamma(\alpha+2)}\left(1+8B^2\operatorname{sech}^2(Bx)-4B^2\tanh^2(Bx)\right)\right),$$

$$v_1(x,t) = -3iB\operatorname{sech}^2(Bx)e^{idx}\frac{t^\beta}{\Gamma(\beta+1)\Gamma(\beta+2)}\left(12B^2 ct\,\Gamma(\beta+1)\operatorname{sech}^2(Bx)\tanh(Bx)\right.$$
$$\left. + \Gamma(\beta+2)\left(-4B^2 + d^2 + 4iBd\tanh(Bx)\right)\right),$$

$$u_2(x,t) = \frac{3}{4}B^2\sec^6(Bx)\,t^\alpha\left(\frac{48Bd\sinh(2Bx)}{\Gamma(1+\alpha+\beta)} + \frac{t^\alpha}{\Gamma(2\alpha+2)}\left(3(-1-40B^2+176B^4)(1+2\alpha)\right.\right.$$
$$-2(1-56B^2+208B^4)(1+2\alpha)\cosh(2Bx) + 2ct(1-4B^2)^2$$
$$(1+2\alpha)\cosh(4Bx)\left((-2-80B^2+928B^4)\sinh(2Bx)\right.$$
$$\left.\left.\left. -(1-4B^2)^2\sinh(4Bx) - 2880B^4\tanh(Bx)\right)\right)\right)$$

$$v_2(x,t) = -\frac{1}{\Gamma(\beta+1)\Gamma(\beta+2)\Gamma(2+\alpha+\beta)\Gamma(3+2\beta)}\Big(3B\,\mathrm{sech}^2(Bx)e^{idx}t^\beta\Big(144B^4c^2t^{\beta+2}$$

$$\Gamma(\beta+1)\Gamma(\beta+3)\Gamma(2+\alpha+\beta)\,\mathrm{sech}^4(Bx)\tanh^2(Bx)+3iB^2t^\alpha\,\Gamma(\beta+1)\Gamma(\beta+2)$$

$$\Gamma(3+2\beta)\,\mathrm{sech}^2(Bx)((-1+4B^2)(1+\alpha+\beta)(-2+3\,\mathrm{sech}^2(Bx)-4ct(1-4B^2$$

$$+12B^2\,\mathrm{sech}^2(Bx))\tanh(Bx))+2t^\beta\Gamma(\beta+2)^2\Gamma(2+\alpha+\beta)(24B^3ct$$

$$\mathrm{sech}^4(Bx)(-id(7+2\beta)+12B\tanh(Bx)+(1+2\beta)(16B^4-24B^2d^2+d^4$$

$$-8iBd(4B^2-d^2)\tanh(Bx))+12B^2\,\mathrm{sech}^2(Bx)(2d(d+2d\beta+2iBct(3+\beta))$$

$$+(cd^2t(2+\beta)-4B^2ct(5+\beta)+2iB(d+2d\beta))\tanh(Bx))\Big)\Big)\Big),$$

and so on.

In this manner, the other components of the homotopy series can be easily obtained by which $u(x,t)$ and $v(x,t)$ can be evaluated in a series form:

$$u(x,t) = u_0(x,t) + u_1(x,t) + u_2(x,t)$$

$$= 6B^2\,\mathrm{sech}^2(Bx) - 12B^2c\,\mathrm{sech}^2(Bx)\tanh(Bx) + 3B^2\,\mathrm{sec}^2(Bx)t^\alpha$$

$$\left(\frac{(-1+4B^2)(-2+\cosh(2Bx))\,\mathrm{sech}^2(Bx)}{\Gamma(\alpha+1)} + \frac{4ct\tanh(Bx)}{\Gamma(\alpha+2)}\right.$$

$$\left(1+8B^2\,\mathrm{sech}^2(Bx)4B^2\tanh^2(Bx)\right)$$

$$+\frac{3}{4}B^2\,\mathrm{sec}^6(Bx)t^\alpha\left(\frac{48B\sinh^2(2Bx)}{\Gamma(1+\alpha+\beta)}\right.$$

$$+\frac{t^\alpha}{\Gamma(2\alpha+2)}\Big(3\big(-1-40B^2+176B^4\big)(1+2\alpha)$$

$$-2(1-56B^2+208B^4)(1+2\alpha)\cosh(2Bx)$$

$$+(1-4B^2)^2(1+2\alpha)\cosh(4Bx)$$

$$2ct((-2-80B^2+928B^4)\sinh(2Bx)$$

$$\left.\left.-(1-4B^2)^2\sinh(4Bx)-2880B^4\tanh(Bx)\Big)\right)\right). \tag{3.95}$$

$$v(x,t) = v_0(x,t) + v_1(x,t) + v_2(x,t)$$

$$= 3B\operatorname{sech}^2(Bx)e^{idx} + 3iB\operatorname{sech}^2(Bx)e^{idx}\frac{t^\beta}{\Gamma(\beta+1)\Gamma(\beta+2)}$$

$$(12B^2ct\,\Gamma(\beta+1)\operatorname{sech}^2(Bx)\tanh(Bx) + \Gamma(\beta+2)(-4B^2+d^2+4iBd\tan h(Bx)))$$

$$-\frac{1}{\Gamma(\beta+1)\Gamma(\beta+2)\Gamma(2+\alpha+\beta)\Gamma(3+2\beta)}\Big(3B\operatorname{sech}^2(Bx)e^{idx}t^\beta\Big(144B^4c^2t^{\beta+2}$$

$$\Gamma(\beta+1)\Gamma(\beta+3)\Gamma(2+\alpha+\beta)\operatorname{sech}^4(Bx)\tanh^2(Bx) + 3iB^2t^\alpha\,\Gamma(\beta+1)\Gamma(\beta+2)$$

$$\Gamma(3+2\beta)\operatorname{sech}^2(Bx)((-1+4B^2)(1+\alpha+\beta)(-2+3\operatorname{sech}^2(Bx) - 4ct(1-4B^2$$

$$+12B^2\operatorname{sech}^2(Bx))\tan h(Bx)) + 2t^\beta\Gamma(\beta+2)^2\,\Gamma(2+\alpha+\beta)(24B^3ct$$

$$\operatorname{sech}^4(Bx)(-id(7+2\beta) + 12B\tanh(Bx) + (1+2\beta)(16B^4 - 24B^2d^2 + d^4$$

$$-8iBd(4B^2 - d^2)\tan h(Bx)) + 12B^2\operatorname{sech}^2(Bx)(2d(d+2d\beta + 2iBct(3+\beta))$$

$$+(cd^2t(2+\beta) - 4B^2ct(5+\beta) + 2iB(d+2d\beta))\tan h(Bx))\Big)\Big).$$

(3.96)

3.4.3 Numerical Results and Discussion

In the present numerical computation, we have used HPTM approximate solutions for which we have assumed $B = 0.575$.

3.4.3.1 The Numerical Simulations for HPTM Method

In this present numerical experiment, eqs. (3.95) and (3.96) have been used to draw the graphs as shown in Figure 3.4 to 3.15, respectively. The numerical solutions of coupled K–G–S eqs. (3.10) and (3.11) have been shown in Figures 3.4 to 3.15 with the help of five-term approximations for the homotopy series solutions of $u(x,t)$ and $v(x,t)$, respectively, Figures 3.4 to 3.8.

In Figures 3.4 to 3.7, the HPTM approximate solutions of $u(x, t)$ and $v(x, t)$ are plotted for the intervals $-10 \le x \le 10$ and $0 \le t \le 1$ for the classical order, that is for $\alpha = 2$ and $\beta = 1$. In Figures 3.8 to 3.15, the HPTM approximate solutions of $u(x, t)$ and $v(x, t)$ are plotted for the same intervals for the fractional order, that is for $\alpha = 1.75$, $\beta = 0.75$ and $\alpha = 1.5$, $\beta = 0.5$, respectively. As value of α decreases from 2 to 1.5 and β decreases from 1 to 0.5, the distribution of $u(x,t)$ bifurcated into two waves. Similarly, when α decreases from 2 to 1.5 and β decreases from 1 to 0.5, the distribution of $\operatorname{Re}(v(x,t))$ and $\operatorname{Im}(v(x,t))$ bifurcated into three waves.

New Analytical Approximate Solutions of Fractional Differential Equations

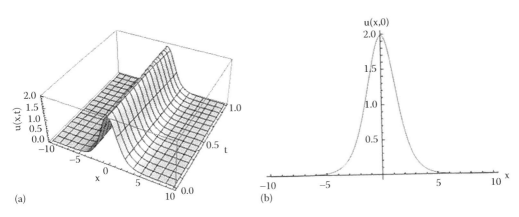

FIGURE 3.4
Case I: For $\alpha = 2$, $\beta = 1$ (Classical order). (a) The HPTM solution for $u(x,t)$, (b) corresponding solution for $u(x,t)$ when $t = 0$.

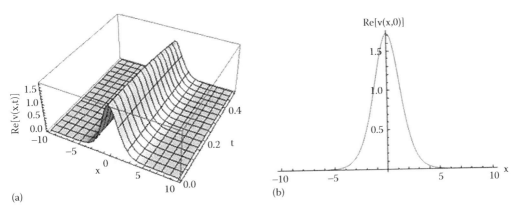

FIGURE 3.5
Case I: For $\alpha = 2$, $\beta = 1$. (a) The HPTM solution for $\text{Re}(v(x,t))$, (b) corresponding solution for $\text{Re}(v(x,t))$ when $t = 0$.

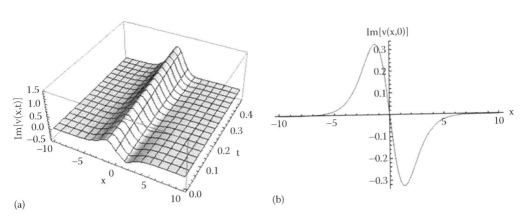

FIGURE 3.6
Case I: For $\alpha = 2$, $\beta = 1$. (a) The HPTM solution for $\text{Im}(v(x,t))$, (b) corresponding solution for $\text{Im}(v(x,t))$ when $t = 0$.

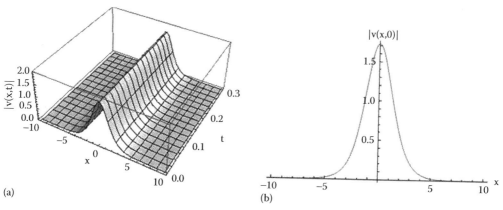

FIGURE 3.7
Case I: For $\alpha = 2$, $\beta = 1$. (a) The HPTM solution for $|v(x,t)|$, (b) corresponding solution for $|v(x,t)|$ when $t = 0$.

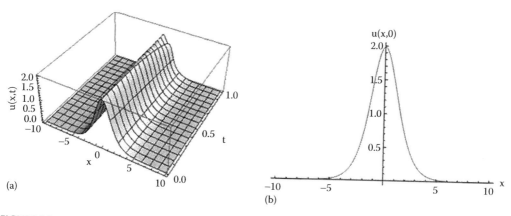

FIGURE 3.8
Case II: For $\alpha = 1.75$, $\beta = 0.75$ (Fractional order). (a) The HPTM solution for $u(x,t)$, (b) corresponding solution for $u(x,t)$ when $t = 0$.

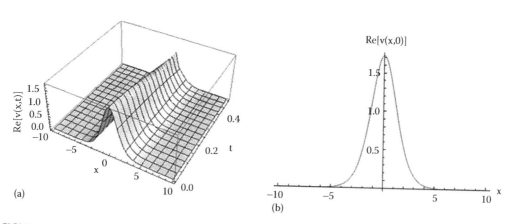

FIGURE 3.9
Case II: For $\alpha = 1.75$, $\beta = 0.75$. (a) The HPTM solution for $\text{Re}(v(x,t))$, (b) corresponding solution for $\text{Re}(v(x,t))$ when $t = 0$.

New Analytical Approximate Solutions of Fractional Differential Equations 75

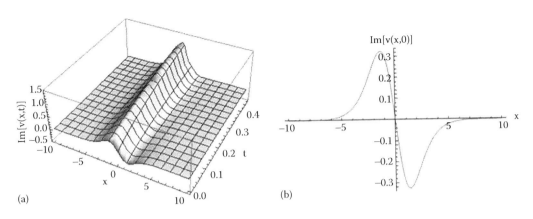

FIGURE 3.10
Case II: For $\alpha = 175$, $\beta = 0.75$. (a) The HPTM solution for $\text{Im}(v(x,t))$, (b) corresponding solution for $\text{Im}(v(x,t))$ when $t = 0$.

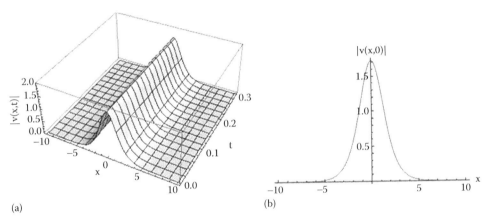

FIGURE 3.11
Case II: For $\alpha = 175$, $\beta = 0.75$. (a) The HPTM solution for $|v(x,t)|$, (b) corresponding solution for $|v(x,t)|$ when $t = 0$.

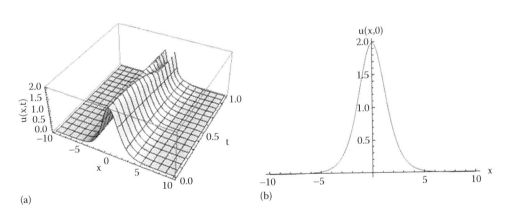

FIGURE 3.12
Case III: For $\alpha = 1.5$, $\beta = 0.5$ (Fractional order). (a) The HPTM solution for $u(x,t)$, (b) corresponding solution for $u(x,t)$ when $t = 0$.

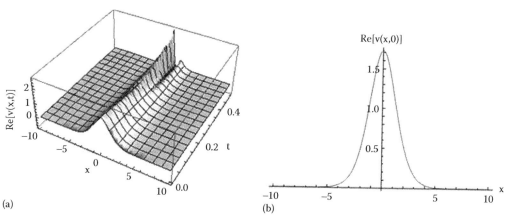

FIGURE 3.13

Case III: For $\alpha = 1.5$, $\beta = 0.5$. (a) The HPTM solution for $\mathrm{Re}(v(x,t))$, (b) corresponding solution for $\mathrm{Re}(v(x,t))$ when $t = 0$.

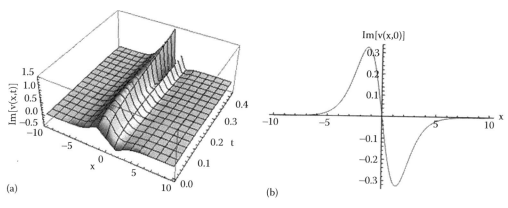

FIGURE 3.14

Case III: For $\alpha = 1.5$, $\beta = 0.5$. (a) The HPTM solution for $\mathrm{Im}(v(x,t))$, (b) corresponding solution for $\mathrm{Im}(v(x,t))$ when $t = 0$.

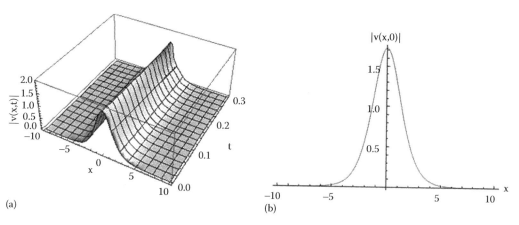

FIGURE 3.15

Case III: For $\alpha = 1.5$, $\beta = 0.5$. (a) The HPTM solution for $|v(x,t)|$, (b) corresponding solution for $|v(x,t)|$ when $t = 0$.

3.4.3.2 The Numerical Simulations for Absolute Errors in HPM and HPTM Solutions

In this section, we present Figures 3.16 to 3.19, citing the numerical simulations for comparison of absolute errors in solutions of $u(x,t)$ and $v(x,t)$ obtained by the HPM and HPTM at $x = 0.3$.

Although both the methods are reliable and efficient, but Figures 3.16 to 3.19 assure plausibility to consider the HPTM provides more accurate solutions than the HPM solutions for coupled fractional K–G–S equations.

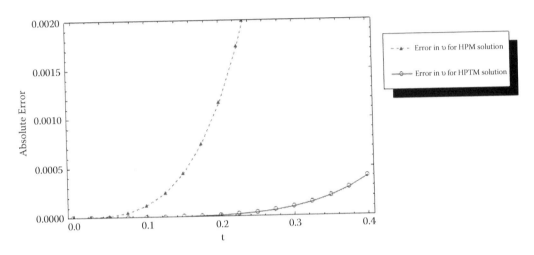

FIGURE 3.16
Graphical comparison of absolute errors in the solution of $u(x,t)$ for HPM and HPTM, when $\alpha = 2, \beta = 1$.

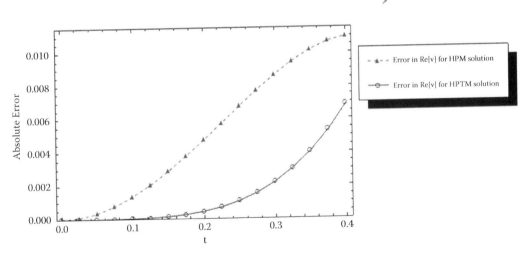

FIGURE 3.17
Graphical comparison of absolute errors in the solution of $\text{Re}(v(x,t))$ for HPM and HPTM, when $\alpha = 2, \beta = 1$.

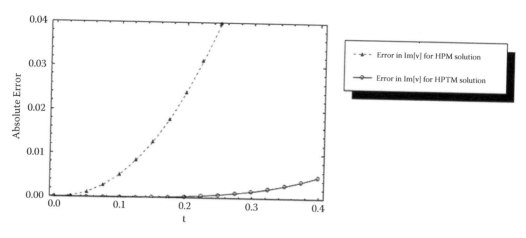

FIGURE 3.18
Graphical comparison of absolute errors in the solution of $\text{Im}(v(x,t))$ for HPM and HPTM, when $\alpha = 2, \beta = 1$.

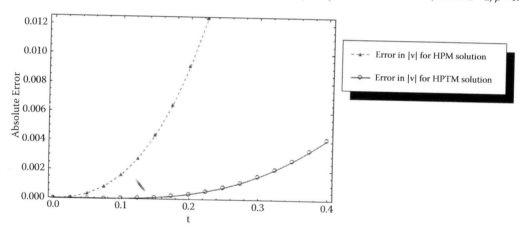

FIGURE 3.19
Graphical comparison of absolute errors in the solution of $|v(x,t)|$ for HPM and HPTM, when $\alpha = 2, \beta = 1$.

3.4.3.3 Comparison of HPM and HPTM Solutions with Regard to Exact Solutions

In the case of integer order, exact solutions of eqs. (3.10) and (3.11) are given by [99]:

$$u(x,t) = 6B^2 \text{sech}^2(Bx+ct), \tag{3.97}$$

$$v(x,t) = 3B\,\text{sech}^2(Bx+ct)\,e^{i(dx+(4B^2-d^2)t)}. \tag{3.98}$$

where $B(\geq 1/2)$, c and d are arbitrary constants with $c = \sqrt{4B^2-1}/2$, $d = -c/2B$ for the fractional coupled K–G–S eqs. (3.10) and (3.11).

In this present analysis, we present the comparison for the solutions of the HPM with HPTM. Here, we demonstrate the absolute and relative errors by taking different values of t with respect to some fixed value of x.

New Analytical Approximate Solutions of Fractional Differential Equations 79

TABLE 3.1

Comparison of Absolute Errors between Four Term HPM and HPTM Solutions with Regard to Exact Solutions for Different Values of t Respectively When $x = 1.5$ in Case of $\alpha = 2$ and $\beta = 1$

				$x = 1.5$				
	Absolute Error between Exact Solution and Four Term HPM Solution				**Absolute Error between Exact Solution and Four Term HPTM Solution**			
t	$u(x,t)$	$Re(v(x,t))$	$Im(v(x,t))$	$v(x,t)$	$u(x,t)$	$Re(v(x,t))$	$Im(v(x,t))$	$|v(x,t)|$
0.2	2.464E-5	1.192E-2	1.002E-2	1.3071E-2	4.107E-8	2.09E-4	1.693E-4	1.877E-4
0.4	3.79E-4	2.597E-2	5.46E-2	2.05E-2	4.56E-6	3.358E-3	2.926E-3	3.716E-3
0.6	1.734E-3	8.084E-3	1.585E-1	3.487E-2	5.157E-5	1.701E-2	1.581E-2	2.16E-2
0.8	4.599E-3	7.651E-2	3.496E-1	1.999E-1	2.645E-4	5.37E-2	5.295E-2	7.4637E-2
1.0	8.418E-3	2.619E-1	6.594E-1	3.403E-1	8.99E-4	1.31E-1	1.362E-1	1.8816E-1

TABLE 3.2

Comparison of Relative Errors between Four Term HPM and HPTM Solutions with Regard to Exact Solutions for Different Values of t Respectively When $x = 1.5$ in Case of $\alpha = 2$ and $\beta = 1$

				$x = 1.5$				
	Relative Error between Exact Solution and Four Term HPM Solution				**Relative Error between Exact Solution and Four Term HPTM Solution**			
t	$u(x,t)$	$Re(v(x,t))$	$Im(v(x,t))$	$v(x,t)$	$u(x,t)$	$Re(v(x,t))$	$Im(v(x,t))$	$|v(x,t)|$
0.2	8.05E-4	7.17E-2	2.43E-2	7.05E-2	4.33E-8	2.57E-4	1.76E-3	2.29E-4
0.4	6.31E-3	3.03E-1	1.01E-1	3.009E-1	5.08E-6	4.51E-3	2.91E-2	4.94E-3
0.6	2.06E-2	7.67E-1	2.22E-2	7.009E-1	5.82E-5	2.66E-2	6.08E-2	3.14E-2
0.8	4.63E-2	6.52E-1	3.36E-1	1.27	2.79E-4	1.06E-1	1.40E-1	1.18E-1
1.0	8.35E-2	3.462	5.05E-1	2.03	7.65E-4	3.61E-1	3.01E-1	3.24E-1

In the case of $\alpha = 2$ and $\beta = 1$, Table 3.1 represents comparison of absolute errors four term HPM and HPTM solutions with respect to exact solutions when $x = 1.5$, and Table 3.2 represents comparison of relative errors four term HPM and HPTM solutions with respect to exact solutions when $x = 1.5$. Comparison results in Tables 3.1 and 3.2 exhibit that there is a good agreement between the HPM and HPTM solutions with exact solutions. Although, the HPTM provides more accurate solutions compared to the HPM solutions, which can be easily verified from Tables 3.1 and 3.2.

3.5 Application of HPTM and MHAM Methods for the Solutions of Fractional Coupled K–G–Z Equations

In this section, two reliable methods, viz., the HPTM and MHAM have been applied for getting analytical approximate solutions of fractional coupled K–G–Z equations.

3.5.1 Implementation of HPTM Method for the Solutions of Fractional Coupled K–G–Z Equations

In this section, we first consider the application of the HPTM for the solution of fractional coupled K–G–Z eqs. (3.12) and (3.13) with given initial conditions:

$$u(x,0) = A\operatorname{sech}(Bx)\exp\left(\frac{ix}{B}\right), \tag{3.99}$$

$$u_t(x,0) = A\operatorname{sech}(Bx)\tanh(Bx)\exp\left(\frac{ix}{B}\right) - A\operatorname{sech}(Bx)\exp\left(\frac{ix}{B}\right),$$

$$v(x,0) = -2\operatorname{sech}^2(Bx),$$

$$v_t(x,0) = -4\operatorname{sech}^2(Bx)\tanh(Bx),$$

where $A = \sqrt{10} - \sqrt{2}/2$ and $B = \sqrt{1+\sqrt{5}/2}$ for fractional coupled K–G–Z eqs. (3.12) and (3.13).

Applying Laplace transform on both sides of eqs. (3.12) and (3.13), respectively, we get:

$$\mathcal{L}[u(x,t)] = \frac{u(x,0)}{s} + \frac{u_t(x,0)}{s^2} + \frac{1}{s^\alpha}\mathcal{L}[u_{xx}] - \frac{1}{s^\alpha}\mathcal{L}[u] - \frac{1}{s^\alpha}\mathcal{L}[uv] - \frac{1}{s^\alpha}\mathcal{L}[|u|^2 u], \tag{3.100}$$

$$\mathcal{L}[v(x,t)] = \frac{v(x,0)}{s} + \frac{v_t(x,0)}{s^2} + \frac{1}{s^\beta}\mathcal{L}[v_{xx}] + \frac{1}{s^\beta}\mathcal{L}\left[\left(|u|^2\right)_{xx}\right]. \tag{3.101}$$

Then applying inverse Laplace transform on both sides of eqs. (3.100) and (3.101), respectively, we get:

$$u(x,t) = u(x,0) + tu_t(x,0) + \mathcal{L}^{-1}\left[\frac{1}{s^\alpha}\mathcal{L}\left[u_{xx} - u - uv - |u|^2 u\right]\right], \tag{3.102}$$

$$v(x,t) = v(x,0) + tv_t(x,0) + \mathcal{L}^{-1}\left[\frac{1}{s^\beta}\mathcal{L}\left[v_{xx} + \left(|u|^2\right)_{xx}\right]\right]. \tag{3.103}$$

By using the homotopy perturbation method, we will construct the homotopy of eqs. (3.102) and (3.103) as:

$$u(x,t) = u(x,0) + tu_t(x,0) + p\left(\mathcal{L}^{-1}\left[\frac{1}{s^\alpha}\mathcal{L}\left[u_{xx} - u - uv - |u|^2 u\right]\right]\right), \tag{3.104}$$

$$v(x,t) = v(x,0) + tv_t(x,0) + p\left(\mathcal{L}^{-1}\left[\frac{1}{s^\beta}\mathcal{L}\left[v_{xx} + \left(|u|^2\right)_{xx}\right]\right]\right). \tag{3.105}$$

New Analytical Approximate Solutions of Fractional Differential Equations 81

By substituting $u(x,t) = \sum_{n=0}^{\infty} p^n u_n(x,t)$ and $v(x,t) = \sum_{n=0}^{\infty} p^n v_n(x,t)$ in eqs. (3.104) and (3.105), we get:

$$
\begin{aligned}
\sum_{n=0}^{\infty} p^n u_n(x,t) = u(x,0) + t u_t(x,0) + p\Bigg(&\mathcal{L}^{-1}\Bigg[\frac{1}{s^\alpha}\mathcal{L}\Bigg(\Bigg(\sum_{n=0}^{\infty} p^n u_n(x,t)\Bigg)_{xx} \\
&- \sum_{n=0}^{\infty} p^n u_n(x,t) - \Bigg(\Bigg(\sum_{n=0}^{\infty} p^n u_n(x,t)\Bigg)\Bigg(\sum_{n=0}^{\infty} p^n v_n(x,t)\Bigg)\Bigg) \\
&- \Bigg(\Bigg(\sum_{n=0}^{\infty} p^n u_n(x,t)\Bigg)^2\Bigg(\sum_{n=0}^{\infty} p^n \bar{u}_n(x,t)\Bigg)\Bigg)\Bigg]\Bigg),
\end{aligned}
$$
(3.106)

$$
\begin{aligned}
\sum_{n=0}^{\infty} p^n v_n(x,t) = v(x,0) + t v_t(x,0) + p\Bigg(&\mathcal{L}^{-1}\Bigg[\frac{1}{s^\alpha}\mathcal{L}\Bigg(\Bigg(\sum_{n=0}^{\infty} p^n v_n(x,t)\Bigg)_{xx} \\
&+ \Bigg(\Bigg(\sum_{n=0}^{\infty} p^n u_n(x,t)\Bigg)\Bigg(\sum_{n=0}^{\infty} p^n \bar{u}_n(x,t)\Bigg)\Bigg)_{xx}\Bigg)\Bigg]\Bigg).
\end{aligned}
$$
(3.107)

Comparing the coefficients of like powers in p for eqs. (3.106) and (3.107), we have the following system of fractional differential equations:

Coefficients of $p^0 : u_0 = u(x,0) + t u_t(x,0)$, $v_0 = v(x,0) + t v_t(x,0)$. $\qquad$ (3.108)

Coefficients of $p : u_1 = \mathcal{L}^{-1}\Bigg[\dfrac{1}{s^\alpha}\mathcal{L}\Bigg[\dfrac{\partial^2 u_0}{\partial x^2} - u_0 - u_0 v_0 - u_0^2 \bar{u}_0\Bigg]\Bigg]$, $v_1 = \mathcal{L}^{-1}\Bigg[\dfrac{1}{s^\beta}\mathcal{L}\Bigg[\dfrac{\partial^2 v_0}{\partial x^2} + (u_0 \bar{u}_0)_{xx}\Bigg]\Bigg]$. $\quad$ (3.109)

Coefficients of $p^2 : u_2 = \mathcal{L}^{-1}\Bigg[\dfrac{1}{s^\alpha}\mathcal{L}\Bigg[\dfrac{\partial^2 u_1}{\partial x^2} - u_1 - (u_0 v_1 + v_0 u_1) - (u_0^2 \bar{u}_1 + 2 u_0 u_1 \bar{u}_0)\Bigg]\Bigg]$, $\qquad$ (3.110)

$$
v_2 = \mathcal{L}^{-1}\Bigg[\frac{1}{s^\beta}\mathcal{L}\Bigg[\frac{\partial^2 v_1}{\partial x^2} + (u_0 \bar{u}_1 + \bar{u}_0 u_1)_{xx}\Bigg]\Bigg],
$$

and so on.

By putting the initial conditions in eq. (3.99) into eqs. (3.108) to (3.110) and solving them, we obtain:

$$
u_0(x,t) = A\operatorname{sech}(Bx)\exp\left(\frac{ix}{B}\right) + t\left(A\operatorname{sech}(Bx)\tanh(Bx)\exp\left(\frac{ix}{B}\right) - A\operatorname{sech}(Bx)\exp\left(\frac{ix}{B}\right)\right),
$$

$$
v_0(x,t) = -2\operatorname{sech}^2(Bx) - t\big(4\operatorname{sech}^2(Bx)\tanh(Bx)\big),
$$

$$u_1(x,t) = \frac{1}{16B^2\Gamma(\alpha+1)\Gamma(\alpha+2)\Gamma(\alpha+3)\Gamma(\alpha+4)}\left(\exp\left(\frac{ix}{B}\right)At^\alpha\sec^6(Bx)\right.$$

$$\left(it\Gamma(\alpha+1)\left(4\cosh^2(Bx)\Gamma(\alpha+3)\Gamma(\alpha+4)\left(\left(3+5B^4+B^2\left(5+4A^2\right)\right)\cos(Bx)\right.\right.$$

$$-\left(-1+B^2+B^4\right)\cosh(3Bx)-2i\left(-1-11B^4+B^2\left(9-6A^2\right)+\left(-1-3B^2+B^4\right)\cosh(2Bx)\right)$$

$$\sinh(Bx)\right)+16B^2t\Gamma(\alpha+2)\left(\cosh(Bx)+\sinh(Bx)\right)\left(6A^2t\cosh(2Bx)\Gamma(\alpha+3)\right.$$

$$+2\cosh(Bx)\Gamma(\alpha+4)\left(iA^2\cosh(Bx)+\left(-4+3A^2\right)\sinh(Bx)\right)\right)\right)+8\cosh^3(Bx)$$

$$\Gamma(\alpha+2)\Gamma(\alpha+3)\Gamma(\alpha+4)\left(-1+3B^2-3B^4-2B^2A^2+\left(-1-B^2+B^4\right)\right.$$

$$\left.\left.\cosh(2Bx)2iB^2\sinh(2Bx)\right)\right)\right),$$

$$v_1(x,t) = 2B^2\operatorname{sech}^4(Bx)\frac{t^\beta}{\Gamma(\beta+1)\Gamma(\beta+2)\Gamma(\beta+3)}\left(2t\left(-2+A^2\right)\left(-5+\cosh(2Bx)\Gamma(\beta+1)\right.\right.$$

$$\Gamma(\beta+3)\tanh(Bx)+\Gamma(\beta+2)\left(\left(-2+A^2\right)\left(-2+\cosh(2Bx)\right)\Gamma(\beta+3)\right.$$

$$\left.\left.+4A^2t^2\left(-4+\cosh(2Bx)\right)\Gamma(\beta+1)\tan^2(Bx)\right),\right.$$

and so on.

In this manner, the other components of the homotopy series can be easily obtained by which $u(x,t)$ and $v(x,t)$ can be evaluated in a series form:

$$u(x,t) = u_0(x,t)+u_1(x,t) = A\operatorname{sech}(Bx)\exp\left(\frac{ix}{B}\right)$$

$$+t\left(A\operatorname{sech}(Bx)\tanh(Bx)\exp\left(\frac{ix}{B}\right)-A\operatorname{sech}(Bx)\exp\left(\frac{ix}{B}\right)\right)$$

$$+\frac{1}{16B^2\Gamma(\alpha+1)\Gamma(\alpha+2)\Gamma(\alpha+3)\Gamma(\alpha+4)}\left(\exp\left(\frac{ix}{B}\right)At^\alpha\sec^6(Bx)\right.$$

$$\left(it\Gamma(\alpha+1)\left(4\cosh^2(Bx)\Gamma(\alpha+3)\Gamma(\alpha+4)\left(\left(3+5B^4+B^2\left(5+4A^2\right)\right)\cos(Bx)\right.\right.$$

$$-\left(-1+B^2+B^4\right)\cosh(3Bx)-2i\left(-1-11B^4+B^2\left(9-6A^2\right)+\left(-1-3B^2+B^4\right)\cosh(2Bx)\right)$$

$$\sinh(Bx)\right)+16B^2t\Gamma(\alpha+2)\left(\cosh(Bx)+\sinh(Bx)\right)\left(6A^2t\cosh(2Bx)\Gamma(\alpha+3)\right.$$

$$+2\cosh(Bx)\Gamma(\alpha+4)\left(iA^2\cosh(Bx)+\left(-4+3A^2\right)\sinh(Bx)\right)\right)\right)+8\cosh^3(Bx)$$

$$\Gamma(\alpha+2)\Gamma(\alpha+3)\Gamma(\alpha+4)\left(-1+3B^2-3B^4-2B^2A^2+\left(-1-B^2+B^4\right)\right.$$

$$\left.\left.\cosh(2Bx)2iB^2\sinh(2Bx)\right)\right)\right),$$

(3.111)

New Analytical Approximate Solutions of Fractional Differential Equations

$$v(x,t) = v_0(x,t) + v_1(x,t)$$

$$= -2\operatorname{sech}^2(Bx) - t\left(4\operatorname{sech}^2(Bx)\tanh(Bx)\right) + 2B^2\operatorname{sech}^4(Bx)\frac{t^\beta}{\Gamma(\beta+1)\Gamma(\beta+2)\Gamma(\beta+3)}$$

$$\left(2t\left(-2+A^2\right)\left(-5+\cosh(2Bx)\Gamma(\beta+1)\Gamma(\beta+3)\tanh(Bx)+\Gamma(\beta+2)\right)\right) \tag{3.112}$$

$$\left(\left(-2+A^2\right)\left(-2+\cosh(2Bx)\right)\Gamma(\beta+3)+4A^2t^2\left(-4+\cosh(2Bx)\right)\Gamma(\beta+1)\tan^2(Bx)\right).$$

3.5.2 Implementation of the MHAM for Approximate Solutions to the K–G–Z Equations

In this section, we first consider the application of the MHAM for the solution of fractional coupled K–G–Z equations (3.12) and (3.13).

Expanding $\phi_i(x,t;p)$ in a Taylor series with respect to p for $i = 1, 2, 3$, we have:

$$\phi_1(x,t;p) = u_0(x,t) + \sum_{m=1}^{+\infty} p^m u_m(x,t),$$

$$\phi_2(x,t;p) = v_0(x,t) + \sum_{m=1}^{+\infty} p^m v_m(x,t), \tag{3.113}$$

$$\phi_3(x,t;p) = \bar{u}_0(x,t) + \sum_{m=1}^{+\infty} p^m \bar{u}_m(x,t),$$

where:

$$u_m(x,t) = \frac{1}{m!}\frac{\partial^m \phi_1(x,t;p)}{\partial p^m}\bigg|_{p=0},$$

$$v_m(x,t) = \frac{1}{m!}\frac{\partial^m \phi_2(x,t;p)}{\partial p^m}\bigg|_{p=0}, \tag{3.114}$$

$$\bar{u}_m(x,t) = \frac{1}{m!}\frac{\partial^m \phi_3(x,t;p)}{\partial p^m}\bigg|_{p=0}.$$

To obtain the approximate solution of the K–G–Z equations [102, 103], we choose the linear operators:

$$L_1\left[\phi_1(x,t)\right] = D_t^\alpha \phi_1(x,t;p) \text{ and } L_2\left[\phi_2(x,t)\right] = D_t^\beta \phi_2(x,t;p). \tag{3.115}$$

From equations (3.12) and (3.13), we define a system of nonlinear operators as:

$$N_1\left[\phi_1(x,t;p),\phi_2(x,t;p),\phi_3(x,t;p)\right]$$
$$= D_t^\alpha\left[\phi_1(x,t;p)\right]-\left(\phi_1(x,t;p)\right)_{xx}+\phi_1(x,t;p)+\phi_1(x,t;p)\phi_2(x,t;p)$$
$$+\left(\phi_1(x,t;p)\right)^2\phi_3(x,t;p), \tag{3.116}$$

$$N_2\left[\phi_1(x,t;p),\phi_2(x,t;p),\phi_3(x,t;p)\right]$$
$$= D_t^\beta\left[\phi_2(x,t;p)\right]-\left(\phi_2(x,t;p)\right)_{xx}-\left(\phi_1(x,t;p)\phi_3(x,t;p)\right)_{xx},$$

where the non-linear terms $\phi_1(x,t;p)\phi_2(x,t;p)$, $\left(\phi_1(x,t;p)\right)^2\phi_3(x,t;p)$ and $\left(\phi_1(x,t;p)\phi_3(x,t;p)\right)_{xx}$ are expanded in Adomian like polynomials.

The first non-linear term $\phi_1(x,t;p)\phi_2(x,t;p)$ has been taken as:

$$\phi_1(x,t;p)\phi_2(x,t;p)=\sum_{n=0}^{\infty}p^n A_{1,n},$$

where:

$$A_{1,n}=\frac{1}{n!}\frac{\partial^n}{\partial p^n}\left(\phi_1(x,t;p)\phi_2(x,t;p)\right)\bigg|_{p=0}$$
$$=\frac{1}{n!}\frac{\partial^n}{\partial p^n}\left(\left(u_0(x,t)+\sum_{n=1}^{+\infty}p^n u_n(x,t)\right)\left(v_0(x,t)+\sum_{n=1}^{+\infty}p^n v_n(x,t)\right)\right)\bigg|_{p=0},\ n\ge 0. \tag{3.117}$$

Similarly, the second non-linear term $\left(\phi_1(x,t;p)\right)^2\phi_3(x,t;p)$ has been taken as:

$$\left(\phi_1(x,t;p)\right)^2\phi_3(x,t;p)=\sum_{n=0}^{\infty}p^n A_{2,n},$$

where:

$$A_{2,n}=\frac{1}{n!}\frac{\partial^n}{\partial p^n}\left(\left(\phi_1(x,t;p)\right)^2\phi_3(x,t;p)\right)\bigg|_{p=0}$$
$$=\frac{1}{n!}\frac{\partial^n}{\partial p^n}\left(\left(u_0(x,t)+\sum_{n=1}^{+\infty}p^n u_n(x,t)\right)^2\left(\bar{u}_0(x,t)+\sum_{n=1}^{+\infty}p^n \bar{u}_n(x,t)\right)\right)\bigg|_{p=0}. \tag{3.118}$$

Finally, the third non-linear term $\left(\phi_1(x,t;p)\phi_3(x,t;p)\right)_{xx}$ has been taken as:

$$\left(\phi_1(x,t;p)\phi_3(x,t;p)\right)_{xx}=\sum_{n=0}^{\infty}p^n A_{3,n},$$

New Analytical Approximate Solutions of Fractional Differential Equations

where:

$$A_{3,n} = \frac{1}{n!} \frac{\partial^n}{\partial p^n} \left(\left(\phi_1(x,t;p) \phi_3(x,t;p) \right)_{xx} \right) \Big|_{p=0}$$

$$= \frac{1}{n!} \frac{\partial^n}{\partial p^n} \left(\left(\left(u_0(x,t) + \sum_{n=1}^{+\infty} p^n u_n(x,t) \right) \left(\bar{u}_0(x,t) + \sum_{n=1}^{+\infty} p^n \bar{u}_n(x,t) \right) \right)_{xx} \right) \Big|_{p=0}, n \geq 0.$$

(3.119)

Using the MHAM, we construct the so-called zeroth-order deformation equations:

$$(1-p)L_1 \left[\phi_1(x,t;p) - u_0(x,t) \right] = p\hbar N_1 \left[\phi_1(x,t;p), \phi_2(x,t;p), \phi_3(x,t;p) \right],$$

$$(1-p)L_2 \left[\phi_2(x,t;p) - u_0(x,t) \right] = p\hbar N_2 \left[\phi_1(x,t;p), \phi_2(x,t;p), \phi_3(x,t;p) \right].$$

(3.120)

Obviously, when $p = 0$ and $p = 1$, eq. (3.113) yields:

$$\phi_1(x,t;0) = u(x,0), \phi_1(x,t;1) = u(x,t),$$

$$\phi_2(x,t;0) = v(x,0), \phi_2(x,t;1) = v(x,t),$$

$$\phi_3(x,t;0) = \bar{u}(x,0), \phi_3(x,t;1) = \bar{u}(x,t).$$

Therefore, as the embedding parameter p increases from 0 to 1, $\phi_i(x,t;p)$ varies from the initial guess to the exact solutions $u(x,t)$, $v(x,t)$, and $\bar{u}(x,t)$, for $i = 1, 2, 3$, respectively.

If the auxiliary linear operator, the initial guess, and the auxiliary parameters $\hbar$ are so properly chosen, the above series in eq. (3.113) converge at $p = 1$, and we obtain:

$$u(x,t) = u_0(x,t) + \sum_{m=1}^{+\infty} u_m(x,t), v(x,t) = v_0(x,t) + \sum_{m=1}^{+\infty} v_m(x,t).$$

(3.121)

According to eq. (2.23) of Chapter 2, we have the mth-order deformation equations:

$$L_1 \left[u_m(x,t) - \chi_m u_{m-1}(x,t) \right] = \hbar \Re_{1,m} \left(u_0, u_1, ..., u_{m-1}, v_0, v_1, ..., v_{m-1}, \bar{u}_0, \bar{u}_1, ..., \bar{u}_{m-1} \right), m \geq 1,$$

$$L_2 \left[v_m(x,t) - \chi_m v_{m-1}(x,t) \right] = \hbar \Re_{2,m} \left(u_0, u_1, ..., u_{m-1}, v_0, v_1, ..., v_{m-1}, \bar{u}_0, \bar{u}_1, ..., \bar{u}_{m-1} \right), m \geq 1,$$

(3.122)

where:

$$\Re_{1,m} \left(u_0, u_1, ..., u_{m-1}, v_0, v_1, ..., v_{m-1}, \bar{u}_0, \bar{u}_1, ..., \bar{u}_{m-1} \right)$$

$$= \frac{1}{(m-1)!} \frac{\partial^{m-1}}{\partial p^{m-1}} \left(N_1[\phi_1(x,t;p), \phi_2(x,t;p), \phi_3(x,t;p)] \right) \Big|_{p=0},$$

$$= D_t^\alpha u_{m-1}(x,t) - \left(u_{m-1}(x,t) \right)_{xx} + u_{m-1}(x,t) + A_{1,m-1} + A_{2,m-1}$$

and

$$\Re_{2,m}\left(u_0,u_1,...,u_{m-1},v_0,v_1,...,v_{m-1},\bar{u}_0,\bar{u}_1,...,\bar{u}_{m-1}\right)_{p=0}$$

$$= \frac{1}{(m-1)!}\frac{\partial^{m-1}}{\partial p^{m-1}}\left(N_2[\phi_1(x,t;p),\phi_2(x,t;p),\phi_3(x,t;p)]\right)\Bigg|_{p=0} \quad (3.123)$$

$$= D_t^\beta v_{m-1}(x,t) - \left(v_{m-1}(x,t)\right)_{xx} - A_{3,m-1}.$$

Now, the solutions of the mth-order deformation eqs. (3.122) for $m \geq 1$ becomes

$$u_m(x,t) = \chi_m u_{m-1}(x,t) + \hbar J_t^\alpha\left[\Re_{1,m}\left(u_0,u_1,...,u_{m-1},v_0,v_1,...,v_{m-1},\bar{u}_0,\bar{u}_1,...,\bar{u}_{m-1}\right)\right],$$

$$v_m(x,t) = \chi_m v_{m-1}(x,t) + \hbar J_t^\beta\left[\Re_{2,m}\left(u_0,u_1,...,u_{m-1},v_0,v_1,...,v_{m-1},\bar{u}_0,\bar{u}_1,...,\bar{u}_{m-1}\right)\right]. \quad (3.124)$$

By putting the initial conditions in eq. (3.99) into eq. (3.124) and then solving them, we now successively obtain:

$$u_0(x,t) = A\operatorname{sech}(Bx)\exp\left(\frac{ix}{B}\right) + t\left(A\operatorname{sech}(Bx)\tanh(Bx)\exp\left(\frac{ix}{B}\right) - A\operatorname{sech}(Bx)\exp\left(\frac{ix}{B}\right)\right)$$

$$u_1(x,t) = \frac{1}{4}\exp\left(\frac{ix}{B}\right)hAt^\alpha\operatorname{sech}^3(Bx)\left(\frac{1}{B^2\Gamma(\alpha+2)}t\operatorname{sech}(Bx)\left(-i\left(3+5B^4+B^2\left(5+4A^2\right)\right)\cosh(Bx)\right.\right.$$

$$+ i\left(-1+B^4+B^2\right)\cosh(3Bx) - 2\left(-1-11B^4+B^2\left(9-6A^2\right)+\left(-1-3B^2+3B^4\right)\cosh(2Bx)\right)$$

$$\sinh(Bx) + \frac{2\left(1-3B^2+3B^4+2B^2A^2+\left(1+B^2-B^4\right)\cosh(2Bx)2iB^2\sinh(2Bx)\right)}{B^2\Gamma(\alpha+1)}$$

$$\frac{24A^2t^3\left(-i+\tanh(Bx)\right)^2\left(i+\tanh(Bx)\right)}{\Gamma(\alpha+4)} + \frac{8t^2\left(-i+\tanh(Bx)\right)\left(iA^2+\left(-4+3A^2\right)\tanh(Bx)\right)}{\Gamma(\alpha+3)}$$

$$+ ...$$

$$v_0(x,t) = -2\operatorname{sech}^2(Bx) - t\left(4\operatorname{sech}^2(Bx)\tanh(Bx)\right)$$

$$v_1(x,t) = -\frac{2B^2h\left(-2+A^2\right)t^\beta\left(-2+\cosh(2Bx)\right)\operatorname{sech}^4(Bx)}{\Gamma(\beta+1)} - \frac{1}{\Gamma(\beta+2)}\left(4B^2h\left(-2+A^2\right)t^{\beta+1}\right.$$

$$\left(-5+\cosh(2Bx)\right)\operatorname{sech}^4(Bx)\tanh(Bx)\right) - \frac{8B^2hA^2t^{\beta+2}\left(-4+\cosh(2Bx)\right)\operatorname{sech}^4(Bx)\tanh^2(Bx)}{\Gamma(\beta+3)},$$

and so on.

New Analytical Approximate Solutions of Fractional Differential Equations

Finally, we obtain third-order approximate solutions given as:

$$
u(x,t) = u_0(x,t) + u_1(x,t) + u_2(x,t)
$$

$$
= A\operatorname{sech}(Bx)\exp\left(\frac{ix}{B}\right) + t\left(A\operatorname{sech}(Bx)\tanh(Bx)\exp\left(\frac{ix}{B}\right) - A\operatorname{sech}(Bx)\exp\left(\frac{ix}{B}\right)\right)
$$

$$
+ \frac{1}{4}\exp\left(\frac{ix}{B}\right)hAt^{\alpha}\operatorname{sech}^3(Bx)\left(\frac{1}{B^2\Gamma(\alpha+2)}t\operatorname{sech}(Bx)\left(-i\left(3+5B^4+B^2\left(5+4A^2\right)\right)\cosh(Bx)\right.\right.
$$

$$
+ i\left(-1+B^4+B^2\right)\cosh(3Bx) - 2\left(-1-11B^4+B^2\left(9-6A^2\right)+\left(-1-3B^2+3B^4\right)\cosh(2Bx)\right)
$$

$$
\operatorname{sinh}(Bx) + \frac{2\left(1-3B^2+3B^4+2B^2A^2+\left(1+B^2-B^4\right)\cosh(2Bx)2iB^2\operatorname{sinh}(2Bx)\right)}{B^2\Gamma(\alpha+1)}
$$

$$
\frac{24A^2t^3\left(-i+\tanh(Bx)\right)^2\left(i+\tanh(Bx)\right)}{\Gamma(\alpha+4)} + \frac{8t^2\left(-i+\tanh(Bx)\right)\left(iA^2+\left(-4+3A^2\right)\tanh(Bx)\right)}{\Gamma(\alpha+3)}\right), \quad (3.125)
$$

$$+\ldots$$

$$
v(x,t) = v_0(x,t) + v_1(x,t) + v_2(x,t)
$$

$$
= -2\operatorname{sech}^2(Bx) - t\left(4\operatorname{sech}^2(Bx)\tanh(Bx)\right)
$$

$$
- \frac{2t^{\beta}B^2h\left(-2+A^2\right)\left(-2+\cosh(2Bx)\right)\operatorname{sech}^4(Bx)}{\Gamma(\beta+1)}
$$

$$
- \frac{1}{\Gamma(\beta+2)}\left(4B^2h\left(-2+A^2\right)t^{\beta+1}\right.
$$

$$
\left(-5+\cosh(2Bx)\right)\operatorname{sech}^4(Bx)\tanh(Bx)\right)
$$

$$
- \frac{8B^2hA^2t^{\beta+2}\left(-4+\cosh(2Bx)\right)\operatorname{sech}^4(Bx)\tanh^2(Bx)}{\Gamma(\beta+3)} + \ldots
$$

3.5.3 Numerical Results and Discussion

In this section, we analyze the numerical solutions obtained by the MHAM method and also present an analysis for the comparison of errors between the HPTM and MHAM methods.

3.5.3.1 *The Numerical Simulations for MHAM*

In this present numerical experiment, eqs. (3.125) and (3.126) have been used to draw the graphs as shown in Figures 3.20 to 3.27 for different fractional order values of α and β, respectively. The numerical solutions of coupled K–G–Z equations (3.12) and (3.13) have been shown in Figures 3.20 to 3.27 with the help of three-term approximations for the homotopy series solutions of $u(x,t)$ and $v(x,t)$, respectively.

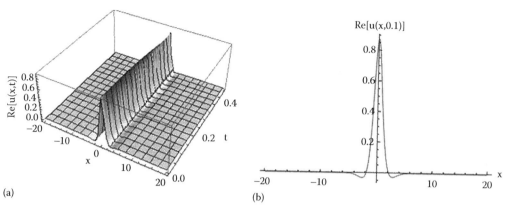

FIGURE 3.20
Case I: For $\alpha = 2, \beta = 2$. (a) The MHAM solution for $\text{Re}\big[u(x,t)\big]$, (b) corresponding solution for $\text{Re}\big[u(x,t)\big]$ when $t = 0.1$.

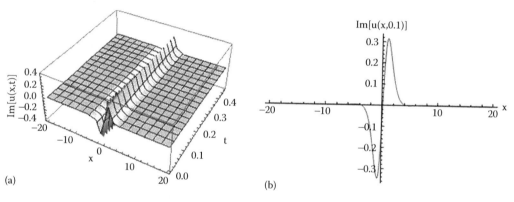

FIGURE 3.21
Case I: For $\alpha = 2, \beta = 2$. (a) The MHAM solution for $\text{Im}\big[u(x,t)\big]$, (b) corresponding solution for $\text{Im}\big[u(x,t)\big]$ when $t = 0.1$.

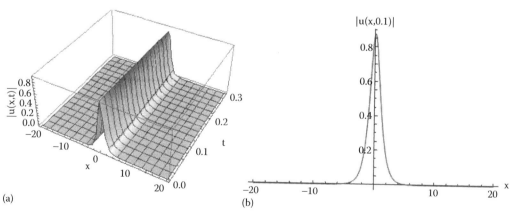

FIGURE 3.22
Case I: For $\alpha = 2, \beta = 2$. (a) The MHAM solution for $|u(x,t)|$, (b) corresponding solution for $|u(x,t)|$ when $t = 0.1$.

New Analytical Approximate Solutions of Fractional Differential Equations

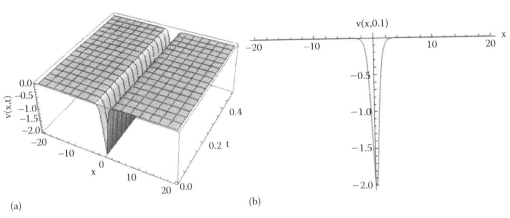

FIGURE 3.23
Case I: For $\alpha = 2$, $\beta = 2$. (a) The MHAM solution for $v(x,t)$, (b) corresponding solution for $v(x,t)$ when $t = 0.1$.

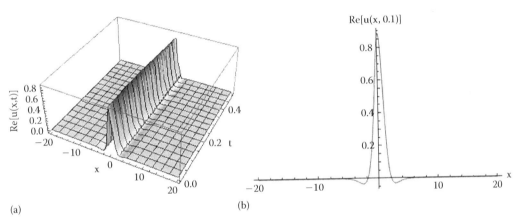

FIGURE 3.24
Case II: For $\alpha = 1.75$, $\beta = 1.5$. (a) The MHAM solution for $\text{Re}\big[u(x,t)\big]$, (b) corresponding solution for $\text{Re}\big[u(x,t)\big]$ when $t = 0.1$.

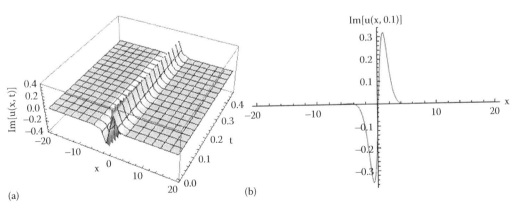

FIGURE 3.25
Case II: For $\alpha = 1.75$, $\beta = 1.5$. (a) The MHAM solution for $\text{Im}\big[u(x,t)\big]$, (b) corresponding solution for $\text{Im}\big[u(x,t)\big]$ when $t = 0.1$.

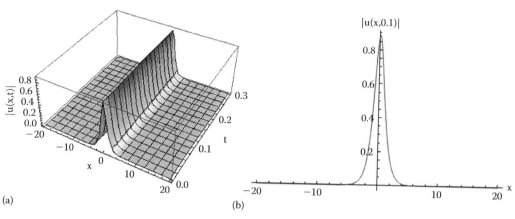

FIGURE 3.26
Case II: For $\alpha = 1.75, \beta = 1.5$. (a) The MHAM solution for $|u(x,t)|$, (b) corresponding solution for $|u(x,t)|$ when $t = 0.1$.

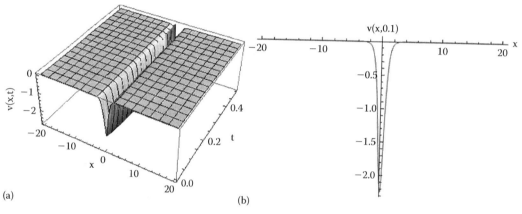

FIGURE 3.27
Case II: For $\alpha = 1.75, \beta = 1.5$. (a) The MHAM solution for $v(x,t)$, (b) corresponding solution for $v(x,t)$ when $t = 0.1$.

3.5.3.2 The ℏ Graph and the Numerical Simulations for u(x, t) and v(x, t) in MHAM

In the case of integer order, exact solutions of eqs. (3.12) and (3.13) are given by [102,103]:

$$u(x,t) = A \operatorname{sech}(Bx - t) \exp\left[i\left(\frac{x}{B} - t\right)\right], \quad (3.127)$$

$$v(x,t) = -2 \operatorname{sech}^2(Bx - t). \quad (3.128)$$

As pointed out by Liao [42] in general, by means of the so-called ℏ-curve, it is straightforward to choose a proper value of ℏ which ensures that the solution series is convergent.

To investigate the influence of ℏ on the solution series, we plot the so called ℏ-curve of partial derivatives of $u(x,t)$ and $v(x,t)$ at (0,0) obtained from the third-order MHAM solutions as shown in Figures 3.28 to 3.30.

In this way, it is found that our series converge when ℏ = −1.01. We also compare here the solutions curve in Figures 3.31 to 3.34, for different values of ℏ. It is observed that the valid region for ℏ is $-1.01 \leq ℏ \leq -1$.

New Analytical Approximate Solutions of Fractional Differential Equations 91

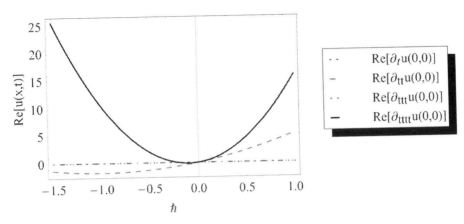

FIGURE 3.28
The $\hbar$-curve for partial derivatives of $\text{Re}\left[u(x,t)\right]$ at (0,0) for the third-order MHAM solution when $\alpha = 2, \beta = 2$.

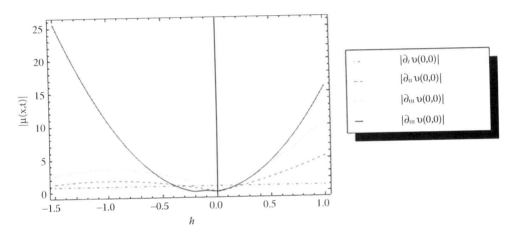

FIGURE 3.29
The $\hbar$-curve for partial derivatives of $|u(x,t)|$ at (0,0) for the third-order MHAM solution when $\alpha = 2, \beta = 2$.

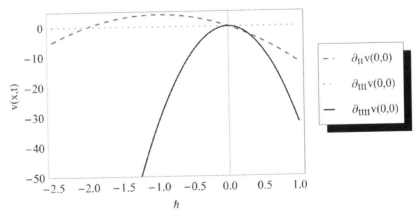

FIGURE 3.30
The $\hbar$-curve for partial derivatives of $v(x,t)$ at (0,0) for the third-order MHAM solution when $\alpha = 2, \beta = 2$.

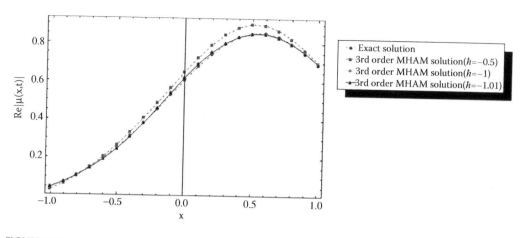

FIGURE 3.31
The result obtained by the MHAM for various $\hbar$ by third-order MHAM approximate solution for $\alpha = 2, \beta = 2$ in comparison with the exact solution when $-1 < x < 1$ and $t = 0.6$.

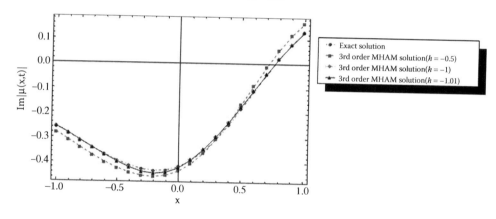

FIGURE 3.32
The result obtained by the MHAM for various $\hbar$ by third-order MHAM approximate solution for $\alpha = 2, \beta = 2$ in comparison with the exact solution when $-1 < x < 1$ and $t = 0.6$.

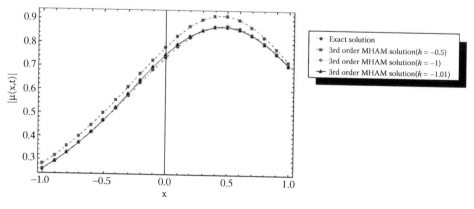

FIGURE 3.33
The result obtained by the MHAM for various $\hbar$ by third-order MHAM approximate solution for $\alpha = 2, \beta = 2$ in comparison with the exact solution when $-1 < x < 1$ and $t = 0.6$.

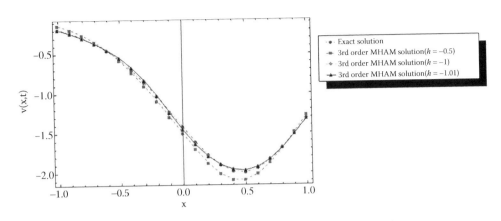

FIGURE 3.34
The result obtained by the MHAM for various $\hbar$ by third-order MHAM approximate solution for $\alpha = 2, \beta = 2$ in comparison with the exact solution when $-1 < x < 1$ and $t = 0.6$.

3.5.3.3 To Analogize the Solutions of K–G–Z Equations by MHAM with Exact Solution for Different Values of $\hbar$

In this section, to examine the accuracy and reliability of the MHAM for the K–G–Z equation, we compare the third-order approximate solution with an exact solutions in eqs. (3.127) and (3.128). The following Figures 3.35 to 3.38 can help us to express the comparison between the third-order approximate solutions and an exact solutions in eqs. (3.127) and (3.128). In Figures 3.35 to 3.38, the absolute errors for the K–G–Z equations obtained by the third-order MHAM solutions are given by $|u_{MHAM} - u_{Exact}|$ and $|v_{MHAM} - v_{Exact}|$ when $\hbar = -0.5, \hbar = -1, \hbar = -1.01$ and $t = 0.6$.

3.5.3.4 Comparison of Absolute Errors for MHAM and HPTM Solutions

In this section, we present Figures 3.39 to 3.42, citing the numerical simulations for comparison of absolute errors in the solutions of $u(x,t)$ and $v(x,t)$ obtained by the MHAM and HPTM at $x = 0.3$.

Although both the present methods are reliable and efficient, but Figures 3.39 to 3.42 assure plausibility to consider the MHAM provides more accurate solutions than the HPTM solutions for coupled fractional K–G–Z equations.

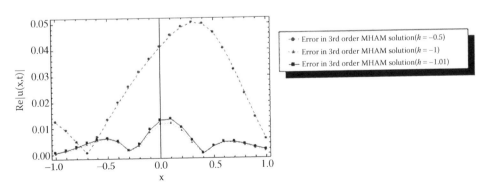

FIGURE 3.35
The absolute errors for the K–G–Z equation by the third-order MHAM approximation of $\text{Re}[u(x,t)]$.

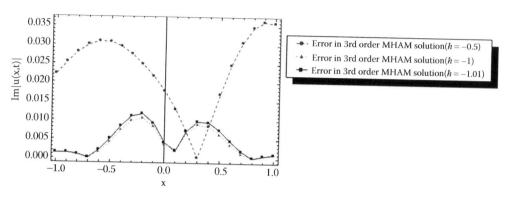

FIGURE 3.36
The absolute errors for the K–G–Z equation by the third-order MHAM approximation of $\mathrm{Im}[u(x,t)]$.

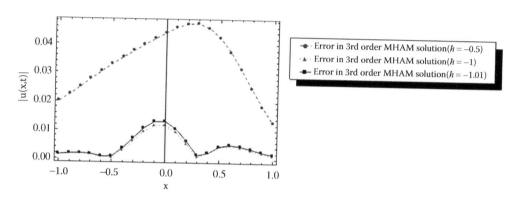

FIGURE 3.37
The absolute errors for the K–G–Z equation by the third-order MHAM approximation of $|u(x,t)|$.

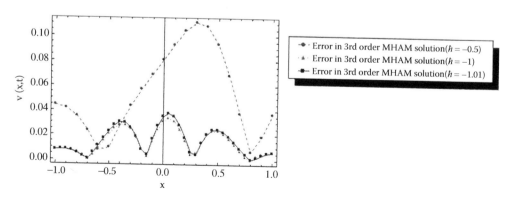

FIGURE 3.38
The absolute errors for the K–G–Z equation by the third-order MHAM approximation of $v(x,t)$.

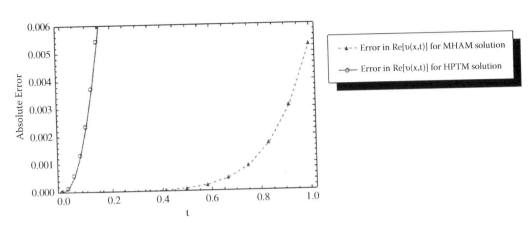

FIGURE 3.39
Graphical comparison of absolute errors in the solution of Re($u(x,t)$) for (a) MHAM and (b) HPTM, when $\alpha = 2$, $\beta = 2$.

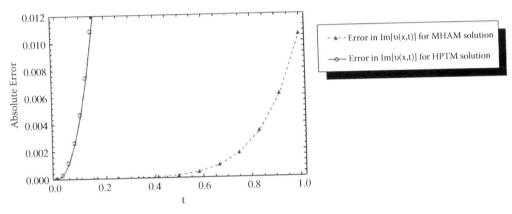

FIGURE 3.40
Graphical comparison of absolute errors in the solution of Im($u(x,t)$) for (a) MHAM and (b) HPTM, when $\alpha = 2$, $\beta = 2$.

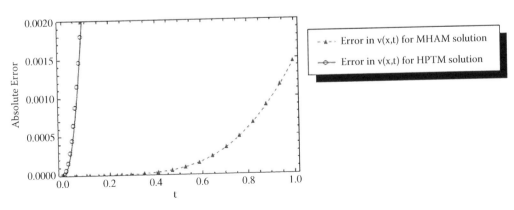

FIGURE 3.41
Graphical comparison of absolute errors in the solution of $|u(x,t)|$ for (a) MHAM and (b) HPTM, when $\alpha = 2, \beta = 2$.

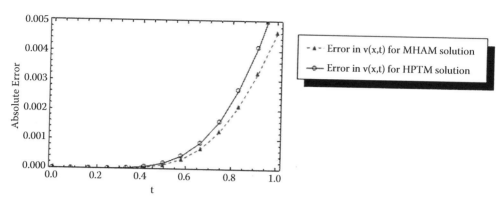

FIGURE 3.42
Graphical comparison of absolute errors in the solution of $v(x,t)$ for (a) MHAM and (b) HPTM, when $\alpha = 2$, $\beta = 2$.

3.6 Application of the MHAM and HPTM for Approximate Solutions of Fractional Coupled S-G Equations

In this section, two reliable methods, viz., the HPTM and MHAM have been applied for getting analytical approximate solutions of fractional coupled S-G equations.

3.6.1 Implementation of the MHAM for Approximate Solutions of Fractional Coupled S-G Equations

In this section, we first consider the application of the MHAM for the solution of fractional sine-Gordon equations (3.14) and (3.15) with given initial conditions [108]:

$$u(x,0) = A\cos(kx),$$
$$u_t(x,0) = 0,$$
$$v(x,0) = 0, \qquad (3.129)$$
$$v_t(x,0) = 0.$$

Expanding $\phi_i(x,t;p)$ in a Taylor series with respect to p for $i = 1, 2$, we have:

$$\phi_1(x,t;p) = u_0(x,t) + \sum_{m=1}^{+\infty} p^m u_m(x,t),$$
$$\phi_2(x,t;p) = v_0(x,t) + \sum_{m=1}^{+\infty} p^m v_m(x,t), \qquad (3.130)$$

New Analytical Approximate Solutions of Fractional Differential Equations 97

where:

$$u_m(x,t) = \frac{1}{m!} \frac{\partial^m \phi_1(x,t;p)}{\partial p^m}\bigg|_{p=0},$$ (3.131)

$$v_m(x,t) = \frac{1}{m!} \frac{\partial^m \phi_2(x,t;p)}{\partial p^m}\bigg|_{p=0}.$$

To obtain the approximate solution of the fractional coupled S-G eqs. (3.14) and (3.15), we choose the linear operators:

$$L_1[\phi_1(x,t)] = D_t^\alpha \phi_1(x,t;p) \text{ and } L_2[\phi_2(x,t)] = D_t^\beta \phi_2(x,t;p),$$ (3.132)

where D_t^α and D_t^β are Caputo fractional differential operators.

From eqs. (3.14) and (3.15), we define a system of non-linear operators as:

$$N_1[\phi(x,t;p)\phi_2(x,t;p),\phi_3(x,t;p)]$$

$$= D_t^\alpha[\phi_1(x,t;p)] - (\phi_1(x,t;p))_{xx} + \delta^2 \sin(\phi_1(x,t;p) - \phi_2(x,t;p)),$$

$$N_2[\phi_1(x,t;p),\phi_2(x,t;p),\phi_3(x,t;p)]$$ (3.133)

$$= D_t^\beta[\phi_2(x,t;p)] - c^2(\phi_2(x,t;p))_{xx} - \sin(\phi_1(x,t;p) - \phi_2(x,t;p)),$$

where the non-linear term $\sin(\phi_1(x,t;p) - \phi_2(x,t;p))$ is expressed in Adomian-like polynomials. The non-linear term $\sin(\phi_1(x,t;p) - \phi_2(x,t;p))$ has been taken as:

$$\sin(\phi_1(x,t;p) - \phi_2(x,t;p)) = \sum_{n=0}^{\infty} p^n A_n,$$

where:

$$A_n = \frac{1}{n!} \frac{\partial^n}{\partial p^n} \left(\sin(\phi_1(x,t;p) - \phi_2(x,t;p)) \right)\bigg|_{p=0}$$

$$= \frac{1}{n!} \frac{\partial^n}{\partial p^n} \left(\sin\left(\left(u_0(x,t) + \sum_{n=1}^{+\infty} p^n u_n(x,t) \right) - \left(v_0(x,t) + \sum_{n=1}^{+\infty} p^n v_n(x,t) \right) \right) \right)\bigg|_{p=0}, n \geq 0.$$ (3.134)

Using the definition given in the above section, we construct the so-called zeroth-order deformation equations:

$$(1-p)L_1[\phi_1(x,t;p) - u_0(x,t)] = p\hbar N_1[\phi_1(x,t;p),\phi_2(x,t;p)],$$

$$(1-p)L_2[\phi_2(x,t;p) - u_0(x,t)] = p\hbar N_2[\phi(x,t;p),\phi_2(x,t;p)].$$ (3.135)

Obviously, when $p = 0$ and $p = 1$, eq. (3.130) yields:

$$\phi_1(x,t;0) = u(x,0), \; \phi_1(x,t;1) = u(x,t),$$

$$\phi_2(x,t;0) = v(x,0), \; \phi_2(x,t;1) = v(x,t).$$

Therefore, as the embedding parameter p increases from 0 to 1, $\phi_i(x,t;p)$ varies from the initial guess to the exact solutions $u(x,t)$ and $v(x,t)$, respectively.

If the auxiliary linear operator, the initial guess, and the auxiliary parameters $\hbar$ are so properly chosen, the above series in eq. (3.130) converge at $p = 1$, and we obtain:

$$u(x,t) = u_0(x,t) + \sum_{m=1}^{+\infty} p^m u_m(x,t),$$

$$v(x,t) = v_0(x,t) + \sum_{m=1}^{+\infty} p^m v_m(x,t). \tag{3.136}$$

According to eq. (2.23) of Chapter 2, we have the mth-order deformation equations:

$$L_1\big[u_m(x,t) - \chi_m u_{m-1}(x,t)\big] = \hbar \Re_{1,m}\big(u_0, u_1, ..., u_{m-1}, v_0, v_1, ..., v_{m-1}\big),$$

$$L_2\big[v_m(x,t) - \chi_m v_{m-1}(x,t)\big] = \hbar \Re_{2,m}\big(u_0, u_1, ..., u_{m-1}, v_0, v_1, ..., v_{m-1}\big), \; m \geq 1, \tag{3.137}$$

where:

$$\Re_{1,m}\big(u_0, u_1, ..., u_{m-1}, v_0, v_1, ..., v_{m-1}\big) = \frac{1}{(m-1)!} \frac{\partial^{m-1}}{\partial p^{m-1}} \Big(N_1\big[\phi_1(x,t;p), \phi_2(x,t;p)\big]\Big)\Big|_{p=0},$$

$$= D_t^\alpha u_{m-1}(x,t) - \big(u_{m-1}(x,t)\big)_{xx} + \delta^2 A_{m-1}$$

$$\Re_{2,m}\big(u_0, u_1, ..., u_{m-1}, v_0, v_1, ..., v_{m-1}\big) = \frac{1}{(m-1)!} \frac{\partial^{m-1}}{\partial p^{m-1}} \Big(N_2\big[\phi_1(x,t;p), \phi_2(x,t;p)\big]\Big)\Big|_{p=0}.$$

$$= D_t^\beta v_{m-1}(x,t) - c^2 \big(v_{m-1}(x,t)\big)_{xx} + A_{m-1}, \tag{3.138}$$

Now, the solutions of the mth-order deformation eq. (3.137) for $m \geq 1$ become:

$$u_m(x,t) = \chi_m u_{m-1}(x,t) + \hbar J_t^\alpha \Big[\Re_{1,m}\big(u_0, u_1, ..., u_{m-1}, v_0, v_1, ..., v_{m-1}\big)\Big],$$

$$v_m(x,t) = \chi_m v_{m-1}(x,t) + \hbar J_t^\beta \Big[\Re_{2,m}\big(u_0, u_1, ..., u_{m-1}, v_0, v_1, ..., v_{m-1}\big)\Big], \tag{3.139}$$

where J_t^α and J_t^β are Riemann–Liouville integral operators.

By putting the initial conditions in eq. (3.129) into eq. (3.139) and solving them, we now successively obtain:

$$u_0(x,t) = A\cos(kx),$$

$$v_0(x,t) = 0,$$

New Analytical Approximate Solutions of Fractional Differential Equations

$$u_1(x,t) = \frac{\hbar t^\alpha}{\Gamma(\alpha+1)}\left(Ak^2\cos(kx)+\delta^2\sin\left(A\cos(kx)\right)\right),$$

$$v_1(x,t) = \frac{\hbar t^\beta \sin\left(A\cos(kx)\right)}{\Gamma(\beta+1)},$$

$$u_2(x,t) = \frac{\hbar t^\alpha}{2\Gamma(\alpha+1)\Gamma(\alpha+3)}\left(2(1+\hbar)\Gamma(\alpha+3)\left(Ak^2\cos(kx)+\delta^2\sin\left(A\cos(kx)\right)\right)\right.$$

$$+ t^\alpha \hbar\Gamma(\alpha+1)\left(2Ak^2\cos(kx)\left(k^2+2\delta^2\sin\left(A\cos(kx)\right)\right)\right)$$

$$\left.+ \delta^2\left(2A^2k^2\sin^2(kx)\sin\left(A\cos(kx)\right)+\left(1+\delta^2\right)\sin\left(2A\cos(kx)\right)\right)\right),$$

$$v_2(x,t) = -\frac{\hbar t^\beta}{2\Gamma(\beta+1)\Gamma(\beta+3)}\left(2(1+\hbar)\Gamma(\beta+3)\sin\left(A\cos(kx)\right)+\hbar t^2\Gamma(\beta+1)\right.$$

$$\left(2A(1+c^2)k^2\cos(kx)\cos\left(A\cos(kx)\right)+2A^2c^2k^2\sin^2(kx)\right.$$

$$\left.\left.\sin\left(A\cos(kx)\right)+\left(1+\delta^2\right)\sin\left(2A\cos(kx)\right)\right)\right),$$

and so on.

Finally, we obtain fourth-order approximate solutions:

$$u(x,t) = u_0(x,t)+u_1(x,t)+u_2(x,t)+u_3(x,t)$$

$$= A\cos(kx)+\frac{\hbar t^\alpha}{\Gamma(\alpha+1)}\left(Ak^2\cos(kx)+\delta^2\sin\left(A\cos(kx)\right)\right)$$

$$+\frac{\hbar t^\alpha}{2\Gamma(\alpha+1)\Gamma(\alpha+3)}\left(2(1+\hbar)\Gamma(\alpha+3)\left(Ak^2\cos(kx)+\delta^2\sin\left(A\cos(kx)\right)\right)\right. \quad (3.140)$$

$$+ t^\alpha \hbar\Gamma(\alpha+1)\left(2Ak^2\cos(kx)\left(k^2+2\delta^2\sin\left(A\cos(kx)\right)\right)\right)$$

$$\left.+ \delta^2\left(2A^2k^2\sin^2(kx)\sin\left(A\cos(kx)\right)+\left(1+\delta^2\right)\sin\left(2A\cos(kx)\right)\right)\right)+\cdots,$$

$$v(x,t) = v_0(x,t)+v_1(x,t)+v_2(x,t)+v_3(x,t)$$

$$= \frac{\hbar t^\beta \sin\left(A\cos(kx)\right)}{\Gamma(\beta+1)}-\frac{\hbar t^\beta}{2\Gamma(\beta+1)\Gamma(\beta+3)}\left(2(1+\hbar)\Gamma(\beta+3)\sin\left(A\cos(kx)\right)\right.$$

$$+ \hbar t^2\Gamma(\beta+1)\left(2A(1+c^2)k^2\cos(kx)\cos\left(A\cos(kx)\right)+2A^2c^2k^2\sin^2(kx)\right. \quad (3.141)$$

$$\left.\left.\sin\left(A\cos(kx)\right)+\left(1+\delta^2\right)\sin\left(2A\cos(kx)\right)\right)\right)+\cdots,$$

3.6.2 Implementation of the HPTM for Approximate Solutions of Fractional Coupled S-G Equations

In this section, we first consider the application of the HPTM for the solution of fractional sine-Gordon equations (3.14) and (3.15).

Applying Laplace transform on both sides of eqs. (3.28) and (3.29), we get

$$\mathcal{L}[u(x,t)] = \frac{u(x,0)}{s} + \frac{u_t(x,0)}{s^2} + \frac{1}{s^\alpha}\mathcal{L}[u_{xx}] + \frac{1}{s^\alpha}\mathcal{L}[-\delta^2 \sin(u-v)], \tag{3.142}$$

$$\mathcal{L}[v(x,t)] = \frac{v(x,0)}{s} + \frac{v_t(x,0)}{s^2} + \frac{1}{s^\alpha}\mathcal{L}[c^2 v_{xx}] + \frac{1}{s^\alpha}\mathcal{L}[\sin(u-v)]. \tag{3.143}$$

Then, applying inverse Laplace transform on both sides of eqs. (3.142) and (3.143), we get:

$$u(x,t) = u(x,0) + t u_t(x,0) + \mathcal{L}^{-1}\left[\frac{1}{s^\alpha}\mathcal{L}[u_{xx} - \delta^2 \sin(u-v)]\right], \tag{3.144}$$

$$v(x,t) = v(x,0) + t v_t(x,0) + \mathcal{L}^{-1}\left[\frac{1}{s^\alpha}\mathcal{L}[c^2 v_{xx} + \sin(u-v)]\right]. \tag{3.145}$$

By using the homotopy perturbation method, we will construct the homotopy of eqs. (3.144) and (3.145):

$$u(x,t) = u(x,0) + t u_t(x,0) + p\left(\mathcal{L}^{-1}\left[\frac{1}{s^\alpha}\mathcal{L}[u_{xx} - \delta^2 \sin(u-v)]\right]\right), \tag{3.146}$$

$$v(x,t) = v(x,0) + t v_t(x,0) + p\left(\mathcal{L}^{-1}\left[\frac{1}{s^\alpha}\mathcal{L}[c^2 v_{xx} + \sin(u-v)]\right]\right). \tag{3.147}$$

By substituting $u(x,t) = u_0(x,t) + \sum_{n=1}^{\infty} p^n u_n(x,t)$ and $v(x,t) = v_0(x,t) + \sum_{n=0}^{\infty} p^n v_n(x,t)$ in eqs. (3.146) and (3.147), we get:

$$u_0(x,t) + \sum_{n=1}^{\infty} p^n u_n(x,t) = u(x,0) + t u_t(x,0) + p\left(\mathcal{L}^{-1}\left[\frac{1}{s^\alpha}\mathcal{L}\left(\left(u_0(x,t) + \sum_{n=1}^{\infty} p^n u_n(x,t)\right)_{xx}\right.\right.\right.$$

$$\left.\left.\left. - \delta^2 \sin\left(\left(u_0(x,t) + \sum_{n=1}^{\infty} p^n u_n(x,t)\right) - \left(v_0(x,t) + \sum_{n=1}^{\infty} p^n v_n(x,t)\right)\right)\right)\right]\right), \tag{3.148}$$

New Analytical Approximate Solutions of Fractional Differential Equations

101

and

$$v_0(x,t) + \sum_{n=1}^{\infty} p^n v_n(x,t) = v(x,0) + t v_t(x,0) + p\left(\mathcal{L}^{-1}\left[\frac{1}{s^\alpha} \mathcal{L}\left(c^2\left(v_0(x,t) + \sum_{n=1}^{\infty} p^n v_n(x,t) \right) \right) \right]_{xx} \right) \quad (3.149)$$

Comparing the coefficients of like powers in p for eq. (3.148) and (3.149), we have the following system of equations:

Coefficients of p^0: $u_0(x,t) = u(x,0) + t u_t(x,0)$, $v_0(x,t) = v(x,0) + t v_t(x,0)$. $\quad (3.150)$

$$p : u_1(x,t) = \mathcal{L}^{-1}\left[\frac{1}{s^\alpha} \mathcal{L}\left[\left((u_0)_{xx} - \delta^2\left(\sin\left(u_0 - v_0 \right) \right) \right) \right] \right],$$

Coefficients of

$$v_1(x,t) = \mathcal{L}^{-1}\left[\frac{1}{s^\alpha} \mathcal{L}\left[c^2\left((v_0)_{xx} + \left(\sin\left(u_0 - v_0 \right) \right) \right) \right] \right]. \quad (3.151)$$

Coefficients of p^2: $u_2(x,t) = \mathcal{L}^{-1}\left[\frac{1}{s^\alpha} \mathcal{L}\left[\left((u_1)_{xx} - \delta^2\left(\cos\left(u_0 - v_0 \right)\left(u_1 - v_1 \right) \right) \right) \right] \right],$

$$v_2(x,t) = \mathcal{L}^{-1}\left[\frac{1}{s^\alpha} \mathcal{L}\left[c^2\left((v_1)_{xx} + \left(\cos\left(u_0 - v_0 \right)\left(u_1 - v_1 \right) \right) \right) \right] \right], \quad (3.152)$$

and so on.

By putting the initial conditions in eq. (3.129) into eqs. (3.150) through (3.152) and solving them, we obtain:

$$u_0(x,t) = A\cos(kx),$$

$$v_0(x,t) = 0,$$

$$u_1(x,t) = \frac{t^\alpha}{\Gamma(1+\alpha)}\left[-Ak^2\cos(kx) - \delta^2\sin\left(A\cos(kx) \right) \right],$$

$$v_1(x,t) = \frac{t^\beta\sin\left(A\cos(kx) \right)}{\Gamma(1+\beta)},$$

$$u_2(x,t) = \frac{t^\alpha}{2\Gamma(1+2\alpha)\Gamma(1+\alpha+\beta)}\left(t^\beta\delta^2\Gamma(1+2\alpha)\sin\left(2A\cos(kx) \right) \right.$$

$$+ t^\alpha\Gamma(1+\alpha+\beta)\left(2Ak^2\delta^2\sin^2(kx)\sin\left(A\cos(kx) \right) \right.$$

$$\left. + 2Ak^2\delta^2\sin^2(kx)\sin\left(A\cos(kx) \right) + \delta^4\sin\left(2A\cos(kx) \right) \right),$$

$$v_2(x,t) = -\frac{t^\beta}{\Gamma(1+2\beta)\Gamma(1+\alpha+\beta)}\left(t^\alpha A\cos(kx)\Gamma(1+2\beta)\right.$$

$$\left(Ak^2\cos(kx)+\partial^2\sin(A\cos(kx))\right)+\frac{1}{2}t^\beta\Gamma(1+\alpha+\beta)2Ack^2\cos(kx)$$

$$\cos(A\cos(kx))+2Ak^2c^2\sin^2(kx)\sin(A\cos(kx))+\sin(2A\cos(kx))\big),$$

and so on.

The solution of eqs. (3.14) and (3.15) is approximated as:

$$u(x,t) = u_0(x,t)+u_1(x,t)+u_2(x,t)$$

$$= A\cos(kx)+\frac{t^\alpha}{\Gamma(1+\alpha)}\left[-Ak^2\cos(kx)-\delta^2\sin(A\cos(kx))\right]$$

$$+\frac{t^\alpha}{2\Gamma(1+2\alpha)\Gamma(1+\alpha+\beta)}\left(t^\beta\delta^2\Gamma(1+2\alpha)\sin(2A\cos(kx))\right. \tag{3.153}$$

$$+t^\alpha\Gamma(1+\alpha+\beta)\left(2Ak^2\delta^2\sin^2(kx)\sin(A\cos(kx))\right.$$

$$+2Ak^2\delta^2\sin^2(kx)\sin(A\cos(kx))+\delta^4\sin(2A\cos(kx)))\big),$$

$$v(x,t) = v_0(x,t)+v_1(x,t)+v_2(x,t)$$

$$= \frac{t^\beta\sin(A\cos(kx))}{\Gamma(1+\beta)}-\frac{t^\beta}{\Gamma(1+2\beta)\Gamma(1+\alpha+\beta)}\left(t^\alpha A\cos(kx)\Gamma(1+2\beta)\right.$$

$$\left(Ak^2\cos(kx)+\partial^2\sin(A\cos(kx))\right)+\frac{1}{2}t^\beta\Gamma(1+\alpha+\beta)2Ack^2\cos(kx) \tag{3.154}$$

$$\cos(A\cos(kx))+2Ak^2c^2\sin^2(kx)\sin(A\cos(kx))+\sin(2A\cos(kx))\big).$$

3.6.3 Numerical Results and Discussion

In this section, we analyze the numerical solutions of the MHAM and also present an analysis for the comparison of errors between the MHAM and HPTM solutions.

3.6.3.1 The ℏ Graph and the Numerical Simulations for u(x, t) and v(x, t) in MHAM

As pointed out by Liao [42] in general, by means of the so-called ℏ-curve, it is straightforward to choose a proper value of ℏ which ensures that the solution series is convergent. To investigate the influence of ℏ on the solution series, we plot the so called ℏ-curve of partial derivatives of $u(x,t)$ and $v(x,t)$ at $(0,0)$ obtained from the fourth-order MHAM solutions as shown in Figures 3.43 and 3.44.

New Analytical Approximate Solutions of Fractional Differential Equations 103

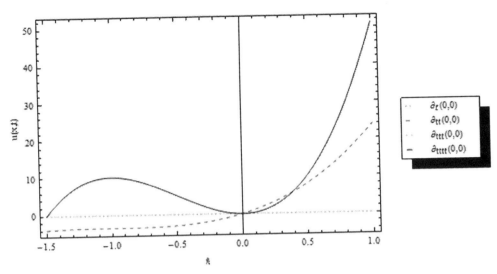

FIGURE 3.43
The $\hbar$-curve for partial derivatives of $u(x,t)$ at $(0,0)$ for the fourth-order MHAM solution when $\alpha = 2, \beta = 2$.

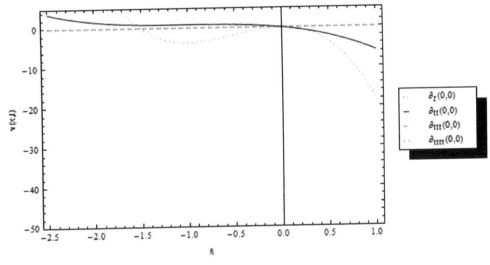

FIGURE 3.44
The $\hbar$-curve for partial derivatives of $v(x,t)$ at $(0,0)$ for the fourth-order MHAM solution when $\alpha = 2, \beta = 2$.

In this way, it is found that our series converge when $\hbar = -1.1$.

3.6.3.2 The Numerical Simulations for MHAM

In this present numerical experiment, eqs. (3.140) and (3.141) obtained by the MHAM have been used to draw the graphs as shown in Figures 3.45 to 3.48 for different fractional order values of α and β, respectively. The numerical solutions of fractional coupled S-G equations (3.14) and (3.15) have been shown in Figures 3.45 to 3.48 with the help of fourth-term approximations for the homotopy series solutions of $u(x,t)$ and $v(x,t)$, respectively.

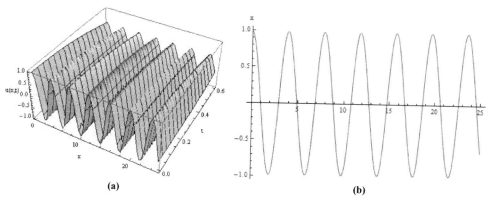

FIGURE 3.45
Case I: For $\alpha = 2, \beta = 2$. (a) The MHAM solution for $u(x,t)$, (b) corresponding solution for $u(x,t)$ when $t = 0.1$.

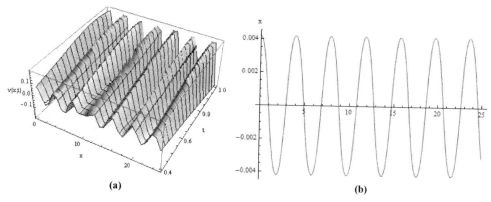

FIGURE 3.46
Case I: For $\alpha = 2, \beta = 2$. (a) The MHAM solution for $v(x,t)$, (b) corresponding solution for $v(x,t)$ when $t = 0.1$.

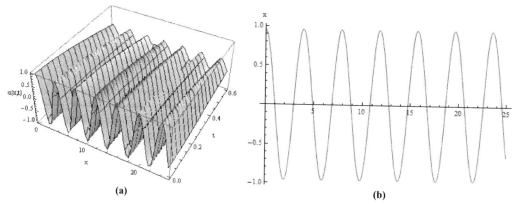

FIGURE 3.47
Case II: For $\alpha = 1.75, \beta = 1.5$. (a) The MHAM solution for $u(x,t)$, (b) corresponding solution for $u(x,t)$ when $t = 0.1$.

New Analytical Approximate Solutions of Fractional Differential Equations 105

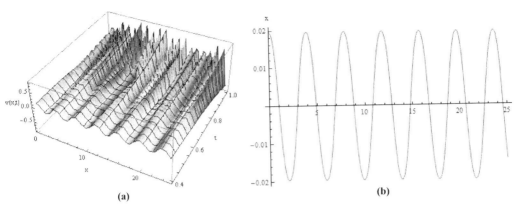

FIGURE 3.48
Case II: For $\alpha = 1.75$, $\beta = 1.5$. (a) The MHAM solution for $v(x,t)$, (b) corresponding solution for $v(x,t)$ when $t = 0.1$.

3.6.3.3 Comparison of Absolute Errors for MHAM and HPTM Solutions

In this section, we present Figures 3.49 and 3.50, citing the numerical simulations for comparison of absolute errors in solutions of $u(x,t)$ and $v(x,t)$ obtained by the MHAM and HPTM compared to the modified decomposition method [108] solutions at $x = 0.3$. In the following comparisons, the solutions obtained by the MDM [108] have been considered as classical solutions for coupled S-G equations.

Although both the present methods are reliable and efficient, but Figures 3.49 and 3.50 assure plausibility to consider the MHAM provides more accurate solutions than the HPTM solutions for fractional coupled S-G equations.

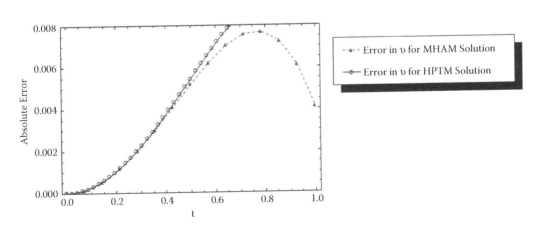

FIGURE 3.49
Graphical comparison of absolute errors in the solution of $u(x,t)$ for MHAM and HPTM, when $\alpha = 2$, $\beta = 2$.

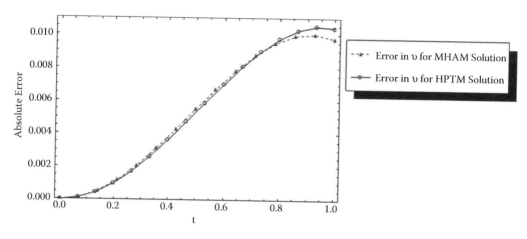

FIGURE 3.50
Graphical comparison of absolute errors in the solution of $v(x,t)$ for MHAM and HPTM, when $\alpha = 2, \beta = 2$.

3.7 Conclusion

In this chapter, we have presented schemes in order to obtain the approximate solutions of a class of time-fractional Lotka–Volterra equations of prey–predator model by using the homotopy perturbation method. Moreover, the homotopy perturbation method and homotopy perturbation transform method have been applied for finding the solutions for fractional coupled K–G–S equations with initial conditions. Also, the homotopy perturbation transform method and modified homotopy analysis method are applied to obtain the approximate solution of the fractional coupled K–G–Z equations and fractional coupled sine-Gordon equations. With appropriate initial conditions, the approximate solutions are obtained in series by means of the homotopy perturbation method, homotopy perturbation transform method, and modified homotopy analysis method.

The analytical method HPM is used since it does not require small parameter assumption in the equations so that the limitation of the traditional perturbation methods can be eliminated. The method is effective and convenient to solve fractional non-linear problems like continuous fractional order interacting population models. The local stability and the behaviors of the systems for different dynamical conditions have also been investigated for the continuous fractional order interacting population models.

Also, the approximate solutions of the fractional coupled K–G–S equations have been calculated by using the homotopy perturbation method and homotopy perturbation transform method. The homotopy perturbation method does not need any transformation techniques and linearization of the equations, whereas in the case of the homotopy perturbation transform method, the Laplace transformation techniques have been used. Additionally, the homotopy perturbation transform method does not need any discretization method to have numerical solutions. This method thus eliminates the difficulties and massive computation work. The above perturbation methods are straightforward, without restrictive assumptions, and the components of the series solution can be easily computed using any mathematical symbolic software. Moreover, these methods do not change the problem into a convenient one for the use of linear theory. It is obvious to see that the

New Analytical Approximate Solutions of Fractional Differential Equations 107

homotopy perturbation transform method is more accurate, easy, and efficient technique for solving fractional coupled K–G–S equations.

Also, in this chapter, the homotopy perturbation transform method and the modified homotopy analysis method are applied to obtain the approximate solution of the fractional coupled K–G–Z equation and sine-Gordon equations. The fractional coupled K–G–Z equations and sine-Gordon equations have been first time solved by the homotopy perturbation transform method and the modified homotopy analysis method in order to justify applicability of the above methods. The approximate solutions obtained by the MHAM provide us with a convenient way to control the convergence of approximate series solution, which is a fundamental qualitative difference in analysis between the homotopy perturbation transform method and the modified homotopy analysis method. The MHAM solves the problem without any need to discretization of the variables.

To control the convergence of the solution, we can choose the proper values of $\hbar$. From the absolute errors between the approximate series solutions and the exact solutions of K–G–Z equations, we can see that our approximate solutions agree well with the exact solutions. In comparing the solutions obtained by the MHAM and HPTM, we have pre-assumed the solutions obtained by the modified decomposition method as the classical solutions for fractional coupled S-G equations. The comparison results confirm that the MHAM provides more accurate solutions than HPTM. The MHAM also does not require a large computer memory and discretization of variables t and x. Moreover, the MHAM provides more accurate solution than the HPTM, which can be observed from the comparison of absolute errors in the HPTM and MHAM solutions. The proposed MHAM is more efficient and accurate for solving coupled fractional K–G–Z equations and fractional coupled S-G equations.

As mentioned, these methods avoid linearization and physically unrealistic assumptions. The above proposed methods are straightforward, without restrictive assumptions, and the components of the series solution can be easily computed using any mathematical symbolic software. Moreover, these methods do not change the problem into a convenient one for the use of linear theory. It is obvious to see that these methods are more accurate, easy, and efficient technique for solving fractional partial differential equations.

4

New Analytical Approximate Solutions
of Riesz Fractional Differential Equations

4.1 Introduction

In this chapter, we consider the analytical solutions of fractional partial differential equations with Riesz space fractional derivatives (FPDEs-RSFDs) on a finite domain. Here, we considered the Riesz fractional diffusion equation (RFDE), Riesz fractional advection–dispersion equation (RFADE), and Riesz time-fractional Camassa–Holm equation.

The RFDE is obtained from the standard diffusion equation by replacing the second-order space derivative with the Riesz fractional derivative of order $\alpha \in (1, 2]$. The RFADE is obtained from the standard advection–dispersion equation by replacing the first-order and second-order space derivatives with the Riesz fractional derivatives of order $\beta \in (0, 1]$ and of order $\alpha \in (1, 2]$, respectively. Here, the analytic solutions of the RFDE, RFADE, and Riesz time-fractional Camassa–Holm equation are derived by using the modified homotopy analysis method with Fourier transform. Then, we analyze the results through numerical simulations, which demonstrate the simplicity and effectiveness of the present method. Here, the space fractional derivatives are defined as Riesz fractional derivatives.

4.2 Outline of Present Study

In this chapter, the derived analytical solutions are based on the modified homotopy analysis method with Fourier transform. In this present analysis, we employ a new technique, such as applying the Fourier transforms followed by the homotopy analysis method, which enables derivation of the analytical solutions for the RFDE, the RFADE, and Riesz time-fractional Camassa–Holm equation.

4.2.1 Riesz Fractional Diffusion Equation and Riesz Fractional Advection–Dispersion Equation

We have considered here the RFDE and the RFADE in the following form [114]:

$$\frac{\partial}{\partial t} u(x,t) = K_\alpha \frac{\partial^\alpha}{\partial |x|^\alpha} u(x,t), \; t > 0, 0 < x < \pi, 1 < \alpha \leq 2, \tag{4.1}$$

and

$$\frac{\partial}{\partial t} u(x,t) = K_\alpha \frac{\partial^\alpha}{\partial |x|^\alpha} u(x,t) + K_\beta \frac{\partial^\beta}{\partial |x|^\beta} u(x,t), \, t > 0, 0 < x < \pi, 1 < \alpha \le 2, 0 < \beta \le 1, \quad (4.2)$$

respectively. Here, $u(x,t)$ is a solute concentration, K_α and K_β represent the dispersion coefficient and the average fluid velocity, respectively.

The RFADE with a symmetric fractional derivative, namely, the Riesz fractional derivative, was derived from the kinetics of chaotic dynamics by Saichev and Zaslavsky [115] and summarized by Zaslavsky [116]. Fractional diffusion equations model phenomena exhibiting anomalous diffusion that cannot be modeled accurately by second-order diffusion equations. Fractional advection–dispersion equations are used in groundwater hydrology to model the transport of a passive tracer carried by fluid flow in a porous medium [117,118].

In recent past, Liu et al. [118] transformed the space-fractional advection–diffusion equation into a system of ordinary differential equations that was then solved using backward differentiation formulas. Saha Ray [119] derived the exact solutions for time-fractional diffusion equations by using the decomposition method. Shen et al. [120] also considered a space-time fractional advection–diffusion equation with a Caputo time-fractional derivative and Riemann–Liouville space fractional derivatives. Shen et al. [13] discussed the fundamental solution and discrete random walk model for the time-space Riesz fractional advection–dispersion equation.

4.2.2 Camassa–Holm Equation

The Camassa–Holm equation [16] with Riesz-fractional time derivative is presented as follows

$$\begin{aligned} {}_0^R D_t^\alpha u(x,t) + 2ku_x(x,t) - \frac{2}{3} u_{xx}(x,t) + 3u(x,t)u_x(x,t) \\ + u_x(x,t)u_{xx}(x,t) + \frac{1}{2} u_x(x,t)u_{xxx}(x,t) = 0, \end{aligned} \quad (4.3)$$

where $0 < \alpha \le 1$ and k is an arbitrary constant. ${}_0^R D_t^\alpha$ is the Riesz fractional derivative.

Recently, the Camassa–Holm (CH) equation has been of great research interest due to its shallow water wave nature and multiple soliton nature as pointed by Boyd [121]. Previously, Camassa and Holm [122] derived a completely integrable dispersive shallow water equation, for example, the Camassa–Holm equation and obtained the solitary wave solution of the form $u = ce^{-|x-ct|}$. The solitary wave obtained by Camassa and Holm is called peakon wave due to the discontinuity of the first derivative at the wave peak, and it has been studied in [121–128]. Camassa et al. [129] presented numerical solutions of the time-dependent form and discussed the CH equation as a Hamiltonian system. For general k, Cooper and Shepard [130] derived the variational approximations to the solitary wave solution of the CH equation. He et al. [131] obtained some exact traveling wave solutions by using the integral bifurcation method. The exact traveling wave solutions for the CH equation are studied by Wazwaz [132,133] and others [134,135]. Yomba [136,137] applied the sub-ordinary differential equation (sub-ODE) method and the generalized auxiliary equation method for obtaining the exact solutions of the CH equation. The coexistence of multifarious solutions has been studied by Liu and Pan [138]. The explicit non-linear wave solutions of the CH equation were proposed by Liu and Liang [139] and Parkes and Vakhnenko [140].

New Analytical Approximate Solutions of Riesz Fractional Differential Equations   111

Here, the derived analytical solutions are based on the homotopy analysis method [42,141,142] with some modification. In this present analysis, we employ a new approach involving the homotopy analysis method along with Adomian's polynomials. It enables successful derivation of the analytical solutions for the Riesz time-fractional Camassa–Holm equation. By taking the third order modified homotopy analysis method (MHAM), approximate solutions of two-dimensional and three-dimensional graphs have been plotted. Then, we compare the MHAM results with the solutions obtained by the variational iteration method (VIM) [16].

4.3 Implementation of the MHAM-FT for Approximate Solution of Riesz Fractional Diffusion Equation

In this section, we first consider the application of the MHAM-Fourier transform (FT) for the solution of the RFDE of eq. (4.1) with given initial and boundary conditions [114]:

$$u(x,0) = \sin(4x) \text{ and } u(0,t) = u(\pi,t) = 0. \tag{4.4}$$

Then, by applying Fourier transform and using eq. (1.37) of Lemma 1.2.5.1 in Chapter 1, on eqs. (4.1) and (4.4), we get:

$$\frac{\partial}{\partial t}\hat{u}(k,t) = -K_\alpha |k|^\alpha \hat{u}(k,t), \tag{4.5}$$

with initial condition

$$\hat{u}(k,0) = i\sqrt{\frac{\pi}{2}}\delta(-4+k) - i\sqrt{\frac{\pi}{2}}\delta(4+k), \tag{4.6}$$

where k is called the transform parameter of Fourier transform and $\delta(.)$ denotes the Dirac delta function.

Expanding $\phi(k,t;p)$ in a Taylor series with respect to p, we have:

$$\phi(k,t;p) = \hat{u}_0(k,t) + \sum_{m=1}^{+\infty} p^m \hat{u}_m(k,t), \tag{4.7}$$

where:

$$\hat{u}_m(k,t) = \frac{1}{m!}\frac{\partial^m \phi(k,t;p)}{\partial p^m}\bigg|_{p=0.}$$

To obtain the approximate solution of the RFDE of eq. (4.5), we choose the linear operator:

$$L[\phi(k,t;p)] = \frac{\partial}{\partial t}\phi(k,t;p). \tag{4.8}$$

According to eq. (2.26) of Chapter 2, we define a non-linear operator as:

$$N\left[\phi\left(k,t;p\right)\right] = \frac{\partial}{\partial t}\phi\left(k,t;p\right) + K_\alpha |k|^\alpha \phi\left(k,t;p\right). \tag{4.9}$$

By using eq. (2.27) of Chapter 2, we construct the so-called zeroth-order deformation equation:

$$\left(1-p\right)L\left[\phi\left(k,t;p\right) - \hat{u}_0\left(k,t\right)\right] = p\hbar N\left[\phi\left(k,t;p\right)\right]. \tag{4.10}$$

Obviously, when $p = 0$ and $p = 1$, eq. (4.10) yields:

$$\phi\left(k,t;0\right) = \hat{u}_0\left(k,t\right); \phi\left(k,t;1\right) = \hat{u}\left(k,t\right)$$

Therefore, as the embedding parameter p increases from 0 to 1, $\phi(k,t;p)$ varies from the initial guess to the exact solution $\hat{u}(k,t)$.

If the auxiliary linear operator, the initial guess, and the auxiliary parameter $\hbar$ are so properly chosen, the above series in eq. (4.7) converge at $p = 1$, and we obtain:

$$\hat{u}\left(k,t\right) = \phi\left(k,t;1\right) = \hat{u}_0\left(k,t\right) + \sum_{m=1}^{+\infty} \hat{u}_m\left(k,t\right). \tag{4.11}$$

According to eq. (2.30) of Chapter 2, we have the following mth-order deformation equation:

$$L\left[\hat{u}_m\left(k,t\right) - \chi_m\hat{u}_{m-1}\left(k,t\right)\right] = \hbar\Re_m\left(\hat{u}_0,\hat{u}_1,...,\hat{u}_{m-1}\right), m \geq 1, \tag{4.12}$$

where:

$$\Re_m\left(\hat{u}_0,\hat{u}_1,...,\hat{u}_{m-1}\right) = \frac{1}{\left(m-1\right)!} \frac{\partial^{m-1}}{\partial p^{m-1}} N\left[\phi\left(k,t;p\right)\right]\Bigg|_{p=0.}$$

$$= \frac{\partial}{\partial t}\hat{u}_{m-1}\left(k,t\right) + K_\alpha |k|^\alpha \hat{u}_{m-1}\left(k,t\right) \tag{4.13}$$

Now, the solution of the mth-order deformation eq. (4.12) for $m \geq 1$ becomes:

$$\hat{u}_m(k,t) = \chi_m\hat{u}_{m-1}(k,t) + \hbar L^{-1}\left[\Re_m\left(\hat{u}_0,\hat{u}_1,...,\hat{u}_{m-1}\right)\right]. \tag{4.14}$$

New Analytical Approximate Solutions of Riesz Fractional Differential Equations 113

By putting the initial condition in eq. (4.6) into eq. (4.14) and solving them, we now successively obtain:

$$\hat{u}_0(k,t) = i\sqrt{\frac{\pi}{2}}\delta(-4+k) - i\sqrt{\frac{\pi}{2}}\delta(4+k), \tag{4.15}$$

$$\hat{u}_1(k,t) = K_\alpha t\hbar|k|^\alpha \, i\left(\sqrt{\frac{\pi}{2}}\delta(-4+k) - \sqrt{\frac{\pi}{2}}\delta(4+k)\right), \tag{4.16}$$

$$\begin{aligned}\hat{u}_2(k,t) = K_\alpha t\hbar|k|^\alpha \left(i\sqrt{\frac{\pi}{2}}\delta(-4+k) - i\sqrt{\frac{\pi}{2}}\delta(4+k)\right) \\ +\hbar\left(iK_\alpha t\hbar|k|^\alpha \sqrt{\frac{\pi}{2}}\delta(-4+k) + \frac{1}{2}iK_\alpha^2 t^2\hbar|k|^\alpha \sqrt{\frac{\pi}{2}}\delta(-4+k)\right. \\ \left. - iK_\alpha t\hbar|k|^\alpha \sqrt{\frac{\pi}{2}}\delta(4+k) - \frac{1}{2}iK_\alpha^2 t^2\hbar|k|^\alpha \sqrt{\frac{\pi}{2}}\delta(4+k)\right),\end{aligned} \tag{4.17}$$

and so on.

Then, by applying inverse Fourier transform of aforementioned eqs. (4.15) to (4.17), we have:

$$u_0(x,t) = \sin(4x),$$

$$u_1(x,t) = 4^\alpha K_\alpha t\hbar \sin(4x),$$

$$u_2(x,t) = 2^{-1+2\alpha} K_\alpha \hbar\left(2 + (2+4^\alpha K_\alpha t)\hbar\right)\sin(4x)t,$$

and so on.

In this manner, the other components of the homotopy series can be easily obtained by which $u(x,t)$ can be evaluated in a series form as:

$$\begin{aligned}u(x,t) &= u_0(x,t) + u_1(x,t) + u_2(x,t) + \dots \\ &= \frac{1}{6}\left(6 + 4^\alpha K_\alpha t\hbar(18 + \hbar(6(3+h) + 4^\alpha K_\alpha t(9 + (6+4^\alpha K_\alpha t)\hbar))))\right)\sin(4x) + \dots\end{aligned} \tag{4.18}$$

4.3.1 The $\hbar$ Graph and Numerical Simulations for MHAM-FT and Discussions

As pointed out by Liao [42] in general, by means of the so-called $\hbar$-curve, it is straightforward to choose a proper value of $\hbar$ which ensures that the solution series is convergent.

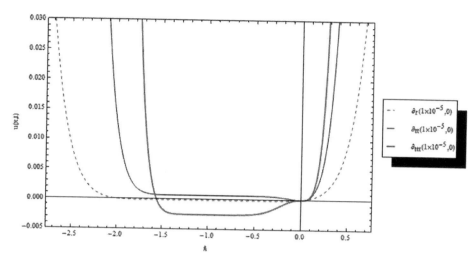

FIGURE 4.1
The $\hbar$-curve for partial derivatives of $u(x,t)$ at $(1\times 10^{-5},0)$ for the 11th-order MHAM-FT solution when $\alpha = 2$.

To investigate the influence of $\hbar$ on the solution series, we plot the so-called $\hbar$-curve of partial derivatives of $u(x,t)$ at $(1\times 10^{-5},0)$ obtained from the 11th-order MHAM-FT solutions as shown in Figure 4.1. In this way, it is found that our series converge when $\hbar = -1$.

In this present numerical experiment, eq. (4.18) obtained by the MHAM-FT has been used to draw the graphs as shown in Figures 4.2 and 4.3 for different fractional order values of α. The numerical solutions of the RFDE in eq. (4.1) has been shown in Figures 4.2 and 4.3 with the help of 11-term approximations for the homotopy series solutions of $u(x,t)$, when $\hbar = -1$ and $K_\alpha = 0.25$.

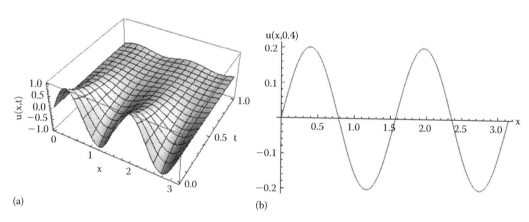

(a) (b)

FIGURE 4.2
Case I: For $a = 2$. (a) The MHAM-FT solution for $u(x,t)$, (b) corresponding solution for $u(x,t)$ when $t = 0.4$.

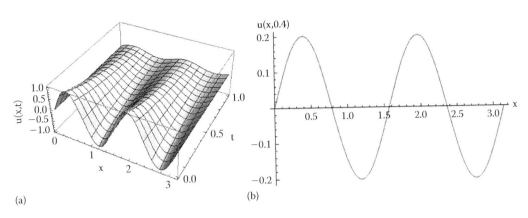

FIGURE 4.3

Case II: For $a = 1.5$. (a) The MHAM-FT solution for $u(x,t)$, (b) corresponding solution for $u(x,t)$ when $t = 0.4$.

4.4 Implementation of the MHAM-FT for Approximate Solution of Riesz Fractional Advection–Dispersion Equation

In this section, we first consider the application of the MHAM-FT for the solution of the RFADE of eq. (4.2) with given initial and boundary conditions in eq. (4.4).

Then applying Fourier transform and using eq. (1.37) of Lemma 1.2.5.1 in Chapter 1 on eqs. (4.2) and (4.4), we get:

$$\frac{\partial}{\partial t}\hat{u}(k,t) = -K_\alpha |k|^\alpha \hat{u}(k,t) - K_\beta |k|^\beta \hat{u}(k,t), \qquad (4.19)$$

with initial condition:

$$\hat{u}(k,0) = i\sqrt{\frac{\pi}{2}}\delta(-4+k) - i\sqrt{\frac{\pi}{2}}\delta(4+k), \qquad (4.20)$$

where k is called the transform parameter of Fourier transform and $\delta(.)$ denotes the Dirac delta function.

Expanding $\phi(k,t;p)$ in a Taylor series with respect to p, we have:

$$\phi(k,t;p) = \hat{u}_0(k,t) + \sum_{m=1}^{+\infty} p^m \hat{u}_m(k,t), \qquad (4.21)$$

where

$$\hat{u}_m(k,t) = \frac{1}{m!} \left.\frac{\partial^m \phi(k,t;p)}{\partial p^m}\right|_{p=0}.$$

To obtain the approximate solution of the RFADE of eq. (4.19), we choose the linear operator:

$$L\big[\phi(k,t;p)\big] = \frac{\partial}{\partial t}\phi(k,t;p),$$ (4.22)

According to eq. (2.26) of Chapter 2, we define a non-linear operator as:

$$N\big[\phi(k,t;p)\big] = \frac{\partial}{\partial t}\phi(k,t;p) + K_\alpha |k|^\alpha \phi(k,t;p) + K_\beta |k|^\beta \phi(k,t;p).$$ (4.23)

By using eq. (2.27) of Chapter 2, we construct the so-called zeroth-order deformation equation as following:

$$(1-p)L\big[\phi(k,t;p) - \hat{u}_0(k,t)\big] = p\hbar N\big[\phi(k,t;p)\big].$$ (4.24)

Obviously, when $p = 0$ and $p = 1$, eq. (4.24) yields:

$$\phi(k,t;0) = \hat{u}_0(k,t); \phi(k,t;1) = \hat{u}(k,t).$$

Therefore, as the embedding parameter p increases from 0 to 1, $\phi(k,t;p)$ varies from the initial guess to the exact solution $\hat{u}(k,t)$.

If the auxiliary linear operator, the initial guess, and the auxiliary parameters $\hbar$ are so properly chosen, the above series in eq. (4.21) converge at $p = 1$ and we obtain:

$$\hat{u}(k,t) = \phi(k,t;1) = \hat{u}_0(k,t) + \sum_{m=1}^{+\infty} \hat{u}_m(k,t).$$ (4.25)

According to eq. (2.30) of Chapter 2, we have the following mth-order deformation equation:

$$L\big[\hat{u}_m(k,t) - \chi_m\hat{u}_{m-1}(k,t)\big] = \hbar\Re_m\big(\hat{u}_0,\hat{u}_1,...,\hat{u}_{m-1}\big), m \geq 1,$$ (4.26)

where:

$$\Re_m\big(\hat{u}_0,\hat{u}_1,...,\hat{u}_{m-1}\big) = \frac{1}{(m-1)!}\frac{\partial^{m-1}}{\partial p^{m-1}} N\big[\phi(k,t;p)\big]\bigg|_{p=0}$$ (4.27)

$$= \frac{\partial}{\partial t}\hat{u}_{m-1}(k,t) + K_\alpha |k|^\alpha \hat{u}_{m-1}(k,t) + K_\beta |k|^\beta \hat{u}_{m-1}(k,t)$$

Now, the solution of the mth-order deformation eq. (4.26) for $m \geq 1$ becomes:

$$\hat{u}_m(k,t) = \chi_m\hat{u}_{m-1}(k,t) + \hbar L^{-1}\big[\Re_m\big(\hat{u}_0,\hat{u}_1,...,\hat{u}_{m-1}\big)\big].$$ (4.28)

By putting the initial condition in eq. (4.20) into eq. (4.28) and solving them, we now successively obtain:

$$\hat{u}_0(k,t) = i\sqrt{\frac{\pi}{2}}\delta(-4+k) - i\sqrt{\frac{\pi}{2}}\delta(4+k),$$ (4.29)

$$\hat{u}_1(k,t) = t\hbar \left(K_\alpha |k|^\alpha i \left(\sqrt{\frac{\pi}{2}} \delta(-4+k) - \sqrt{\frac{\pi}{2}} \delta(4+k) \right) \right.$$
$$\left. + K_\beta |k|^\beta i \left(\sqrt{\frac{\pi}{2}} \delta(-4+k) - \sqrt{\frac{\pi}{2}} \delta(4+k) \right) \right), \quad (4.30)$$

and so on.

Then, by applying inverse Fourier transform of above eqs. (4.29) and (4.30), we have:
$$u_0(x,t) = \sin(4x),$$
$$u_1(x,t) = (4^\alpha K_\alpha + 4^\beta K_\beta) t\hbar \sin(4x), \quad (4.31)$$

and so on.

In this manner, the other components of the homotopy series can be easily obtained by which $u(x,t)$ can be evaluated in a series form as:
$$u(x,t) = u_0(x,t) + u_1(x,t) + \dots$$
$$= \frac{1}{6} \Big(6 + (4^\alpha K_\alpha + 4^\beta K_\beta) t\hbar (18 + \hbar(6(3+\hbar) \quad (4.32)$$
$$+ 4(4^\alpha K_\alpha + 4^\beta K_\beta) t(9 + (6 + 4(4^\alpha K_\alpha + 4^\beta K_\beta)t)\hbar)))\Big) \sin(4x) + \dots.$$

4.4.1 The ℏ Graph and Numerical Simulations for MHAM-FT Method and Discussions

As pointed out by Liao [42] in general, by means of the so-called ℏ-curve, it is straightforward to choose a proper value of ℏ which ensures that the solution series is convergent.

To investigate the influence of ℏ on the solution series, we plot the so-called ℏ-curve of partial derivatives of $u(x,t)$ at $(1\times 10^{-5}, 0)$ obtained from the 11th-order MHAM-FT solutions as shown in Figure 4.4. In this way, it is found that our series converge when $\hbar = -1$.

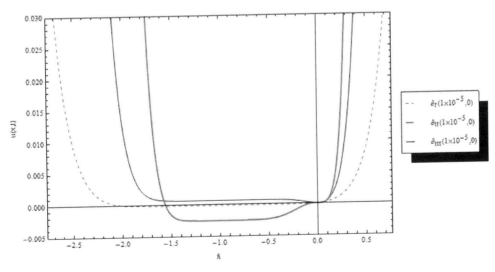

FIGURE 4.4
The ℏ-curve for partial derivatives of $u(x,t)$ at $(1\times 10^{-5}, 0)$ for the 11th-order MHAM-FT solution when $\alpha = 2, \beta = 1$.

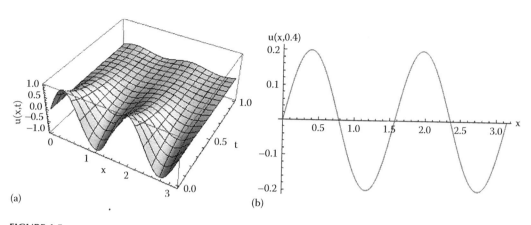

FIGURE 4.5
Case I: For $\alpha = 2$, $\beta = 1$. (a) The MHAM-FT solution for $u(x,t)$, (b) corresponding solution for $u(x,t)$ when $t = 0.5$.

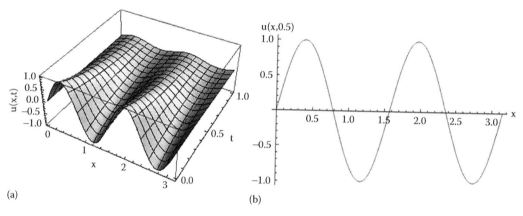

FIGURE 4.6
Case II: For $\alpha = 1.5$, $\beta = 0.7$. (a) The MHAM-FT solution for $u(x,t)$, (b) corresponding solution for $u(x,t)$ when $t = 0.5$.

In this present numerical experiment, eq. (4.32) obtained by the MHAM-FT has been used to draw the graphs as shown in Figures 4.5 and 4.6 for different fractional order values of α. The numerical solutions of the RFADE of eq. (4.2) has been shown in Figures 4.5 and 4.6 with the help of 11-term approximations for the homotopy series solutions of $u(x,t)$, when $\hbar = -1$, and $K_\alpha = 0.2$ and $K_\beta = 0.2$.

4.5 Implementation of the MHAM for Approximate Solution of Riesz Time-Fractional Camassa–Holm Equation

In this section, we first consider the application of the MHAM for the solution of the Riesz time-fractional Camassa–Holm of eq. (4.3) with given initial condition [16]:

$$u(x,0) = -k + k\sin(x). \tag{4.33}$$

By applying ${}_0^R D_t^{1-\alpha}$ and by Remark 1.2.5.2 of Chapter 1, the eq. (4.3) can be written in the following form as:

New Analytical Approximate Solutions of Riesz Fractional Differential Equations   119

$$u_t(x,t) + {}_0^R D_t^{1-\alpha}\left(2ku_x(x,t) - \frac{2}{3}u_{xx}(x,t) + 3u(x,t)u_x(x,t) \right.$$

$$\left. + u_x(x,t)u_{xx}(x,t) + \frac{1}{2}u_x(x,t)u_{xxxx}(x,t) \right) = 0 \tag{4.34}$$

Expanding $\phi(x,t;p)$ in a Taylor series with respect to p, we have:

$$\phi(x,t;p) = u_0(x,t) + \sum_{m=1}^{+\infty} p^m u_m(x,t), \tag{4.35}$$

where:

$$u_m(x,t) = \frac{1}{m!}\frac{\partial^m \phi(x,t;p)}{\partial p^m}\bigg|_{p=0}.$$

To obtain the approximate solution of the eq. (4.3), we choose the linear operators:

$$L(\phi(x,t;p)) = \frac{\partial}{\partial t}\phi(x,t;p). \tag{4.36}$$

According to eq. (2.26) of Chapter 2, the non-linear term can be defined as following:

$$N\big[\phi(x,t;p)\big] = \frac{\partial}{\partial t}\phi(x,t;p) + {}_0^R D_t^{1-\alpha}\left(2k\phi_x(x,t;p) - \frac{2}{3}\phi_{xx}(x,t;p) + 3\phi(x,t;p)\phi_x(x,t;p) \right.$$

$$\left. + \phi_x(x,t;p)\phi_{xx}(x,t;p) + \frac{1}{2}\phi_x(x,t;p)\phi_{xxxx}(x,t;p) \right) = 0. \tag{4.37}$$

Using eq. (2.27) of Chapter 2, the zeroth-order deformation equation can be presented as follows:

$$(1-p)L\big[\phi(x,t;p) - u_0(x,t)\big] = p\hbar N\big[\phi(x,t;p)\big]. \tag{4.38}$$

Obviously, when $p = 0$ and $p = 1$, eq. (4.38) yields:

$$\phi(x,t;0) = u(x,0);\ \phi(x,t;1) = u(x,t).$$

Therefore, as the embedding parameter p increases from 0 to 1, $\phi(x,t;p)$ varies from the initial guess to the exact solution $u(x,t)$.

If the auxiliary linear operator, the initial guess, and the auxiliary parameters $\hbar$ are so properly chosen, the above series in eq. (4.35) converges at $p = 1$, and we obtain:

$$\phi(x,t;p) = \hat{u}_0(x,t) + \sum_{m=1}^{+\infty} u_m(x,t). \tag{4.39}$$

According to eq. (2.30) of Chapter 2, the mth-order deformation equation can be written as:

$$L\left[u_m(x,t)-\chi_m u_{m-1}(x,t)\right]=\hbar\mathfrak{R}_m\left(\hat{u}_0,\hat{u}_1,\ldots,\hat{u}_{m-1}\right),m\geq 1, \tag{4.40}$$

where:

$$\mathfrak{R}_m\left(\hat{u}_0,\hat{u}_1,\ldots,\hat{u}_{m-1}\right)=\frac{1}{(m-1)!}\frac{\partial^{m-1}}{\partial p^{m-1}}N\left[\phi(x,t;p)\right]\Bigg|_{p=0}$$

$$=\frac{\partial}{\partial t}u_{m-1}(x,t)+{}_0^R D_t^{1-\alpha}\left(2k(u_{m-1})_x(x,t)-\frac{2}{3}(u_{m-1})_{xx}(x,t)+3u_{m-1}(x,t)(u_{m-1})_x(x,t)\right. \tag{4.41}$$

$$\left.+(u_{m-1})_x(x,t)(u_{m-1})_{xx}(x,t)+\frac{1}{2}(u_{m-1})_x(x,t)(u_{m-1})_{xxxx}(x,t)\right).$$

Now, the solutions of the mth-order deformation eq. (4.40) for $m\geq 1$ becomes:

$$u_m(x,t)=\chi_m u_{m-1}(x,t)+\hbar L^{-1}[\mathfrak{R}_m(\hat{u}_0,\hat{u}_1,\ldots,\hat{u}_{m-1})]. \tag{4.42}$$

NOTE: In view of the right-hand side, Riemann–Liouville fractional derivative is interpreted as a future state of the process in physics. For this reason, the right-derivative is usually neglected in applications, when the present state of the process does not depend on the results of the future development, and so the right-derivative is used equal to zero in the following calculations.

By putting the initial conditions in eq. (4.33) into eq. (4.42) and solving them, we now successively obtain:

$$u_0(x,t)=u(x,0)=-k+k\sinh(x),$$

$$u_1(x,t)=-\frac{3k^2 t^{1-\alpha}\hbar\cosh(x)\sec\left(\dfrac{\alpha\pi}{2}\right)(-1+3\sinh(x))}{4\Gamma(2-\alpha)},$$

$$u_2(x,t)=\frac{1}{28}\hbar k^2 t^{1-2\alpha}\sec\left(\frac{\alpha\pi}{2}\right)$$

$$\left(-\frac{96t^\alpha(1+\hbar)\cosh(x)(-1+3\sinh(x))}{\Gamma(2-\alpha)}+\frac{2\hbar\sec\left(\dfrac{\alpha\pi}{2}\right)(-32(-1+\alpha)\cosh(x))}{\Gamma(3-2\alpha)}\right.$$

$$\left.+\frac{2\hbar\sec\left(\dfrac{\alpha\pi}{2}\right)(24(kt+8(-1+\alpha))\sinh(2x)+3kt(-100\cosh(2x)-27\sinh(x)+105\sinh(3x)))}{\Gamma(3-2\alpha)}\right),$$

and so on.

By the homotopy third order series, the solution of eq. (4.3) is approximated as:

$$u(x,t) = u_0(x,t) + u_1(x,t) + u_2(x,t)$$

$$= -k + k\sinh(x) - \frac{k^2 t^{1-\alpha} \hbar \cosh(x) \sec\left(\frac{\alpha\pi}{2}\right)(-1 + 3\sinh(x))(21 - 96(1+\hbar))}{28\Gamma(2-\alpha)} + \ldots \quad (4.43)$$

4.5.1 The ℏ Graph and Numerical Simulations for MHAM

As pointed out by Liao [42] in general, by means of the so-called ℏ-curve, it is straightforward to choose a proper value of ℏ which ensures that the solution series is convergent.

To investigate the influence of ℏ on the solution series, we plot the so called ℏ-curve of partial derivatives of $u(x,t)$ at $(0,0)$ obtained from the MHAM solutions as shown in Figure 4.7. In this way, it is found that our series converge when $\hbar = -0.4$.

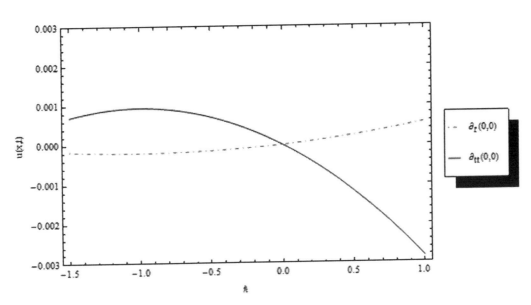

FIGURE 4.7
The ℏ-curve for partial derivatives of $u(x, t)$, at $(0, 0)$ for the MHAM solution.

In this present numerical experiment, eq. (4.43) obtained by the MHAM has been used to draw the graphs as shown in Figures 4.8 and 4.9 for different fractional order values of α. The numerical solutions of eq. (4.3) have been shown in Figures 4.8 and 4.9 with the help of the homotopy series solutions of $u(x,t)$, when $\hbar = -0.4$ and $k = 0.01$.

In Figures 4.8 and 4.9, the MHAM approximate solutions graphs 3D and 2D are plotted for the intervals $-1 \leq x \leq 1$ and $-1 \leq t \leq 1$ using different fractional orders that are for $\alpha = 0.5$ and $\alpha = 0.75$, respectively.

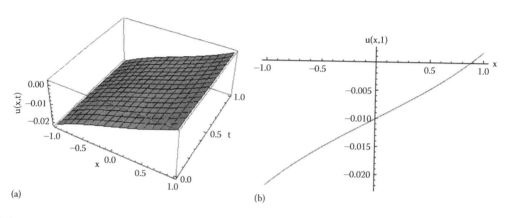

FIGURE 4.8
Case I: For $\alpha = 0.5$. (a) The MHAM travelling wave solution for $u(x,t)$, (b) corresponding 2-D solution for $u(x,t)$ when $t = 1$.

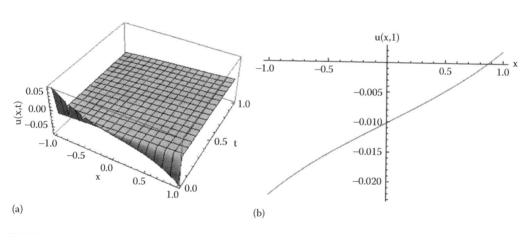

FIGURE 4.9
Case II: For $\alpha = 0.75$. (a) The MHAM travelling wave solution for $u(x,t)$, (b) corresponding 2-D solution for $u(x,t)$ when $t = 1$.

4.5.2 Comparison of Present MHAM Solution with Regard to VIM Solution

In this present analysis, we examine the comparison for the solutions of the MHAM with VIM [16]. Here, we tabulate the solutions for eq. (4.3) using different values of x and t in Table 4.1.

In order to compare the solutions obtained by the present method with regard to those obtained by the VIM [16], L_2 and L_∞ error norms have been also presented in Table 4.2. It may be observed that there is a good agreement between the present MHAM solution and VIM solution.

TABLE 4.1

Comparison of the Solutions between Third Order MHAM and VIM Solutions for Different Values of x and t When $\alpha = 0.5$

| | Comparison of the Solutions between Third Order MHAM and VIM Solutions | | | | | | | | | |
| | 0.1 | | 0.3 | | 0.5 | | 0.7 | | 0.9 | |
x \ t	MHAM	VIM	MHAM	VIM	MHAM	VIM	MHAM	VIM	MHAM	VIM
0.1	−0.00901	−0090186	−0.009028	−0.00903	−0.009036	−0.009044	−0.00904	−0.009053	−0.009050	−0.009061
0.3	−0.00695	−0.006957	−0.006958	−0.00695	−0.006959	−0.006960	−0.00696	−0.006961	−0.006961	−0.006963
0.5	−0.00477	−0.004770	−0.004762	−0.00475	−0.004754	−0.004746	−0.00474	−0.004738	−0.004742	−0.004731
0.7	−0.00237	−0.002367	−0.002346	−0.00233	−0.002325	−0.002307	−0.00230	−0.002286	−0.002295	−0.002267
0.9	0.000338	0.0003522	0.0003922	0.000419	0.0004299	0.0004666	0.000460	0.0005063	0.0004877	0.000541

TABLE 4.2

The L_2 and L_∞ Errors for Third Order MHAM Solutions with Regard to VIM Solutions for Different Values of x When $\alpha = 0.5$

| | Comparison of MHAM Solution with Regard to VIM Solution | |
| | $\alpha = 0.5$ | |
x	L_2	L_∞
0.1	8.02236E-6	9.57534E-6
0.3	1.06901E-6	7.74385E-7
0.5	7.40666E-6	8.84726E-6
0.7	1.94153E-5	7.44980E-6
0.9	3.79915E-5	5.35954E-5

4.6 Conclusion

In this chapter, we have proposed a new analytical technique, the MHAM-FT, to obtain the approximate solution of the RFDE, RFADE, and Riesz time-fractional Camassa–Holm equation. The RFDE, RFADE, and Riesz time-fractional Camassa–Holm equation have been first-time solved by the MHAM-FT in order to justify applicability of the MHAM-FT. The MHAM-FT provides us with a convenient way to control the convergence of approximate series solution and solves the problem without any need for discretization of the variables. To control the convergence of the solution, we can choose the proper values of $\hbar$. The proposed MHAM-FT is very simple and efficient for solving Riesz fractional differential equations.

5

New Exact Solutions of Fractional Differential Equations by Fractional Sub-Equation and Improved Fractional Sub-Equation Method

5.1 Introduction

The study of physical phenomena by means of mathematical models is an essential element in both theoretical and applied sciences. Often such models lead to non-linear systems in a surprisingly large number of cases, to certain prototypical equations. For this reason, the analysis for getting exact solutions plays an important role for analyzing their physical nature of the non-linear equations.

Here, in this chapter, two reliable methods, viz., the fractional sub-equation method and improved fractional sub-equation method have been used for getting new exact solutions for fractional non-linear partial differential equations.

5.2 Outline of Present Study

In the chapter, we construct the analytical exact solutions of some non-linear evolution equations in mathematical physics; namely, the space-time fractional Zakharov–Kuznetsov (ZK) and fractional modified Zakharov–Kuznetsov (mZK) equations by using the fractional sub-equation method. Also, we have used the improved fractional sub-equation method for getting the exact solutions for the time-fractional Korteweg–de Vries (KdV)–Zakharov–Kuznetsov (KdV–ZK) and space-time fractional modified KdV–Zakharov–Kuznetsov (mKdV–ZK) equations. As a result, new types of exact analytical solutions are obtained.

5.2.1 Space-Time Fractional ZK Equation

Let us consider the space-time fractional ZK [143,144] equation:

$$_0D_t^\alpha u + au\, _0D_x^\alpha u + b\, _0D_x^{3\alpha} u + c\, _0D_x^\alpha D_y^{2\alpha} u = 0, \tag{5.1}$$

where $0 < \alpha \leq 1$ and a, b, c are arbitrary constants.

The ZK equation was first derived for analyzing weakly non-linear ion acoustic waves in heavily magnetized lossless plasma and geophysical flows in two dimensions [145]. The ZK equation is one of the two well-established canonical two-dimensional extensions

125

of the Korteweg–de Vries equation [146]. There are some analytical methods like the variational iteration method (VIM) [147,148] by which the fractional ZK equation is solved. The fractional ZK has been examined formerly by using the VIM [149] and HPM [150]. The ZK equation determines the behavior of infirmly non-linear ion-acoustic waves in an incorporation of hot isothermal electrons and cold ions in the existence of a uniform magnetic field [151,152]. A traveling wave analysis is disposed in [153] for the ZK equation.

5.2.2 Space-Time Fractional Modified ZK Equation

Let us consider the space-time fractional mZK [154,155] equation:

$$_0D_t^\alpha u - \beta u^2 \, _0D_x^\alpha u - _0D_x^{3\alpha} u - _0D_x^\alpha \, _0D_y^{2\alpha} u = 0, \tag{5.2}$$

where $0 < \alpha \leq 1$ and β is an arbitrary constant.

When the electron plasma or ion does not fulfill the Boltzmann distribution, Munro and Parkes obtain the mZK equation, and they also examined the stability of traveling wave solutions in isolation and planar periodic two-dimensional long wave perturbation wave solutions [156,157]. The modified ZK equation interprets an anisotropic two-dimensional generalization of the KdV equation and can be analyzed in magnetized plasma for a tiny amplitude Alfven wave at a critical angle to the uninterrupted magnetic field. The mZK equation is efficaciously applied to identify the evolution of several solitary waves in isothermal multicomponent magnetized plasma, likewise, the analysis of stableness of solitary waves of the mZK equation has been described in [158]. Due to the above reason, the mZK equation has enticed the attention for researchers in the past few years.

5.2.3 (3 + 1)-Dimensional Time-Fractional KdV–ZK Equation

Let us consider the (3 + 1)-dimensional time-fractional KdV–ZK equation [159]:

$$D_t^\alpha u + auu_x + u_{xxx} + c(u_{yyx} + u_{zzx}) = 0, \tag{5.3}$$

where $0 < \alpha \leq 1$ and a, c are arbitrary constants.

It is well known that the KdV equation arises as an model for one-dimensional long wavelength surface waves propagating in weakly non-linear dispersive media, as well as the evolution of weakly non-linear ion acoustic waves in plasmas [160]. There are several weakly two-dimensional variations on the KdV equation. The ZK equation is one of two well-studied canonical two-dimensional extensions of the Korteweg–de Vries equation [146]. In recent past, Guo et al [161] derived the (3 + 1)-dimensional variable-coefficient cylindrical KdV) equation describing the non-linear propagation of dust acoustic waves (DAWs). By considering this, the KdV–ZK equation is derived for a plasma comprised of cool and hot electrons and a species of fluid ions [159].

5.2.4 (3 + 1)-Dimensional Space-Time Fractional Modified KdV–ZK

Let us consider the (3 + 1)-dimensional space-time fractional mKdV–ZK equation [162–164]:

$$D_t^\alpha u + \delta u^2 D_x^\alpha u + D_x^{3\alpha} u + D_x^\alpha D_y^{2\alpha} u + D_x^\alpha D_z^{2\alpha} u = 0, \tag{5.4}$$

where $0 < \alpha \leq 1$ and a, c, δ are arbitrary constants.

New Exact Solutions of FDEs

127

The mKdV can be derived as a model for the evolution of ion acoustic perturbations in a plasma with two negative ion components of different temperatures [165]. As in the case of the KdV equation, the mZK equation occurs naturally as weakly two-dimensional variations of the mKdV equation. Exact hyperbolic, trigonometric, and rational solutions of the fractional differential-difference equation related to the discrete mKdV equation has been studied by Aslan [166].

5.3 Application of the Fractional Sub-Equation Method for the Solution of Space-Time Fractional ZK Equation

In this section, we apply the fractional sub-equation method to determine the exact solutions for the space-time fractional ZK equation (5.1).

By considering the traveling wave transformation (2.33) of Chapter 2, eq. (5.1) can be reduced to the following non-linear fractional ordinary differential equation (ODE):

$$v^{\alpha}{}_{0}D_{\xi}^{\alpha}\Phi(\xi) + ak_1^{\alpha}\Phi(\xi){}_{0}D_{\xi}^{\alpha}\Phi(\xi) + bk_1^{3\alpha}{}_{0}D_{\xi}^{3\alpha}\Phi(\xi) + ck_1^{\alpha}k_2^{2\alpha}{}_{0}D_{\xi}^{3\alpha}\Phi(\xi) = 0. \tag{5.5}$$

By balancing the highest order derivative term and non-linear term in eq. (5.5), the value of n can be determined, which is $n = 2$ in this problem.

Thus, the eq. (5.5) has the following formal solution:

$$\Phi(\xi) = a_0 + a_1\varphi(\xi) + a_2\varphi^2(\xi), \tag{5.6}$$

where $\varphi(\xi)$ satisfies the fractional Riccati equation (2.36) of Chapter 2.

Substituting eq. (5.6) along with eq. (2.36) of Chapter 2 into eq. (5.5) and then equating each coefficient of $\varphi^i(\xi)$ ($i = 0, 1, 2, ...$) to zero, we can obtain a set of algebraic equations for $k_1, k_2, v, a_0, a_1, a_2, \sigma$ as follows:

$$\varphi^0 : v^{\alpha}\sigma a_1 + a\sigma a_0 a_1 k_1^{\alpha} + 2\sigma^2 a_1 (bk_1^{3\alpha} + ck_1^{\alpha}k_2^{2\alpha}) = 0,$$

$$\varphi^1 : 2v^{\alpha}\sigma a_2 + a\sigma a_1^2 k_1^{\alpha} + 2a\sigma a_0 a_2 k_1^{\alpha} + 16\sigma^2 a_2 (bk_1^{3\alpha} + ck_1^{\alpha}k_2^{2\alpha}) = 0,$$

$$\varphi^2 : v^{\alpha}a_1 + aa_0 a_1 k_1^{\alpha} + 3a\sigma a_1 a_2 k_1^{\alpha} + 8\sigma a_1 (bk_1^{3\alpha} + k_1^{\alpha}k_2^{2\alpha}) = 0,$$

$$\varphi^3 : 2v^{\alpha}a_2 + aa_1^2 k_1^{\alpha} + 2aa_0 a_2 k_1^{\alpha} + 2a\sigma a_2^2 k_1^{\alpha} + 40\sigma a_2 (bk_1^{3\alpha} + ck_1^{\alpha}k_2^{2\alpha}) = 0, \tag{5.7}$$

$$\varphi^4 : 3aa_1 a_2 k_1^{\alpha} + 6a_1 (bk_1^{3\alpha} + ck_1^{\alpha}k_2^{2\alpha}) = 0,$$

$$\varphi^5 : 2aa_2^2 k_1^{\alpha} + 24a_2 (bk_1^{3\alpha} + ck_1^{\alpha}k_2^{2\alpha}) = 0.$$

Solving the above algebraic eq. (5.7), we have:

Case I:

In this case, a, b, and c are arbitrary. The set of coefficients for the solution in eq. (5.6) are given as:

$$a_0 = -\frac{k_1^{-\alpha}\left(v^\alpha + 8b\sigma k_1^{3\alpha} + 8c\sigma k_1^\alpha k_2^{2\alpha}\right)}{a}, a_1 = 0, a_2 = -\frac{12\left(bk_1^{2\alpha} + ck_2^{2\alpha}\right)}{a}. \tag{5.8}$$

Therefore, using eqs. (2.37) of Chapter 2, from eqs. (5.6) and (5.8), we obtain three types of exact solutions of eq. (5.5), namely, two generalized hyperbolic function solutions, two generalized trigonometric function solutions, and one rational solution as follows:

$$\Phi_{11}(\xi) = -\frac{k_1^{-\alpha}\left(v^\alpha + 8b\sigma k_1^{3\alpha} + 8c\sigma k_1^\alpha k_2^{2\alpha}\right)}{a} + \frac{12\sigma\left(bk_1^{2\alpha} + ck_2^{2\alpha}\right)}{a}\tanh_\alpha^2\left(\sqrt{-\sigma}\xi\right), \quad \sigma < 0,$$

$$\Phi_{12}(\xi) = -\frac{k_1^{-\alpha}\left(v^\alpha + 8b\sigma k_1^{3\alpha} + 8c\sigma k_1^\alpha k_2^{2\alpha}\right)}{a} + \frac{12\sigma\left(bk_1^{2\alpha} + ck_2^{2\alpha}\right)}{a}\coth_\alpha^2\left(\sqrt{-\sigma}\xi\right), \quad \sigma < 0,$$

$$\Phi_{13}(\xi) = -\frac{k_1^{-\alpha}\left(v^\alpha + 8b\sigma k_1^{3\alpha} + 8c\sigma k_1^\alpha k_2^{2\alpha}\right)}{a} - \frac{12\sigma\left(bk_1^{2\alpha} + ck_2^{2\alpha}\right)}{a}\tan_\alpha^2\left(\sqrt{\sigma}\xi\right), \quad \sigma > 0, \tag{5.9}$$

$$\Phi_{14}(\xi) = -\frac{k_1^{-\alpha}\left(v^\alpha + 8b\sigma k_1^{3\alpha} + 8c\sigma k_1^\alpha k_2^{2\alpha}\right)}{a} - \frac{12\sigma\left(bk_1^{2\alpha} + ck_2^{2\alpha}\right)}{a}\cot_\alpha^2\left(\sqrt{\sigma}\xi\right), \quad \sigma > 0,$$

$$\Phi_{15}(\xi) = -\frac{k_1^{-\alpha}\left(v^\alpha + 8b\sigma k_1^{3\alpha} + 8c\sigma k_1^\alpha k_2^{2\alpha}\right)}{a} - \frac{12\sigma\left(bk_1^{2\alpha} + ck_2^{2\alpha}\right)}{a}\left(\frac{\Gamma(1+\alpha)}{\xi^\alpha + \omega}\right)^2,$$

$$\omega = \text{constant}, \ \sigma = 0.$$

Case II:

In the second case, a, b are arbitrary and $c = 0$. The set of coefficients of the solution in eq. (5.6) are given as:

$$a_0 = -\frac{k_1^{-\alpha}\left(v^\alpha + 8b\sigma k_1^{3\alpha}\right)}{a}, a_1 = 0, a_2 = -\frac{12bk_1^{2\alpha}}{a}. \tag{5.10}$$

Therefore, using eq. (2.37) of Chapter 2, from eqs. (5.6) and (5.10), we obtain three types of exact solutions of eq. (5.5), namely, two generalized hyperbolic function solutions, two generalized trigonometric function solutions, and one rational solution as follows:

$$\Phi_{21}(\xi) = -\frac{k_1^{-\alpha}\left(v^\alpha + 8b\sigma k_1^{3\alpha}\right)}{a} + \frac{12\sigma b k_1^{2\alpha}}{a}\tanh_\alpha^2\left(\sqrt{-\sigma}\xi\right), \quad \sigma < 0,$$

$$\Phi_{22}(\xi) = -\frac{k_1^{-\alpha}\left(v^\alpha + 8b\sigma k_1^{3\alpha}\right)}{a} + \frac{12\sigma b k_1^{2\alpha}}{a}\coth_\alpha^2\left(\sqrt{-\sigma}\xi\right), \quad \sigma < 0,$$

$$\Phi_{23}(\xi) = -\frac{k_1^{-\alpha}\left(v^\alpha + 8b\sigma k_1^{3\alpha}\right)}{a} - \frac{12\sigma b k_1^{2\alpha}}{a}\tan_\alpha^2\left(\sqrt{\sigma}\xi\right), \quad \sigma > 0, \qquad (5.11)$$

$$\Phi_{24}(\xi) = -\frac{k_1^{-\alpha}\left(v^\alpha + 8b\sigma k_1^{3\alpha}\right)}{a} - \frac{12\sigma b k_1^{2\alpha}}{a}\cot_\alpha^2\left(\sqrt{\sigma}\xi\right), \quad \sigma > 0,$$

$$\Phi_{25}(\xi) = -\frac{k_1^{-\alpha}\left(v^\alpha + 8b\sigma k_1^{3\alpha}\right)}{a} - \frac{12\sigma b k_1^{2\alpha}}{a}\left(\frac{\Gamma(1+\alpha)}{\xi^\alpha + \omega}\right)^2, \quad \omega = \text{constant}, \ \sigma = 0.$$

Case III:

In the third case, a, c are arbitrary and $b = 0$. The set of coefficients of the solution in eq. (5.6) are given as:

$$a_0 = -\frac{v^\alpha k_1^{-\alpha} + 8c\sigma k_2^{2\alpha}}{a}, a_1 = 0, a_2 = -\frac{12ck_2^{2\alpha}}{a} \qquad (5.12)$$

Again using eq. (2.37) of Chapter 2, from eqs. (5.6) and (5.12), we obtain three types of exact solutions of eq. (5.5), namely, two generalized hyperbolic function solutions, two generalized trigonometric function solutions, and one rational solution as follows:

$$\Phi_{31}(\xi) = -\frac{v^\alpha k_1^{-\alpha} + 8c\sigma k_2^{2\alpha}}{a} + \frac{12ck_2^{2\alpha}}{a}\tanh_\alpha^2\left(\sqrt{-\sigma}\xi\right), \quad \sigma < 0,$$

$$\Phi_{32}(\xi) = -\frac{v^\alpha k_1^{-\alpha} + 8c\sigma k_2^{2\alpha}}{a} + \frac{12ck_2^{2\alpha}}{a}\coth_\alpha^2\left(\sqrt{-\sigma}\xi\right), \quad \sigma < 0,$$

$$\Phi_{33}(\xi) = -\frac{v^\alpha k_1^{-\alpha} + 8c\sigma k_2^{2\alpha}}{a} - \frac{12ck_2^{2\alpha}}{a}\tan_\alpha^2\left(\sqrt{\sigma}\xi\right), \quad \sigma > 0, \qquad (5.13)$$

$$\Phi_{34}(\xi) = -\frac{v^\alpha k_1^{-\alpha} + 8c\sigma k_2^{2\alpha}}{a} - \frac{12ck_2^{2\alpha}}{a}\cot_\alpha^2\left(\sqrt{\sigma}\xi\right), \quad \sigma > 0,$$

$$\Phi_{35}(\xi) = -\frac{v^\alpha k_1^{-\alpha} + 8c\sigma k_2^{2\alpha}}{a} - \frac{12ck_2^{2\alpha}}{a}\left(\frac{\Gamma(1+\alpha)}{\xi^\alpha + \omega}\right)^2, \quad \omega = \text{constant}, \ \sigma = 0.$$

5.3.1 Numerical Simulations of Space-Time Fractional ZK Equation

In this section, the solutions obtained by the fractional sub-equation method have been used to draw the solution graphs for ZK equations.

The numerical solutions of the space-time fractional ZK equation in eq. (5.1) have been presented in Figures 5.1 to 5.4 with the help of the fractional sub-equation method, when: $k_1 = 0.5, k_2 = 1.5, \alpha = 1, a = 1, b = 1/8, c = 1/8, v = 1$, and $t = 0.5$.

In this numerical simulation, Figures 5.1 to 5.4 show solitary wave solutions for $\Phi_{11}(\xi)$, $\Phi_{12}(\xi), \Phi_{13}(\xi)$, and $\Phi_{14}(\xi)$ from eq. (5.9) in the three-dimensional figures, respectively. These solution surfaces have been drawn when $t = 0.5$.

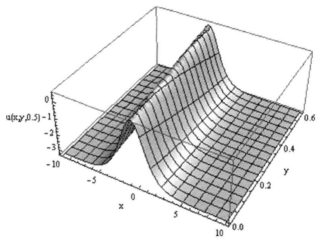

FIGURE 5.1
The bell shaped solitary wave solution for $\Phi_{11}(\xi)$ obtained from eq. (5.9), when $\sigma = -1$.

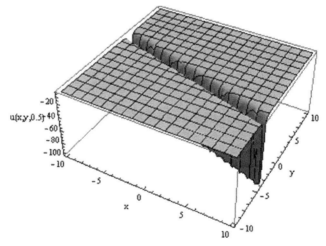

FIGURE 5.2
The solitary wave solution for $\Phi_{12}(\xi)$ obtained from eq. (5.9), when $\sigma = -1$.

New Exact Solutions of FDEs

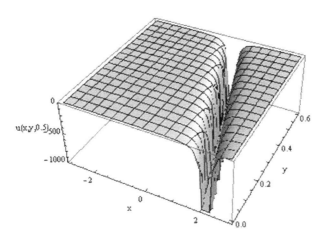

FIGURE 5.3
The solitary wave solution for $\Phi_{13}(\xi)$ obtained from eq. (5.9), when $\sigma = 1$.

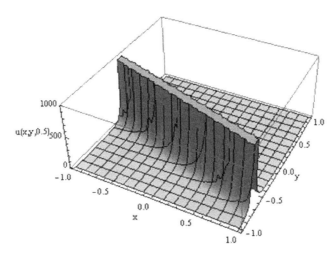

FIGURE 5.4
The solitary wave solution for $\Phi_{14}(\xi)$ obtained from eq. (5.9), when $\sigma = 1$.

5.3.2 Influence of Fractional Order Derivative on Solitary Wave Solutions of Fractional ZK Equation

In this present analysis, we have examined the solitary wave solutions by plotting two-dimensional figures of the fractional Zakharov–Kuznetsov equation for different values of α.

In this numerical simulation, Figure 5.5 shows solitary wave solution for the fractional ZK for different values of α. It may be observed from Figure 5.5 that when the fractional derivative parameter α increases, the peak of the solitary wave nature increases.

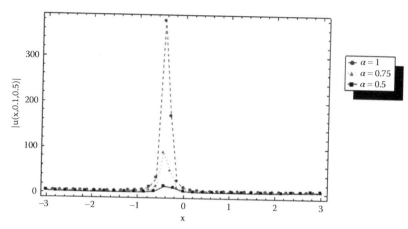

FIGURE 5.5
The solitary wave solution for $|\Phi_{14}(\xi)|$ obtained from eq. (5.9) for the ZK equation for different values of α.

5.4 Implementation of Fractional Sub-Equation Method to the Space-Time Fractional mZK Equation

In this section, we apply the fractional sub-equation method to construct the exact solutions for space-time fractional mZK equation (5.2).

By considering the traveling wave transformation eq. (2.33) of Chapter 2, eq. (5.2) can be reduced to the following non-linear fractional ODE:

$$v^\alpha {}_0D_\xi^\alpha \Phi(\xi) - \beta k_1^\alpha \Phi(\xi)^2 {}_0D_\xi^\alpha \Phi(\xi) - k_1^{3\alpha} {}_0D_\xi^{3\alpha} \Phi(\xi) - k_1^\alpha k_2^{2\alpha} {}_0D_\xi^{3\alpha} \Phi(\xi) = 0. \tag{5.14}$$

By balancing the highest order derivative term and non-linear term in eq. (5.14), we have $n = 1$.
Thus, the eq. (5.14) has the following formal solution:

$$\Phi(\xi) = a_0 + a_1 \varphi(\xi), \tag{5.15}$$

where $\varphi(\xi)$ satisfies the fractional Riccati equation (2.36) of Chapter 2.

Substituting eq. (5.15) along with the Riccati equation (2.36) of Chapter 2 into eq. (5.14), and then equating each coefficient of $\varphi^i(\xi) (i = 0, 1, 2,...)$ to zero, we can obtain a set of algebraic equations for $k_1, k_2, v, a_0, a_1, \sigma$ as follows:

$$\varphi^0 : a_1\left(v^\alpha \sigma - \beta \sigma a_0^2 k_1^\alpha - 2\sigma^2\left(k_1^{3\alpha} + k_1^\alpha k_2^{2\alpha}\right)\right) = 0,$$

$$\varphi^1 : -2\beta\sigma a_0 a_1^2 k_1^\alpha = 0,$$

$$\varphi^2 : a_1\left(v^\alpha - \beta a_0^2 k_1^\alpha - \beta \sigma a_1^2 k_1^\alpha - 8\sigma\left(k_1^{3\alpha} + k_1^\alpha k_2^{2\alpha}\right)\right) = 0, \tag{5.16}$$

$$\varphi^3 : -2\beta a_0 a_1^2 k_1^\alpha = 0,$$

$$\varphi^4 : a_1\left(-\beta a_1^2 k_1^\alpha - 6\left(k_1^{3\alpha} + k_1^\alpha k_2^{2\alpha}\right)\right) = 0.$$

New Exact Solutions of FDEs

Solving the above algebraic equations (5.16), we have:

Case I:

$$\sigma = \frac{v^\alpha k_1^{-\alpha}}{2(k_1^{2\alpha} + k_2^{2\alpha})}, a_0 = 0, a_1 = -\sqrt{\frac{-6\left(k_1^{2\alpha} + k_2^{2\alpha}\right)}{\beta}}. \tag{5.17}$$

Therefore, using eq. (2.37) of Chapter 2, from eqs. (5.15) and (5.17), we obtain two types of exact solutions of eq. (5.14), namely, two generalized hyperbolic function solutions, two generalized trigonometric function solutions as follows:

$$\Phi_{11}(\xi) = \sqrt{\frac{3v^\alpha k_1^{-\alpha}}{\beta}} \tanh_\alpha\left(\sqrt{\frac{-v^\alpha k_1^{-\alpha}}{2(k_1^{2\alpha} + k_2^{2\alpha})}}\xi\right), \qquad \sigma < 0,$$

$$\Phi_{12}(\xi) = \sqrt{\frac{3v^\alpha k_1^{-\alpha}}{\beta}} \coth_\alpha\left(\sqrt{\frac{-v^\alpha k_1^{-\alpha}}{2(k_1^{2\alpha} + k_2^{2\alpha})}}\xi\right), \qquad \sigma < 0,$$

$$\Phi_{13}(\xi) = -i\sqrt{\frac{3v^\alpha k_1^{-\alpha}}{\beta}} \tan_\alpha\left(\sqrt{\frac{v^\alpha k_1^{-\alpha}}{2(k_1^{2\alpha} + k_2^{2\alpha})}}\xi\right), \qquad \sigma > 0, \tag{5.18}$$

$$\Phi_{14}(\xi) = i\sqrt{\frac{3v^\alpha k_1^{-\alpha}}{\beta}} \cot_\alpha\left(\sqrt{\frac{v^\alpha k_1^{-\alpha}}{2(k_1^{2\alpha} + k_2^{2\alpha})}}\xi\right), \qquad \sigma > 0.$$

Case II:

$$\sigma = \frac{v^\alpha k_1^{-\alpha}}{2(k_1^{2\alpha} + k_2^{2\alpha})}, a_0 = 0, a_1 = \sqrt{\frac{-6\left(k_1^{2\alpha} + k_2^{2\alpha}\right)}{\beta}}. \tag{5.19}$$

Thus, using eq. (2.37) of Chapter 2, from eqs. (5.15) and (5.19), we obtain two types of exact solutions of eq. (5.14), namely, two generalized hyperbolic function solutions, two generalized trigonometric function solutions as follows:

$$\Phi_{21}(\xi) = -\sqrt{\frac{3v^\alpha k_1^{-\alpha}}{\beta}} \tanh_\alpha\left(\sqrt{\frac{-v^\alpha k_1^{-\alpha}}{2(k_1^{2\alpha} + k_2^{2\alpha})}}\xi\right), \qquad \sigma < 0,$$

$$\Phi_{22}(\xi) = -\sqrt{\frac{3v^\alpha k_1^{-\alpha}}{\beta}} \coth_\alpha\left(\sqrt{\frac{-v^\alpha k_1^{-\alpha}}{2(k_1^{2\alpha} + k_2^{2\alpha})}}\xi\right), \qquad \sigma < 0,$$

$$\Phi_{23}(\xi) = i\sqrt{\frac{3v^\alpha k_1^{-\alpha}}{\beta}} \tan_\alpha\left(\sqrt{\frac{v^\alpha k_1^{-\alpha}}{2(k_1^{2\alpha} + k_2^{2\alpha})}}\xi\right), \qquad \sigma > 0, \tag{5.20}$$

$$\Phi_{24}(\xi) = -i\sqrt{\frac{3v^\alpha k_1^{-\alpha}}{\beta}} \cot_\alpha\left(\sqrt{\frac{v^\alpha k_1^{-\alpha}}{2(k_1^{2\alpha} + k_2^{2\alpha})}}\xi\right), \qquad \sigma > 0.$$

5.4.1 Numerical Simulations of Space-Time Fractional mZK Equation

In this section, the solutions obtained by the fractional sub-equation method have been used to draw the solution graphs for the space-time fractional mZK equation.

The numerical solutions of the fractional mZK equation in eq. (5.2) have been shown in Figures 5.6 and 5.7 with the help of the fractional sub-equation method, when $k_1 = 1.5$, $k_2 = 0.5$, $v = 1$, $\alpha = 1$, $\beta = 1$, and $t = 0.5$.

In this numerical simulation, Figures 5.6 and 5.7 show solitary wave solutions for $|\Phi_{13}(\xi)|$ and $|\Phi_{14}(\xi)|$ from eq. (5.18) in the three-dimensional figures, respectively. These solution surfaces have been drawn when $t = 0.5$.

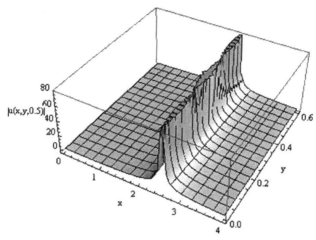

FIGURE 5.6
The solitary wave solution for $|\Phi_{13}(\xi)|$ obtained from eq. (5.18).

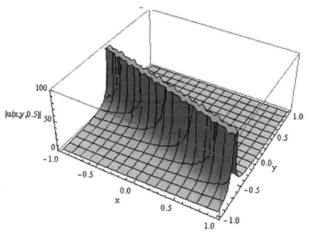

FIGURE 5.7
The solitary wave solution for $|\Phi_{14}(\xi)|$ obtained from eq. (5.18).

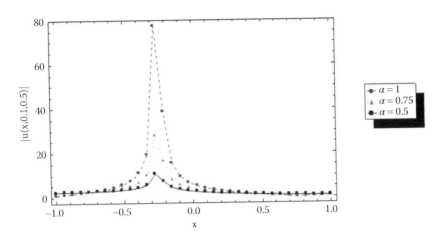

FIGURE 5.8
The solitary wave solution for $|\Phi_{14}(\xi)|$ obtained from eq. (5.18) for the mZK equation for different values of α.

5.4.2 Influence of Fractional Order Derivative on Solitary Wave Solutions of Fractional mZK Equation

In this present analysis, we have examined the solitary wave solutions by plotting two-dimensional figures of the fractional modified Zakharov–Kuznetsov for different values of α.

In this numerical simulation, Figure 5.8 shows solitary wave solution for the fractional mZK equation for different values of α. It may be observed from Figure 5.8 that when the fractional derivative parameter α increases, the peak of the solitary waves nature increases.

5.5 Application of Improved Fractional Sub-Equation Method to the Time-Fractional KdV–ZK Equation

In this section, we apply the improved fractional sub-equation method to determine the exact solutions for the time-fractional KdV–ZK equation (5.3).

By considering the traveling wave transformation eq. (2.40) of Chapter 2, eq. (5.3) can be reduced to the following non-linear fractional ODE:

$$v^\alpha{}_0 D_\xi^\alpha \Phi(\xi) + a k_1^\alpha \Phi(\xi) \Phi'(\xi) + k_1^{3\alpha} \Phi'''(\xi) + c\left(k_1^\alpha k_2^{2\alpha} + k_1^\alpha k_3^{2\alpha}\right) \Phi'''(\xi) = 0. \tag{5.21}$$

By balancing the highest order derivative term and non-linear term in eq. (5.21), the value of n can be determined, which is $n = 2$ in this problem.

Therefore, we have the following ansatz:

$$\Phi(\xi) = a_0 + a_1\left(\frac{D_\xi^\alpha \varphi}{\varphi}\right) + a_2\left(\frac{D_\xi^\alpha \varphi}{\varphi}\right)^2, \tag{5.22}$$

where $\varphi(\xi)$ satisfies the fractional Riccati equation (2.43) of Chapter 2.

Substituting eq. (5.22) along with the Riccati equation (2.43) of Chapter 2 into eq. (5.21), and then equating each coefficient of $(D_\xi^\alpha \varphi/\varphi)^i (i = 0, 1, 2, ...)$ to zero, we can obtain a set of algebraic equations for $k_1, k_2, k_3, v, a_0, a_1, a_2, \sigma$ as follows:

$$\left(\frac{D_\xi^\alpha \varphi}{\varphi}\right)^0 : v^\alpha \sigma a_1 + a\sigma a_0 a_1 k_1^\alpha + 2\sigma^2 a_1 \left(k_1^{3\alpha} + ck_1^\alpha \left(k_2^{2\alpha} + k_3^{2\alpha}\right)\right) = 0,$$

$$\left(\frac{D_\xi^\alpha \varphi}{\varphi}\right)^1 : 2v^\alpha \sigma a_2 + a\sigma a_1^2 k_1^\alpha + 2a\sigma a_0 a_2 k_1^\alpha + 16\sigma^2 a_2 \left(k_1^{3\alpha} + ck_1^\alpha \left(k_2^{2\alpha} + k_3^{2\alpha}\right)\right) = 0,$$

$$\left(\frac{D_\xi^\alpha \varphi}{\varphi}\right)^2 : v^\alpha a_1 + aa_0 a_1 k_1^\alpha + 3a\sigma a_1 a_2 k_1^\alpha + 8\sigma a_1 \left(k_1^{3\alpha} + ck_1^\alpha \left(k_2^{2\alpha} + k_3^{2\alpha}\right)\right) = 0,$$

$$\left(\frac{D_\xi^\alpha \varphi}{\varphi}\right)^3 : 2v^\alpha a_2 + aa_1^2 k_1^\alpha + 2aa_0 a_2 k_1^\alpha + 2a\sigma a_2^2 k_1^\alpha + 40\sigma a_2 \left(k_1^{3\alpha} + ck_1^\alpha \left(k_2^{2\alpha} + k_3^{2\alpha}\right)\right) = 0, \quad (5.23)$$

$$\left(\frac{D_\xi^\alpha \varphi}{\varphi}\right)^4 : 3aa_1 a_2 k_1^\alpha + 6a_1 \left(k_1^{3\alpha} + ck_1^\alpha \left(k_2^{2\alpha} + k_3^{2\alpha}\right)\right) = 0,$$

$$\left(\frac{D_\xi^\alpha \varphi}{\varphi}\right)^5 : 2aa_2^2 k_1^\alpha + 24a_2 \left(k_1^{3\alpha} + ck_1^\alpha \left(k_2^{2\alpha} + k_3^{2\alpha}\right)\right) = 0.$$

Solving the above algebraic eq. (5.23), we have the following cases:

Case I:

In this case, a and c are arbitrary. The set of coefficients for the solution of eq. (5.22) are given as:

$$a_0 = -\frac{k_1^{-\alpha} \left(v^\alpha + 8\sigma k_1^{3\alpha} + 8c\sigma k_1^\alpha \left(k_2^{2\alpha} + k_3^{2\alpha}\right)\right)}{a}, a_1 = 0, a_2 = -\frac{12\left(k_1^{2\alpha} + ck_1^\alpha \left(k_2^{2\alpha} + k_3^{2\alpha}\right)\right)}{a}. \quad (5.24)$$

New Exact Solutions of FDEs

Therefore, using eqs. (2.37) of Chapter 2, from eqs. (5.22) and (5.24), we obtain three types of exact solutions of eq. (5.21), namely, two generalized hyperbolic function solutions, two generalized trigonometric function solutions, and one rational solution as follows:

$$\Phi_{11}(\xi) = -\frac{k_1^{-\alpha}\left(v^\alpha + 8\sigma k_1^{3\alpha} + 8c\sigma k_1^\alpha\left(k_2^{2\alpha} + k_3^{2\alpha}\right)\right)}{a} + \frac{12\sigma\left(k_1^{2\alpha} + ck_1^\alpha\left(k_2^{2\alpha} + k_3^{2\alpha}\right)\right)}{a}$$
$$\tanh_\alpha^2\left(\sqrt{-\sigma}\xi\right), \ \sigma < 0,$$

$$\Phi_{12}(\xi) = -\frac{k_1^{-\alpha}\left(v^\alpha + 8\sigma k_1^{3\alpha} + 8c\sigma k_1^\alpha\left(k_2^{2\alpha} + k_3^{2\alpha}\right)\right)}{a} + \frac{12\sigma\left(k_1^{2\alpha} + ck_1^\alpha\left(k_2^{2\alpha} + k_3^{2\alpha}\right)\right)}{a}$$
$$\coth_\alpha^2\left(\sqrt{-\sigma}\xi\right), \ \sigma < 0,$$

$$\Phi_{13}(\xi) = -\frac{k_1^{-\alpha}\left(v^\alpha + 8\sigma k_1^{3\alpha} + 8c\sigma k_1^\alpha\left(k_2^{2\alpha} + k_3^{2\alpha}\right)\right)}{a} - \frac{12\sigma\left(k_1^{2\alpha} + ck_1^\alpha\left(k_2^{2\alpha} + k_3^{2\alpha}\right)\right)}{a}$$
$$\tan_\alpha^2\left(\sqrt{\sigma}\xi\right), \ \ \sigma > 0,$$

$$\Phi_{14}(\xi) = -\frac{k_1^{-\alpha}\left(v^\alpha + 8\sigma k_1^{3\alpha} + 8c\sigma k_1^\alpha\left(k_2^{2\alpha} + k_3^{2\alpha}\right)\right)}{a} - \frac{12\sigma\left(k_1^{2\alpha} + ck_1^\alpha\left(k_2^{2\alpha} + k_3^{2\alpha}\right)\right)}{a}$$
$$\cot_\alpha^2\left(\sqrt{\sigma}\xi\right), \ \ \sigma > 0,$$

$$\Phi_{15}(\xi) = -\frac{k_1^{-\alpha}\left(v^\alpha + 8\sigma k_1^{3\alpha} + 8c\sigma k_1^\alpha\left(k_2^{2\alpha} + k_3^{2\alpha}\right)\right)}{a} - \frac{12\sigma\left(k_1^{2\alpha} + ck_1^\alpha\left(k_2^{2\alpha} + k_3^{2\alpha}\right)\right)}{a}\left(\frac{\Gamma(1+\alpha)}{\xi^\alpha + \omega}\right)^2,$$

$$\omega = \text{constant}, \quad \sigma = 0. \tag{5.25}$$

Case II:

In the second case, a is arbitrary and $c = 0$. The set of coefficients of the solution of eq. (5.22) are given as:

$$a_0 = -\frac{k_1^{-\alpha}\left(v^\alpha + 8\sigma k_1^{3\alpha}\right)}{a}, a_1 = 0, a_2 = -\frac{12k_1^{2\alpha}}{a} \tag{5.26}$$

Thus, from eqs. (2.37) of Chapter 2, from eqs. (5.22) and (5.26), we obtain three types of exact solutions of eq. (5.21), namely, two generalized hyperbolic function solutions, two generalized trigonometric function solutions, and one rational solution as follows:

$$\Phi_{21}(\xi) = -\frac{k_1^{-\alpha}\left(v^\alpha + 8\sigma k_1^{3\alpha}\right)}{a} + \frac{12\sigma k_1^{2\alpha}}{a}\tanh_\alpha^2\left(\sqrt{-\sigma}\xi\right), \quad \sigma < 0,$$

$$\Phi_{22}(\xi) = -\frac{k_1^{-\alpha}\left(v^\alpha + 8\sigma k_1^{3\alpha}\right)}{a} + \frac{12\sigma k_1^{2\alpha}}{a}\coth_\alpha^2\left(\sqrt{-\sigma}\xi\right), \quad \sigma < 0,$$

$$\Phi_{23}(\xi) = -\frac{k_1^{-\alpha}\left(v^\alpha + 8\sigma k_1^{3\alpha}\right)}{a} - \frac{12\sigma k_1^{2\alpha}}{a}\tan_\alpha^2\left(\sqrt{\sigma}\xi\right), \quad \sigma > 0, \qquad (5.27)$$

$$\Phi_{24}(\xi) = -\frac{k_1^{-\alpha}\left(v^\alpha + 8\sigma k_1^{3\alpha}\right)}{a} - \frac{12\sigma k_1^{2\alpha}}{a}\cot_\alpha^2\left(\sqrt{\sigma}\xi\right), \quad \sigma > 0,$$

$$\Phi_{25}(\xi) = -\frac{k_1^{-\alpha}\left(v^\alpha + 8\sigma k_1^{3\alpha}\right)}{a} - \frac{12\sigma k_1^{2\alpha}}{a}\left(\frac{\Gamma(1+\alpha)}{\xi^\alpha + \omega}\right)^2, \quad \omega = \text{constant}, \sigma = 0.$$

5.5.1 Numerical Simulations of Time-Fractional KdV–ZK Equation

In this section, the solutions obtained by the improved fractional sub-equation method have been used to draw the solution graphs for the time-fractional KdV–ZK equation.

The numerical solutions of the time-fractional KdV–ZK equation in eq. (5.3) have been shown in Figures 5.9 to 5.12 with the help of the improved fractional

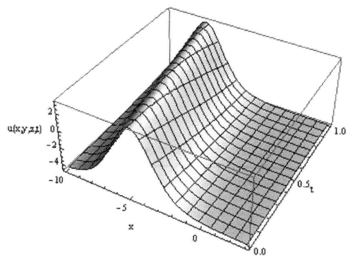

FIGURE 5.9
The bell shaped solitary wave solution for $\Phi_{11}(\xi)$ obtained from eq. (5.25), when $\sigma = -1$, $y = 1$, and $z = 1$.

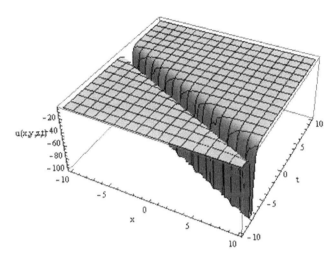

FIGURE 5.10
The solitary wave solution for $\Phi_{12}(\xi)$ obtained from eq. (5.25), when $\sigma = -1$, $y = 1$, and $z = 1$.

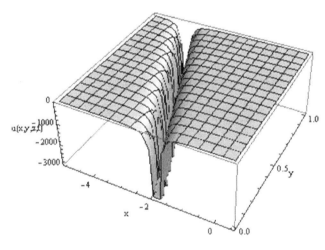

FIGURE 5.11
The solitary wave solution for $\Phi_{13}(\xi)$ obtained from eq. (5.25), when $\sigma = 1$, $y = 1$, and $z = 1$.

sub-equation method, when $k_1 = 0.5$, $k_2 = 1.5$, $k_3 = 1$, $\alpha = 1$, $a = 1$, $c = 1/8$, $v = 1$, $y = 1$, and $z = 1$, respectively.

In this numerical simulation, Figures 5.9 to 5.12 show solitary wave solutions for $\Phi_{11}(\xi)$, $\Phi_{12}(\xi)$, $\Phi_{13}(\xi)$, and $\Phi_{14}(\xi)$ from eq. (5.25) in the three-dimensional figures, respectively. These solution surfaces have been drawn assuming $y = 1$ and $z = 1$, respectively.

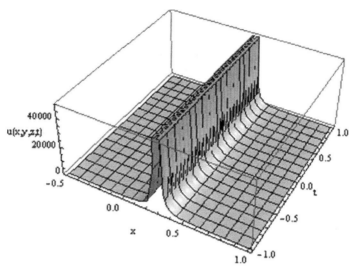

FIGURE 5.12
The solitary wave solution for $\Phi_{14}(\xi)$ obtained from eq. (5.25), when $\sigma = 1$, $y = 1$, and $z = 1$.

5.6 Implementation of Improved Fractional Sub-Equation Method to the Space-Time Fractional Modified mKdV–ZK Equation

In this section, we apply the improved fractional sub-equation method to construct the exact solutions for the space-time fractional mKdV–ZK equation (5.4).

By considering the traveling wave transformation (2.40) of Chapter 2, eq. (5.4) can be reduced to the following non-linear fractional ODE:

$$v^\alpha {}_0D_\xi^\alpha \Phi(\xi) + \delta\, k_1^\alpha \Phi(\xi)^2 D_\xi^\alpha \Phi(\xi) + \left(k_1^{3\alpha} + k_1^\alpha k_2^{2\alpha} + k_1^\alpha k_3^{2\alpha}\right) D_\xi^{3\alpha} \Phi(\xi) = 0. \tag{5.28}$$

By balancing the highest order derivative term and non-linear term in eq. (5.28), the value of n can be determined, which is $n = 1$ in this problem, which yields the following ansatz:

$$\Phi(\xi) = a_0 + a_1 \left(\frac{D_\xi^\alpha \phi}{\phi}\right), \tag{5.29}$$

where $\phi(\xi)$ satisfies the fractional Riccati equation (2.43) of Chapter 2.

Substituting eq. (5.29) along with eq. (2.43) of Chapter 2 into eq. (5.28), and then equating each coefficient of $(D_\xi^\alpha \phi/\phi)^i (i = 0,1,2,...)$ to zero, we can obtain a set of algebraic equations for $k_1, k_2, v, a_0, a_1, \sigma$ as follows:

New Exact Solutions of FDEs

141

$$\left(\frac{D_\xi^\alpha \phi}{\phi}\right)^0 : a_1\left(v^\alpha \sigma - 8\sigma a_0^2 k_1^\alpha + 2\sigma^2\left(k_1^{3\alpha} + k_1^\alpha k_2^{2\alpha} + + k_1^\alpha k_3^{2\alpha}\right)\right) = 0,$$

$$\left(\frac{D_\xi^\alpha \phi}{\phi}\right)^1 : 2\delta\sigma a_0 a_1^2 k_1^\alpha = 0,$$

$$\left(\frac{D_\xi^\alpha \phi}{\phi}\right)^2 : a_1\left(v^\alpha + \delta a_0^2 k_1^\alpha + 8\sigma a_1^2 k_1^\alpha + 8\sigma\left(k_1^{3\alpha} + k_1^\alpha k_2^{2\alpha} + + k_1^\alpha k_3^{2\alpha}\right)\right) = 0, \qquad (5.30)$$

$$\left(\frac{D_\xi^\alpha \phi}{\phi}\right)^3 : 2\delta a_0 a_1^2 k_1^\alpha = 0,$$

$$\left(\frac{D_\xi^\alpha \phi}{\phi}\right)^4 : a_1\left(\delta a_1^2 k_1^\alpha + 6\left(k_1^{3\alpha} + k_1^\alpha k_2^{2\alpha} + + k_1^\alpha k_3^{2\alpha}\right)\right) = 0.$$

Solving the above algebraic eq. (5.30), we have:

Case I:

$$\sigma = -\frac{v^\alpha k_1^{-\alpha}}{2(k_1^{2\alpha} + k_2^{2\alpha} + k_3^{2\alpha})}, a_0 = 0, a_1 = -\sqrt{\frac{-6\left(k_1^{2\alpha} + k_2^{2\alpha} + k_3^{2\alpha}\right)}{\delta}}. \qquad (5.31)$$

Therefore, using eqs. (2.37) of Chapter 2, from eqs. (5.29) and (5.31), we obtain two types of exact solutions of eq. (5.28), namely, two generalized hyperbolic function solutions, two generalized trigonometric function solutions as follows:

$$\Phi_{11}(\xi) = i\sqrt{\frac{3v^\alpha k_1^{-\alpha}}{\delta}} \tanh_\alpha\left(\sqrt{\frac{v^\alpha k_1^{-\alpha}}{2(k_1^{2\alpha} + k_2^{2\alpha} + + k_3^{2\alpha})}}\xi\right), \qquad \sigma < 0,$$

$$\Phi_{12}(\xi) = i\sqrt{\frac{3v^\alpha k_1^{-\alpha}}{\delta}} \coth_\alpha\left(\sqrt{\frac{v^\alpha k_1^{-\alpha}}{2(k_1^{2\alpha} + k_2^{2\alpha} + k_3^{2\alpha})}}\xi\right), \qquad \sigma < 0, \qquad (5.32)$$

$$\Phi_{13}(\xi) = -\sqrt{\frac{3v^\alpha k_1^{-\alpha}}{\delta}} \tan_\alpha\left(\sqrt{\frac{-v^\alpha k_1^{-\alpha}}{2(k_1^{2\alpha} + k_2^{2\alpha} + k_3^{2\alpha})}}\xi\right), \qquad \sigma > 0,$$

$$\Phi_{14}(\xi) = \sqrt{\frac{3v^\alpha k_1^{-\alpha}}{\delta}} \cot_\alpha\left(\sqrt{\frac{-v^\alpha k_1^{-\alpha}}{2(k_1^{2\alpha} + k_2^{2\alpha} + k_3^{2\alpha})}}\xi\right), \qquad \sigma > 0.$$

Case II:

$$\sigma = -\frac{v^\alpha k_1^{-\alpha}}{2(k_1^{2\alpha} + k_2^{2\alpha} + k_3^{2\alpha})}, a_0 = 0, a_1 = \sqrt{\frac{-6\left(k_1^{2\alpha} + k_2^{2\alpha} + k_3^{2\alpha}\right)}{\delta}}. \tag{5.33}$$

Thus, using equation (2.37) of Chapter 2, from eqs. (5.29) and (5.33), we obtain two types of exact solutions of eq. (5.28), namely, two generalized hyperbolic function solutions, two generalized trigonometric function solutions as follows:

$$\Phi_{21}(\xi) = -i\sqrt{\frac{3v^\alpha k_1^{-\alpha}}{\delta}} \tanh_\alpha\left(\sqrt{\frac{v^\alpha k_1^{-\alpha}}{2(k_1^{2\alpha} + k_2^{2\alpha} + +k_3^{2\alpha})}}\xi\right), \quad \sigma < 0,$$

$$\Phi_{22}(\xi) = -i\sqrt{\frac{3v^\alpha k_1^{-\alpha}}{\delta}} \coth_\alpha\left(\sqrt{\frac{v^\alpha k_1^{-\alpha}}{2(k_1^{2\alpha} + k_2^{2\alpha} + k_3^{2\alpha})}}\xi\right), \quad \sigma < 0,$$

$$\tag{5.34}$$

$$\Phi_{23}(\xi) = \sqrt{\frac{3v^\alpha k_1^{-\alpha}}{\delta}} \tan_\alpha\left(\sqrt{\frac{-v^\alpha k_1^{-\alpha}}{2(k_1^{2\alpha} + k_2^{2\alpha} + k_3^{2\alpha})}}\xi\right), \quad \sigma > 0,$$

$$\Phi_{24}(\xi) = -\sqrt{\frac{3v^\alpha k_1^{-\alpha}}{\delta}} \cot_\alpha\left(\sqrt{\frac{-v^\alpha k_1^{-\alpha}}{2(k_1^{2\alpha} + k_2^{2\alpha})}}\xi\right), \quad \sigma > 0.$$

5.6.1 Numerical Simulations of Space-Time Fractional mKdV–ZK Equation

In this section, the solutions obtained by the improved fractional sub-equation method have been used to draw the solution graphs for the space-time fractional mKdV–ZK equation.

The numerical solutions of the mKdV–ZK equation in eq. (5.4) have been shown in Figures 5.13 and 5.14 with the help of the improved fractional sub-equation method, when: $k_1 = 1.5$, $k_2 = 0.5$, $k_3 = 1$, $v = 1$, $\alpha = 1$, $\delta = 1$, $v = 1$, $y = 1$, and $z = 1$, respectively.

In this numerical simulation, Figures 5.13 and 5.14 show solitary wave solution for $|\Phi_{11}(\xi)|$ and $|\Phi_{12}(\xi)|$ from eq. (5.32) in the three-dimensional figures, respectively. These solution surfaces have been drawn using $y = 1$ and $z = 1$, respectively.

New Exact Solutions of FDEs

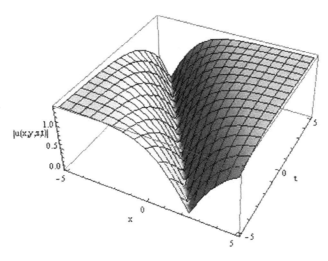

FIGURE 5.13
The solitary wave solution for $|\Phi_{11}(\xi)|$ obtained from eq. (5.32), when $y = 1$ and $z = 1$.

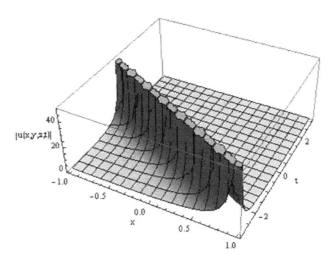

FIGURE 5.14
The solitary wave solution for $|\Phi_{12}(\xi)|$ obtained from eq. (5.32), when $y = 1$ and $z = 1$.

5.7 Conclusion

In this chapter, we have successfully obtained exact solutions for the space-time fractional ZK equation, space-time fractional mZK equation, time-fractional (3 + 1)-dimensional KdV–Zakharov–Kuznetsov equation, and space-time fractional (3 + 1)-dimensional modified KdV–Zakharov–Kuznetsov equations. By using the fractional sub-equation method, three types of exact analytical solutions including the generalized hyperbolic function solutions, generalized trigonometric function solutions, and rational solution for the space-time fractional ZK equation and two types exact analytical solutions including the generalized hyperbolic function solutions, generalized trigonometric function solutions for the space-time fractional mZK equation have been obtained. We have also applied the improved fractional sub-equation method for getting the hyperbolic, trigonometric, and rational function form involving some parameters for time-fractional $(3 + 1)$-dimensional KdV–Zakharov–Kuznetsov and space-time fractional $(3 + 1)$-dimensional modified KdV–Zakharov–Kuznetsov equations.

The proposed methods have many advantages: they are straightforward and concise. From the obtained results, it manifests that the proposed methods are powerful, effective, and convenient for non-linear fractional partial differential equations. Furthermore, these methods provide a powerful mathematical tool to obtain more general explicit exact analytical solutions for non-linear fractional partial differential equations arising in mathematical physics.

6

New Exact Solutions of Fractional Differential Equations by (G′/G)-Expansion Method and Improved (G′/G)-Expansion Method

6.1 Introduction

Non-linear phenomena play a significant role in applied mathematics and physics. Recently, the study of non-linear partial differential equations in modeling physical phenomena has become an important tool. The investigation of the traveling wave solutions plays an important role in non-linear sciences. So in this chapter, we have applied analytical methods like (G'/G)-expansion method and improved (G'/G)-expansion method for getting new exact analytical solutions for some fractional non-linear differential equations.

6.2 Outline of Present Study

In this chapter, we construct the analytical exact solutions of some non-linear evolution equations in mathematical physics; namely, time-fractional modified Kawahara equations, time-fractional coupled Jaulent–Miodek equations, and time-fractional modified Korteweg–de Vries (KdV) equation, by using (G'/G)-expansion method via fractional complex transform. Also time-fractional modified KdV and time-fractional Kaup–Kupershmidt (KK) equation have been solved by using the improved (G'/G)-expansion method. As a result, new types of exact analytical solutions are obtained.

6.2.1 Time-Fractional Fifth Order Modified Kawahara Equation

Let us consider the time-fractional fifth order modified Kawahara equation [167–169]:

$$_0D_t^\alpha u + au^2 u_x + bu_{xxx} - cu_{xxxxx} = 0, \tag{6.1}$$

where $0 < \alpha \le 1$ and a, b, c are arbitrary constants.

The Kawahara equation is a higher order Korteweg–de Vries equation with an additional fifth order derivative term. It was derived by Hasimoto [170] as a model of capillary-gravity waves in an infinitely long canal over a flat bottom in a long wave regime when the bond number is nearly one third. Historically, this type of equation was first found

145

by Kakutani and Ono [171] in analysis of magnet-acoustic waves in a cold collision-free plasma. Then, Hasimoto [170] derived the above equation from capillary-gravity waves. Kawahara [172] studied this type of equation numerically and observed that the equation has both oscillatory and monotone solitary wave solutions. The modified Kawahara equation (6.1) is known as the critical surface-tension model. This equation arises in the modeling of weakly non-linear waves in a wide variety of media [171,172]. A variety of physical phenomena, like magneto acoustic waves in plasma [172], shallow water waves with surface tension [173], and capillary-gravity water waves [174], are described by the modified Kawahara equation (6.1).

The use of fractional calculus for modeling physical systems has been widely considered in the last decades. Due to these, the present chapter is devoted to construct the exact solutions for a time-fractional modified Kawahara equation using a relatively new technique, namely, (G'/G)-expansion method [57–59] via fractional complex transform.

6.2.2 Time-Fractional Coupled JM Equations

Let us consider the non-linear time-fractional coupled Jaulent–Miodek equations [175,176]:

$$
{}_0D_t^\alpha u + u_{xxx} + \frac{3}{2}vv_{xxx} + \frac{9}{2}v_xv_{xx} - 6uu_x - 6uvv_x - \frac{3}{2}u_xv^2 = 0, \tag{6.2}
$$

and

$$
{}_0D_t^\alpha v + v_{xxx} - 6u_xv - 6uv_x - \frac{15}{2}v_xv^2 = 0, \tag{6.3}
$$

where $0 < \alpha \le 1$.

Jaulent and Miodek [177] introduced the Jaulent–Miodek (JM) equation by using inverse scattering transform from the non-linear evolution equation. Matsuno [178] linearized JM equations by hodograph transformation and related it with the Euler–Darboux equation. The JM spectral problem which associates with the JM equation has been studied in [179,180]. By utilizing the Lax pair of the JM spectral problem, Xu [181] constructed the exact solutions of Jaulent–Miodek equations. The symmetries of the JM hierarchy have been studied by Ruan and Lou [182].

Atangana and Baleanu [175] have studied non-linear fractional Jaulent–Miodek equations and presented the series solution by using the Sumudu transform homotopy perturbation method (STHPM). By three-dimensional solution graphs, the nature of the solutions have also been presented by Atangana and Baleanu in their study. By using the two-dimensional Hermite wavelet method, Gupta and Saha Ray [183] have studied the numerical approximation for time-fractional coupled Jaulent–Miodek equations. The fractional JM hierarchy and its super Hamiltonian structure by using fractional supertrace identity have been studied by Wang and Xia [184].

In the past few years, a great deal of attention has been intended by the researchers on the study of non-linear evolution equations [66,70,185,186] that appeared in mathematical physics. Recent past various analytical and numerical methods like the tanh-coth and sech [187] method, extended tanh method [188], Adomian's decomposition method [179], homotopy analysis method [189], (G'/G)-expansion method [190], exp-function method [191,192], and homotopy perturbation method [193] have been used for solving classical JM equations.

In this chapter, an attempt is made to study the solutions of fractional coupled Jaulent–Miodek equations. Hence, it emphasizes on the implementation of a reliable method, (G'/G)-expansion [57–59] method, to find the exact solutions of time-fractional coupled Jaulent–Miodek equations. With a view to exhibit the capability of the method, we employ the method to deal with fractional order coupled Jaulent–Miodek equations. The main objective here is to ascertain additional new general exact solutions of time-fractional coupled Jaulent–Miodek equations by implementing a reliable method, viz., the (G'/G)-expansion method. Some solutions given in this chapter are new solutions which have not been reported yet.

6.2.3 Time-Fractional KK Equation

Let us consider the time-fractional KK equation [194,195]:

$$
{}_0D_t^\alpha u + 45u^2 u_x - \frac{75}{2} u_x u_{xx} - 15 u u_{xxx} + u_{5x} = 0,
\tag{6.4}
$$

where $0 < \alpha \le 1$.

The KK equation has been widely studied over the past three decades. Kaup introduced the equation in 1980 [196] and found a solitary wave solution by using inverse scattering techniques, but its N-soliton solutions have not been found. Date et al. [197] presented the N-soliton solutions of the KK equation in the classification of integrable systems based on the Sato theory and transformation groups and also justified that the KK equation appears as a particular reduction of the C-type sub-hierarchy of the Kadomtsev–Petviashvili (CKP) hierarchy.

In 1997, Hereman and Nuseir [198] constructed only the two- and three-soliton solutions of the KK equation by using the Hirota bilinear method and symbolic computation. Musette et al. [199] and Zait [200] derived the Bäcklund transformation of the KK equation by using singularity analysis based on the Painlevé-Gambier classification. Parker [201,202] introduced the direct approach for converting the KK equation bilinear and arrived at the general N-soliton solutions. Musette and Verhoeven [203] derived the permutability theorem for the KK equation from the Bäcklund transformation. Reyes [204,205] introduced a non-local symmetry for the KK equation.

The fractional differential equations can be described best in discontinuous media and the fractional order is equivalent to its fractional dimensions. Fractal media which is complex, appears in different fields of engineering and physics. In this context, the local fractional calculus theory is very important for modeling problems for fractal mathematics and engineering on Cantorian space in fractal media. Here, an attempt is made successfully to construct the new exact solutions for time-fractional Kaup–Kupershmidt using a relatively new technique, namely, the improved (G'/G)-expansion method [57,65].

6.2.4 Time-Fractional Third Order Time-Fractional Modified KdV Equation

6.2.4.1 Formulation of Fractional mKdV Equation

Let us consider the non-linear classical modified Korteweg–de Vries (mKdV) equation in the following form:

$$
v_t(x,t) + a v^2(x,t) v_x(x,t) + b v_{xxx}(x,t) = 0,
\tag{6.5}
$$

where a, b are arbitrary constants. Here, $v(x,t)$ denotes the amplitude, $x \in \Omega$ is the space variable in the propagation of the field, and $t \in \Gamma$ is the time.

Let $v(x,t)$ be the potential function on the field variable. Putting $v(x,t) = w_x(x,t)$, the eq. (6.5) can be written as following form:

$$w_{xt}(x,t) + aw_x^2(x,t)w_{xx}(x,t) + bw_{xxxx}(x,t) = 0. \tag{6.6}$$

The functional of eq. (6.6) can be represented as:

$$J(w) = \int_\Omega dx \int_\Gamma w(x,t)(c_1 w_{xt}(x,t) + c_2 a w_x^2(x,t)w_{xx}(x,t) + c_3 b w_{xxxx}(x,t))dt, \tag{6.7}$$

where c_1, c_2, and c_3 are the constant Lagrangian multipliers to be determined later on. Integrating eq. (6.7) by parts and taking $w_t(x,t)\big|_\Omega = w_x(x,t)\big|_\Omega = w_{xxx}(x,t)\big|_\Omega = 0$ yield

$$J(w) = \int_\Omega dx \int_\Gamma \left(-c_1 w_t(x,t)w_x(x,t) - \frac{c_2 a}{3}w_x^4(x,t) - c_3 b w_x(x,t)w_{3x}(x,t)\right)dt. \tag{6.8}$$

The Lagrangian multipliers c_1, c_2, and c_3 can be evaluated by taking the variation of the functional (6.8) to make it optimal.

By applying the variation of this functional with respect to $w(x,t)$ leads to:

$$\delta J(w) = \int_\Omega dx \int_\Gamma \left(-c_1(w_x(x,t)\delta w_t(x,t) + w_t(x,t)\delta w_x(x,t))\right.$$

$$-\frac{4c_2 a}{3}w_x^3(x,t)\delta w_x(x,t) \tag{6.9}$$

$$\left. + 2c_3 b w_{xx}(x,t)\delta w_{xx}(x,t). \right.$$

Integrating each term of eq. (6.9) by parts, using $w_t(x,t)\big|_\Omega = w_x(x,t)\big|_\Omega = w_{xxx}(x,t)\big|_\Omega = 0$ and optimizing the variation $J(w) = 0$ yields:

$$c_1 w_{xt}(x,t) + 2c_2 a w_x^2(x,t)w_{xx}(x,t) + c_3 b w_{4x}(x,t) = 0. \tag{6.10}$$

We can determine the Lagrangian multipliers c_1, c_2 and c_3 by comparing the eq. (6.10) with the eq. (6.6). So by comparing the eq. (6.10) with the eq. (6.6), we have here $c_1 = 1$, $c_2 = 1/2$, and $c_3 = 1$. The functional expression given by eq. (6.8) obtains directly the Lagrangian form of the mKdV equation:

$$L(w_t, w_x, w_{xxx}) = -w_t(x,t)w_x(x,t) - \frac{a}{6}w_x^4(x,t) - bw_x(x,t)w_{3x}(x,t). \tag{6.11}$$

Like eq. (6.11), we have the Lagrangian of the time-fractional mKdV equation, which can be written as:

$$F(_0D_t^\alpha w, w_x, w_{xxx}) = -_0D_t^\alpha w(x,t)w_x(x,t) - \frac{a}{6}w_x^4(x,t) - bw_x(x,t)w_{3x}(x,t). \tag{6.12}$$

So the functional of the mKdV equation for eq. (6.12) can be written as:

$$J(w) = \int_{\Omega} dx \int_{\Gamma} F\left({}_0D_t^\alpha w, w_x, w_{xxx}\right) dt. \tag{6.13}$$

The variation of eq. (6.13) with respect to $w(x,t)$ can be written as:

$$\delta J(w) = \int_{\Omega} dx \int_{\Gamma} \left[\frac{\partial F}{\partial_0 D_t^\alpha w} \delta\left({}_0D_t^\alpha w(x,t)\right) + \frac{\partial F}{\partial w_x} \delta w_x(x,t) + \frac{\partial F}{\partial w_{xxx}} \delta w_{xxx}(x,t) \right] dt. \tag{6.14}$$

By Lemma 1.2.5.2 in Chapter 1, eq. (6.14) can be written as:

$$\delta J(w) = \int_{\Omega} dx \int_{\Gamma} \left[{}_tD_\tau^\alpha \left(\frac{\partial F}{\partial_0 D_t^\alpha w} \right) - \frac{\partial}{\partial x}\left(\frac{\partial F}{\partial w_x} \right) - \frac{\partial^3}{\partial x^3}\left(\frac{\partial F}{\partial w_{xxx}} \right) \right] \delta w(x,t) dt. \tag{6.15}$$

By optimizing functional $J(w)$, for example, $\delta J(w) = 0$, the Euler–Lagrange equation for the time-fractional mKdV equation can be written as:

$${}_tD_\tau^\alpha \left(\frac{\partial F}{\partial_0 D_t^\alpha w} \right) - \frac{\partial}{\partial x}\left(\frac{\partial F}{\partial w_x} \right) - \frac{\partial^3}{\partial x^3}\left(\frac{\partial F}{\partial w_{xxx}} \right) = 0. \tag{6.16}$$

Substituting eq. (6.12) into eq. (6.16), we have:

$$- {}_tD_\tau^\alpha w_x(x,t) + {}_0D_t^\alpha w_x(x,t) + 2aw_x^2(x,t)w_{xx}(x,t) + 2bw_{4x}(x,t) = 0. \tag{6.17}$$

Substituting the potential function $w_x(x,t) = v(x,t)$ and taking $a = r/2$, $b = s/2$, we have

$$- {}_tD_\tau^\alpha v(x,t) + {}_0D_t^\alpha v(x,t) + rv^2(x,t)v_x(x,t) + sv_{xxx}(x,t) = 0. \tag{6.18}$$

The right-hand side Riemann–Liouville fractional derivative is interpreted as a future state of the process in physics. For this reason, the right-derivative is usually neglected in applications, when the present state of the process does not depend on the results of the future development. So neglecting the right-derivative, for example, ${}_tD_\tau^\alpha u(x,t)$ from eq. (6.18), we have:

$${}_0D_t^\alpha v(x,t) + rv^2(x,t)v_x(x,t) + sv_{xxx}(x,t) = 0. \tag{6.19}$$

The physical models can be analyzed with the aid of eq. (6.19). The physical significance of this equation has been mentioned in [206].

However, from a mathematical perspective, it is inconvenient to study competition between linear and non-linear effects in the presence of dimensional coefficients. Thus, to make sense, we introduce dimensionless counterparts of the above eq. (6.19).

For making dimensionless, we use the following transformations, which are given as:

$$t' = \frac{Ut^\alpha}{l\Gamma(\alpha+1)}, \, x' = \frac{x}{l}, \, u(x',t') = \frac{v(x,t)}{U},$$

where U is the characteristic amplitude and l is the characteristic spatial scale of the initial assumed field.

Using the above transformations in the fractional mKdV eq. (6.19) with regard to the new dimensionless coordinates x', t' and the dimensionless field $u(x',t')$, we have the following dimensionless version of eq. (6.19):

$$_0D_{t'}^{\alpha}u(x',t') + Uru^2(x',t')u_{x'}(x',t') + \frac{s}{Ul^2}u_{x'x'x'}(x',t') = 0. \tag{6.20}$$

Further, the eq. (6.20) can be written as:

$$_0D_{t'}^{\alpha}u(x',t') + pu^2(x',t')u_{x'}(x',t') + qu_{x'x'x'}(x',t') = 0, \tag{6.21}$$

where $p = Ur$ and $q = s/Ul^2$ are the dimensionless parameters.

For sake of convenience, we have taken $x' = x$ and $t' = t$. Putting it in eq. (6.21), we have the fractional mKdV equation in the dimensionless form, which can be written as [207]:

$$_0D_t^{\alpha}u(x,t) + pu^2(x,t)u_x(x,t) + qu_{xxx}(x,t) = 0, \tag{6.22}$$

where $0 < \alpha \leq 1$, p, q are arbitrary constants.

In the recent past, a great deal of attention has been intended by the researchers on the study of non-linear evolution equations [66,176,208,209] that appeared in mathematical physics. It is widely known that the mKdV equation played an important role in constructing infinitely many conservation laws [210] and the Lax pair for the KdV equation. The Lax pair led to the improvement of the inverse scattering transform (IST) [211] and then the soliton theory. The modified KdV equation is derived by perturbation expansions based on the assumption that the soliton width is small compared with the scale length of the plasma inhomogeneity. In this assumption, soliton maintains all of its identity and its amplitude, width, and speed. The mKdV equation is well known for its special soliton behavior, breathers. Actually, the mKdV equation or its Galilean transformed version, which is usually referred to as the mixed KdV–mKdV equation, appeared in many physical phenomena, such as an harmonic lattices [212], Alfven waves [171], ion acoustic solitons [213], traffic jam [214], Schottky barrier transmission lines [215], and so on. It acquires many remarkable properties, such as Painlevé integrability, Bäcklund transformation, inverse scattering transformation, breather solutions, conservation laws, bilinear transformation, N-solitons, Darboux transformation, and Miura transformation [216–220]. The mKdV equation appears in applications, such as multi-component plasmas and electric circuits, electro-magnetic waves in size-quantized films, electrodynamics, elastic media, and traffic flow [70].

The mKdV equation is used for representing physical models in various physical phenomena, such as to describe the ion-acoustic waves in a magnetized plasma, dipole blocking, study of coastal waves in ocean, and in the issues of atmospheric blocking phenomenon etc. [206,221–227].

With regard to exact solutions, many classical solving methods, such as Hirota's bilinear method [228], the IST [229,230], commutation methods [231], Wronskian technique [232], and reduced differential transform method [233] have been used to solve the mKdV equation.

The fractional differential equations can be described best in discontinuous media and the fractional order is equivalent to its fractional dimensions. Fractal media which is complex, appears in different fields of engineering and physics. In this context, the local fractional calculus theory is very important for modeling problems for fractal mathematics

New Exact Solutions of FDEs by (G'/G)-Expansion and Modified (G'/G)-Expansion Methods 151

and engineering on Cantorian space in fractal media. The present chapter is devoted to construct the exact solutions for the time-fractional modified KdV equation using a relatively new technique, namely, the (G'/G)-expansion method [57–59] and improved (G'/G)-expansion method [57,65]. The fractional complex transform considered here by means of the local fractional derivatives, which can convert fractional differential equations into its equivalent integer-order ordinary differential equations. The fractional complex transform often is used to change fractal time-space to continuous time-space.

6.3 Application of Proposed (G'/G)-Expansion Method to the Time-Fractional Fifth Order Modified Kawahara Equation

In this section, we apply the fractional complex transform and (G'/G)-expansion method to determine the exact solutions for the time-fractional modified Kawahara equation (6.1).

By considering the complex transform (2.45) of Chapter 2, eq. (6.1) can be reduced to the following non-linear fractional ordinary differential equation (ODE):

$$v\Phi'(\xi) + ak\Phi^2(\xi)\Phi'(\xi) + bk^3\Phi'''(\xi) - ck^5\Phi^{\diamond}(\xi) = 0. \tag{6.23}$$

By balancing the highest order derivative term and non-linear term in eq. (6.23), the value of n can be determined, which is $n = 2$ in this problem.

Therefore, we have the following ansatz:

$$\Phi(\xi) = a_0 + a_1\left(\frac{G'}{G}\right) + a_2\left(\frac{G'}{G}\right)^2, \tag{6.24}$$

where G satisfies eq. (2.50) of Chapter 2.

Substituting eq. (6.24) along with eq. (2.51) of Chapter 2 into eq. (6.23), and then equating each coefficient of $(G'/G)(i = 0,1,2,...)$ to zero, we can obtain a set of algebraic equations for $a_i(i = 0,1,2,...,n)$, λ, k, v, and μ as follows:

$$\left(\frac{G'}{G}\right)^0 : -aa_0^2 a_1 k\mu - a_1 v\mu - a_1 bk^3\lambda^2\mu + a_1 ck^5\lambda^4\mu - 2a_1 bk^3\mu^2 - 6a_2 bk^3\lambda\mu^2 + 22a_1 ck^5\lambda^2\mu^2,$$

$$+ 30a_2 ck^5\lambda^3\mu^2 + 16a_1 ck^5\mu^3 + 120a_2 ck^5\lambda\mu^3 = 0,$$

$$\left(\frac{G'}{G}\right)^1 : -\left(aa_0^2 a_1 k\lambda + a_1 v\lambda + a_1 bk^3\lambda^3 - a_1 ck^5\lambda^5 + 2aa_0 a_1^2 k\mu + 2aa_0^2 a_2 k\mu + 2a_2 v\mu + 8a_1 bk^3\lambda\mu \right.$$

$$+ 14a_2 bk^3\lambda^2\mu - 22a_1 ck^5\lambda^3\mu - 62a_2 ck^5\lambda^4\mu + 16a_2 bk^3\mu^2 - 16a_1 ck^5\lambda\mu^2 \tag{6.25}$$

$$\left. - 584a_2 ck^5\lambda^2\mu^2 - 272a_2 ck^5\mu^3 - 30a_1 ck^5\lambda\mu(\lambda^2 + 4\mu)\right) = 0,$$

and so on.

Solving the above algebraic equation (6.25), we have the following cases.

In these cases, a, b, and c are arbitrary. The set of coefficients for the solutions of eq. (6.24) are given as follows:

1. For $\lambda = \lambda$ (arbitrary), we have the following sets of solutions:

Set I:

$$\mu = \frac{\sqrt{15}\sqrt{-b^2c^2k^6 - 10c^3k^5v} + 15c^2k^5\lambda^2}{60c^2k^5}, a_0 = -M_1, a_1 = \frac{6\sqrt{10c}k^2\lambda}{\sqrt{a}}, a_2 = \frac{6\sqrt{10c}}{\sqrt{a}}k^2,$$

$$\text{where} \quad M_1 = \frac{\dfrac{3\sqrt{10}bc^2v^2}{\sqrt{a}} + \dfrac{10\sqrt{6}c^2v^2\sqrt{-c^2k^5(b^2k+10cv)}}{\sqrt{a}k^3} - \dfrac{45\sqrt{10}c^{\frac{7}{2}}k^2v^2\lambda^2}{\sqrt{a}}}{30c^3v^2}. \quad (6.26)$$

Set II:

$$\mu = \frac{\sqrt{15}\sqrt{-b^2c^2k^6 - 10c^3k^5v} + 15c^2k^5\lambda^2}{60c^2k^5}, a_0 = -M_1, a_1 = \frac{6\sqrt{10c}k^2\lambda}{\sqrt{a}}, a_2 = \frac{6\sqrt{10c}}{\sqrt{a}}k^2. \quad (6.27)$$

Set III:

$$\mu = \frac{-\sqrt{15}\sqrt{-b^2c^2k^6 - 10c^3k^5v} + 15c^2k^5\lambda^2}{60c^2k^5}, a_0 = -M_1, a_1 = \frac{6\sqrt{10c}k^2\lambda}{\sqrt{a}}, a_2 = \frac{6\sqrt{10c}}{\sqrt{a}}k^2. \quad (6.28)$$

Set IV:

$$\mu = \frac{\sqrt{15}\sqrt{-b^2c^2k^6 - 10c^3k^5v} + 15c^2k^5\lambda^2}{60c^2k^5}, a_0 = -M_1, a_1 = \frac{6\sqrt{10c}k^2\lambda}{\sqrt{a}}, a_2 = \frac{6\sqrt{10c}}{\sqrt{a}}k^2. \quad (6.29)$$

a. For Set I [eq. (6.26)], we have the following solutions:

Case I:

When $\Delta = \lambda^2 - 4\mu > 0$, we obtain hyperbolic function solution as:

$$\Phi_{11} = M_1 - \frac{6\sqrt{10c}k^2\lambda}{\sqrt{a}}S_1 - \frac{6\sqrt{10c}}{\sqrt{a}}k^2S_1^2,$$

$$\text{where } S_1 = \frac{\sqrt{\Delta}}{2}\left(\frac{C_1\sinh\left(\frac{\sqrt{\Delta}}{2}\xi\right) + C_2\cosh\left(\frac{\sqrt{\Delta}}{2}\xi\right)}{C_1\cosh\left(\frac{\sqrt{\Delta}}{2}\xi\right) + C_2\sinh\left(\frac{\sqrt{\Delta}}{2}\xi\right)}\right) - \frac{\lambda}{2}.$$

Case II:

When $\Delta = \lambda^2 - 4\mu < 0$, we obtain trigonometric function solution as:

$$\Phi_{12} = M_1 - \frac{6\sqrt{10c}k^2\lambda}{\sqrt{a}} S_2 - \frac{6\sqrt{10c}}{\sqrt{a}} k^2 S_2^2.$$

where $S_2 = \left(\frac{\sqrt{-\Delta}}{2} \left(\frac{-C_1 \sin\left(\frac{\sqrt{-\Delta}}{2}\xi\right) + C_2 \cos\left(\frac{\sqrt{-\Delta}}{2}\xi\right)}{C_1 \cos\left(\frac{\sqrt{-\Delta}}{2}\xi\right) + C_2 \sin\left(\frac{\sqrt{-\Delta}}{2}\xi\right)} \right) - \frac{\lambda}{2} \right).$

Case III:

When $\Delta = \lambda^2 - 4\mu = 0$, we obtain the following solution as:

$$\Phi_{13} = M_1 - \frac{6\sqrt{10c}k^2\lambda}{\sqrt{a}} S_3 - \frac{6\sqrt{10c}}{\sqrt{a}} k^2 S_3^2,$$

where $S_3 = \left(\left(\frac{C_2}{C_1 + C_2\xi} \right) - \frac{\lambda}{2} \right)$ and C_1, C_2 are arbitrary constants.

b.   For Set II (eq. 6.27), we have the following solutions:

Case I:

When $\Delta = \lambda^2 - 4\mu > 0$, we obtain hyperbolic function solution as:

$$\Phi_{21} = -M_1 + \frac{6\sqrt{10c}k^2\lambda}{\sqrt{a}} S_1 + \frac{6\sqrt{10c}}{\sqrt{a}} k^2 S_1^2.$$

Case II:

When $\Delta = \lambda^2 - 4\mu < 0$, we obtain trigonometric function solution as:

$$\Phi_{22} = -M_1 + \frac{6\sqrt{10c}k^2\lambda}{\sqrt{a}} S_2 + \frac{6\sqrt{10c}}{\sqrt{a}} k^2 S_2^2.$$

Case III:

When $\Delta = \lambda^2 - 4\mu = 0$, we obtain the following solution as:

$$\Phi_{23} = -M_1 + \frac{6\sqrt{10c}k^2\lambda}{\sqrt{a}} S_3 + \frac{6\sqrt{10c}}{\sqrt{a}} k^2 S_3^2.$$

c. For Set III (eq. 6.28), we have the following solutions:

Case I:

When $\Delta = \lambda^2 - 4\mu > 0$, we obtain hyperbolic function solution as:

$$\Phi_{31} = M_1 - \frac{6\sqrt{10c}k^2\lambda}{\sqrt{a}}S_1 - \frac{6\sqrt{10c}}{\sqrt{a}}k^2S_1^2.$$

Case II:

When $\Delta = \lambda^2 - 4\mu < 0$, we obtain trigonometric function solution as:

$$\Phi_{32} = M_1 - \frac{6\sqrt{10c}k^2\lambda}{\sqrt{a}}S_2 - \frac{6\sqrt{10c}}{\sqrt{a}}k^2S_2^2.$$

Case III:

When $\Delta = \lambda^2 - 4\mu = 0$, we obtain the following solution as:

$$\Phi_{33} = M_1 - \frac{6\sqrt{10c}k^2\lambda}{\sqrt{a}}S_3 - \frac{6\sqrt{10c}}{\sqrt{a}}k^2S_3^2.$$

d. For Set IV (eq. 6.29), we have the following solutions:

Case I:

When $\Delta = \lambda^2 - 4\mu > 0$, we obtain hyperbolic function solution as:

$$\Phi_{41} = -M_1 + \frac{6\sqrt{10c}k^2\lambda}{\sqrt{a}}S_1 + \frac{6\sqrt{10c}}{\sqrt{a}}k^2S_1^2.$$

Case II:

When $\Delta = \lambda^2 - 4\mu < 0$, we obtain trigonometric function solution as:

$$\Phi_{42} = -M_1 + \frac{6\sqrt{10c}k^2\lambda}{\sqrt{a}}S_2 + \frac{6\sqrt{10c}}{\sqrt{a}}k^2S_2^2.$$

Case III:

When $\Delta = \lambda^2 - 4\mu = 0$, we obtain the following solution as:

$$\Phi_{43} = -M_1 + \frac{6\sqrt{10c}k^2\lambda}{\sqrt{a}}S_3 + \frac{6\sqrt{10c}}{\sqrt{a}}k^2S_3^2.$$

2. For $\lambda = 0$, $\mu = 0$. Here, $\lambda^2 - 4\mu = 0$. Therefore, substituting the general solution of eq. (2.50) of Chapter 2 along with eq. (2.52) of Chapter 2 into eq. (2.49) of Chapter 2, we can obtain the sets of solutions of eq. (6.23) as follows:

Set I:

$$a_0 = -\frac{i\sqrt{v}}{\sqrt{ak}}, a_1 = 0, a_2 = \frac{6\sqrt{10c}}{\sqrt{a}} k^2. \tag{6.30}$$

Set II:

$$a_0 = -\frac{i\sqrt{v}}{\sqrt{ak}}, a_1 = 0, a_2 = \frac{6\sqrt{10c}}{\sqrt{a}} k^2. \tag{6.31}$$

Set III:

$$a_0 = -\frac{i\sqrt{v}}{\sqrt{ak}}, a_1 = 0, a_2 = \frac{6\sqrt{10c}}{\sqrt{a}} k^2. \tag{6.32}$$

Set IV:

$$a_0 = \frac{i\sqrt{v}}{\sqrt{ak}}, a_1 = 0, a_2 = \frac{6\sqrt{10c}}{\sqrt{a}} k^2. \tag{6.33}$$

For Set I (eq. 6.30), we have the following solution:

$$\Phi_{11} = -\frac{i\sqrt{v}}{\sqrt{ak}} - \frac{6\sqrt{10c}}{\sqrt{a}} k^2 \left(\frac{C_2}{C_1 + C_2 \xi} \right)^2.$$

For Set II (eq. 6.31), we have the following solution:

$$\Phi_{21} = \frac{i\sqrt{v}}{\sqrt{ak}} - \frac{6\sqrt{10c}}{\sqrt{a}} k^2 \left(\frac{C_2}{C_1 + C_2 \xi} \right)^2.$$

For Set III (eq. 6.32), we have the following solution:

$$\Phi_{31} = -\frac{i\sqrt{v}}{\sqrt{ak}} + \frac{6\sqrt{10c}}{\sqrt{a}} k^2 \left(\frac{C_2}{C_1 + C_2 \xi} \right)^2$$

For Set IV (eq. 6.33), we have the following solution:

$$\Phi_{41} = \frac{i\sqrt{v}}{\sqrt{ak}} + \frac{6\sqrt{10c}}{\sqrt{a}} k^2 \left(\frac{C_2}{C_1 + C_2 \xi} \right)^2,$$

where C_1 and C_2 are arbitrary constants.

3. when $\mu = 0$, we have the following sets of solutions:

Set I:

$$\lambda = \frac{-(-b^2k - 10cv)^{\frac{1}{4}}}{15^{\frac{1}{4}}\sqrt{c}k^{\frac{5}{4}}}, a_0 = \frac{\dfrac{3\sqrt{10}bc^{\frac{5}{2}}}{\sqrt{a}} - \dfrac{5\sqrt{6}c^{\frac{5}{2}}\sqrt{-b^2k - 10cv}}{\sqrt{ak}}}{30c^3},$$

$$a_1 = \frac{2\sqrt{2}3^{\frac{3}{4}}5^{\frac{1}{4}}k^{\frac{3}{4}}(-b^2k - 10cv)^{\frac{1}{4}}}{\sqrt{a}}, a_2 = -\frac{6\sqrt{10c}}{\sqrt{a}}k^2. \tag{6.34}$$

Set II:

$$\lambda = \frac{-(-b^2k - 10cv)^{\frac{1}{4}}}{15^{\frac{1}{4}}\sqrt{c}k^{\frac{5}{4}}}, a_0 = \frac{-\dfrac{3\sqrt{10}bc^{\frac{5}{2}}}{\sqrt{a}} + \dfrac{5\sqrt{6}c^{\frac{5}{2}}\sqrt{-b^2k - 10cv}}{\sqrt{ak}}}{30c^3},$$

$$a_1 = -\frac{2\sqrt{2}3^{\frac{3}{4}}5^{\frac{1}{4}}k^{\frac{3}{4}}(-b^2k - 10cv)^{\frac{1}{4}}}{\sqrt{a}}, a_2 = \frac{6\sqrt{10c}}{\sqrt{a}}k^2. \tag{6.35}$$

Set III:

$$\lambda = \frac{-i(-b^2k - 10cv)^{\frac{1}{4}}}{15^{\frac{1}{4}}\sqrt{c}k^{\frac{5}{4}}}, a_0 = \frac{\dfrac{3\sqrt{10}bc^{\frac{5}{2}}}{\sqrt{a}} + \dfrac{5\sqrt{6}c^{\frac{5}{2}}\sqrt{-b^2k - 10cv}}{\sqrt{ak}}}{30c^3},$$

$$a_1 = \frac{2i\sqrt{2}3^{\frac{3}{4}}5^{\frac{1}{4}}k^{\frac{3}{4}}(-b^2k - 10cv)^{\frac{1}{4}}}{\sqrt{a}}, a_2 = -\frac{6\sqrt{10c}}{\sqrt{a}}k^2. \tag{6.36}$$

Set IV:

$$\lambda = \frac{-i(-b^2k - 10cv)^{\frac{1}{4}}}{15^{\frac{1}{4}}\sqrt{c}k^{\frac{5}{4}}}, a_0 = \frac{-\dfrac{3\sqrt{10}bc^{\frac{5}{2}}}{\sqrt{a}} - \dfrac{5\sqrt{6}c^{\frac{5}{2}}\sqrt{-b^2k - 10cv}}{\sqrt{ak}}}{30c^3},$$

$$a_1 = -\frac{2i\sqrt{2}3^{\frac{3}{4}}5^{\frac{1}{4}}k^{\frac{3}{4}}(-b^2k - 10cv)^{\frac{1}{4}}}{\sqrt{a}}, a_2 = \frac{6\sqrt{10c}}{\sqrt{a}}k^2. \tag{6.37}$$

Set V:

$$\lambda = \frac{i(-b^2k - 10cv)^{\frac{1}{4}}}{15^{\frac{1}{4}}\sqrt{c}k^{\frac{5}{4}}} \,, a_0 = \frac{\dfrac{3\sqrt{10}bc^{\frac{5}{2}}}{\sqrt{a}} + \dfrac{5\sqrt{6}c^{\frac{5}{2}}\sqrt{-b^2k - 10cv}}{\sqrt{a}k}}{30c^3}\,,$$

$$a_1 = -\frac{2i\sqrt{2}\,3^{\frac{3}{4}}5^{\frac{1}{4}}k^{\frac{3}{4}}(-b^2k - 10cv)^{\frac{1}{4}}}{\sqrt{a}} \,, a_2 = -\frac{6\sqrt{10c}}{\sqrt{a}}k^2.$$

(6.38)

Set VI:

$$\lambda = \frac{i(-b^2k - 10cv)^{\frac{1}{4}}}{15^{\frac{1}{4}}\sqrt{c}k^{\frac{5}{4}}} \,, a_0 = \frac{-\dfrac{3\sqrt{10}bc^{\frac{5}{2}}}{\sqrt{a}} - \dfrac{5\sqrt{6}c^{\frac{5}{2}}\sqrt{-b^2k - 10cv}}{\sqrt{a}k}}{30c^3}\,,$$

$$a_1 = \frac{2i\sqrt{2}\,3^{\frac{3}{4}}5^{\frac{1}{4}}k^{\frac{3}{4}}(-b^2k - 10cv)^{\frac{1}{4}}}{\sqrt{a}} \,, a_2 = \frac{6\sqrt{10c}}{\sqrt{a}}k^2.$$

(6.39)

Set VII:

$$\lambda = \frac{(-b^2k - 10cv)^{\frac{1}{4}}}{15^{\frac{1}{4}}\sqrt{c}k^{\frac{5}{4}}} \,, a_0 = \frac{\dfrac{3\sqrt{10}bc^{\frac{5}{2}}}{\sqrt{a}} - \dfrac{5\sqrt{6}c^{\frac{5}{2}}\sqrt{-b^2k - 10cv}}{\sqrt{a}k}}{30c^3}\,,$$

$$a_1 = -\frac{2\sqrt{2}\,3^{\frac{3}{4}}5^{\frac{1}{4}}k^{\frac{3}{4}}(-b^2k - 10cv)^{\frac{1}{4}}}{\sqrt{a}} \,, a_2 = -\frac{6\sqrt{10c}}{\sqrt{a}}k^2.$$

(6.40)

Set VIII:

$$\lambda = \frac{(-b^2k - 10cv)^{\frac{1}{4}}}{15^{\frac{1}{4}}\sqrt{c}k^{\frac{5}{4}}} \,, a_0 = \frac{-\dfrac{3\sqrt{10}bc^{\frac{5}{2}}}{\sqrt{a}} + \dfrac{5\sqrt{6}c^{\frac{5}{2}}\sqrt{-b^2k - 10cv}}{\sqrt{a}k}}{30c^3}\,,$$

$$a_1 = \frac{2\sqrt{2}\,3^{\frac{3}{4}}5^{\frac{1}{4}}k^{\frac{3}{4}}(-b^2k - 10cv)^{\frac{1}{4}}}{\sqrt{a}} \,, a_2 = \frac{6\sqrt{10c}}{\sqrt{a}}k^2.$$

(6.41)

NOTE: Similarly, as obtained in (I) and (II), we can obtain the exact solutions for other sets, those have been omitted here.

6.3.1 The Numerical Simulations for Solutions of Modified Kawahara Equation

In this present numerical experiment, three exact solutions of eqs. (6.1) have been utilized to draw the three-dimensional and two-dimensional solutions graphs for both integer and fractional values of α (Figures 6.1–6.4).

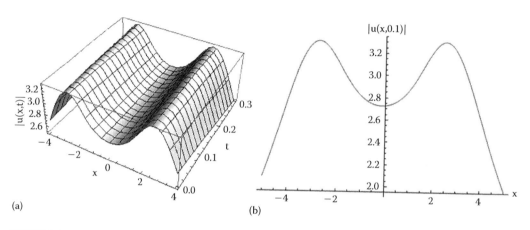

FIGURE 6.1

I: For $\alpha = 1$ (classical order). (a) Three-dimensional solitary wave solution graph for $|u(x,t)|$ appears in Φ_{11} of eq. (6.26) of **Case I**, (b) corresponding two-dimensional solution graph for $|u(x,t)|$ when $t = 0.1$, $a = 1$, $b = 1$, $c = 1$, $k = 1$, $\lambda = 2$, $v = 1$, $C_1 = 1$, $C_2 = 0$, and $\alpha = 1$.

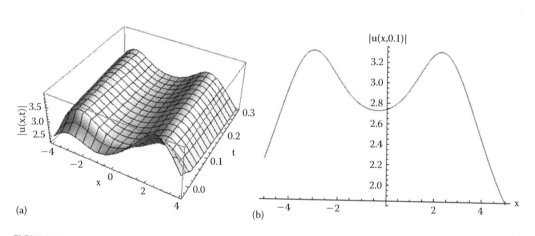

FIGURE 6.2

II: For $\alpha = 0.5$ (fractional order). (a) Three-dimensional solitary wave solution graph for $|u(x,t)|$ appears in Φ_{11} of eq. (6.26) of **Case I**, (b) corresponding two-dimensional solution graph for $|u(x,t)|$ when $t = 0.1$, $a = 1$, $b = 1$, $c = 1$, $k = 1$, $\lambda = 2$, $v = 1$, $C_1 = 1$, $C_2 = 0$, and $\alpha = 0.5$.

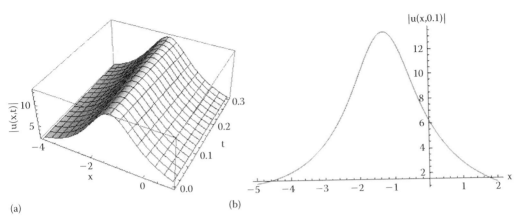

FIGURE 6.3
III: For $\alpha = 1$ (classical order). (a) Three-dimensional solitary wave solution graph for $|u(x,t)|$ appears in Φ_{12} of eq. (6.26) of **Case I**, (b) corresponding two-dimensional solution graph for $|u(x,t)|$ when $t = 0.1$, $a = 1$, $b = 1$, $c = 1$, $k = 1$, $\lambda = 2$, $v = 1$, $C_1 = 1$, $C_2 = 0$, and $\alpha = 1$.

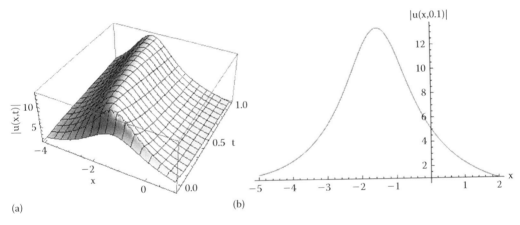

FIGURE 6.4
IV: For $\alpha = 0.5$ (fractional order). (a) Three-dimensional solitary wave solution graph for $|u(x,t)|$ appears in Φ_{12} of eq. (6.26) of **Case I**, (b) corresponding two-dimensional solution graph for $|u(x,t)|$ when $t = 0.1$, $a = 1$, $b = 1$, $c = 1$, $k = 1$, $\lambda = 2$, $v = 1$, $C_1 = 1$, $C_2 = 0$, and $\alpha = 0.5$.

In the present section, we introduce some graphs representing the solutions to visualize the underlying mechanism of the eq. (6.1). Figures 6.1 to 6.4 give a good impression about the change of amplitude and width of the soliton due to the variation of the fractional order of solutions Φ_{11} and Φ_{12}, respectively. Here specifically, three-dimensional graphs describe the behavior of the absolute value of $u(x, t)$ in space x at time t corresponding to different values of the classical order as well as fractional order. The behavior of the solutions graphs indicate that transition of the fractional parameter changes the nature of the solitary wave solution. Therefore, it may be inferred that the fractional-order derivative aids to modulate the shape and nature of the waves in realizing non-linear physical phenomena.

6.4 Implementation of (G'/G)-Expansion Method to the Time-Fractional Coupled JM Equations

In this part, we apply the (G'/G)-expansion method to determine the new exact solutions for time-fractional coupled Jaulent–Miodek equations (6.2) and (6.3).

By applying the following fractional complex transform:

$$u(x,t) = \Phi(\xi),\ v(x,t) = \Psi(\xi),\ \xi = kx + \frac{ct^{\alpha}}{\Gamma(\alpha+1)},$$

the eqs. (6.2) and (6.3) can be reduced to the following non-linear ODE:

$$c\Phi'(\xi) + k^3\Phi'''(\xi) + \frac{3}{2}k^3\Psi(\xi)\Psi'''(\xi) + \frac{9}{2}k^3\Psi'(\xi)\Psi''(\xi),$$

$$-6k\Phi(\xi)\Phi'(\xi) - 6k\Phi(\xi)\Psi(\xi)\Psi'(\xi) - \frac{3}{2}k\Phi'(\xi)\Psi^2(\xi) = 0 \tag{6.42}$$

and

$$c\Psi'(\xi) + k^3\Psi'''(\xi) - 6k\Phi'(\xi)\Psi(\xi) - 6k\Phi(\xi)\Psi'(\xi) - \frac{15}{2}k\Psi'(\xi)\Psi^2(\xi) = 0. \tag{6.43}$$

$$\text{Let } \Phi(\xi) = a_0 + \sum_{i=1}^{n_1} a_i\left(\frac{G'}{G}\right)^i \text{ and } \Psi(\xi) = b_0 + \sum_{i=1}^{n_2} b_i\left(\frac{G'}{G}\right)^i. \tag{6.44}$$

By balancing the highest order derivative term and the non-linear term in eqs. (6.42) and (6.43), the values of n_1 and n_2 can be determined, which are $n_1 = 2$ and $n_2 = 1$ in this problem.

Therefore, by eq. (6.44), we have the following ansatz:

$$\Phi(\xi) = a_0 + a_1\left(\frac{G'}{G}\right) + a_2\left(\frac{G'}{G}\right)^2 \text{ and } \Psi(\xi) = b_0 + b_1\left(\frac{G'}{G}\right), \tag{6.45}$$

where G satisfies eq. (2.50) of Chapter 2.

Substituting eqs. (6.45) along with eq. (2.51) of Chapter 2 into eq. (6.42), then equating each coefficient of $(G'/G)(i=0,1,2,...)$ to zero, we can find a system of algebraic equations for a_i $(i=0,1,2,...,n)$, $b_i(i=0,1,2,...,n)$, λ, k, c, and μ as follows:

$$\left(\frac{G'}{G}\right)^0 : \frac{1}{2}\left(-2c\mu a_1 - 2k^3\lambda^2\mu a_1 - 4k^3\mu^2 a_1 + 12k\mu a_0 a_1 - 12k^3\lambda\mu^2 a_2\right),$$

$$+ 3k\mu a_1 b_0^2 - 3k^3\lambda^2\mu b_0 b_1 - 6k^3\mu^2 b_0 b_1 + 12k\mu a_0 b_0 b_1 - 9k^3\lambda\mu^2 b_1^2\right) = 0,$$

$$\left(\frac{G'}{G}\right)^1 : -\frac{1}{2}\left(2c\lambda a_1 + 2k^3\lambda^3 a_1 + 16k^3\lambda\mu a_1 - 12k\lambda a_0 a_1 - 12k\mu a_1^2 + 4c\mu a_2, \right.$$

$$+ 28k^3\lambda^2\mu a_2 + 32k^3\mu^2 a_2 - 24k\mu b_0 b_1 - 3k\lambda a_1 b_0^2 - 6k\mu a_2 b_0^2,$$

$$+ 3k^3\lambda^3 b_0 b_1 + 24\lambda\mu b_0 b_1 - 12k\lambda a_0 b_0 b_1 - 18k\mu a_1 b_0 b_1,$$

$$\left. + 21k^3\lambda^2\mu b_1^2 + 24k^3\mu^2 b_1^2 - 12k\mu a_0 b_1^2 \right) = 0,$$

$$\left(\frac{G'}{G}\right)^2 : -\frac{1}{2}\left(2c a_1 + 14k^3\lambda^2 a_1 + 16k^3\mu a_1 - 12k a_0 a_1 - 12k\lambda a_1^2 + 4c\lambda a_2, \right.$$

$$+ 16k^3\lambda^3 a_2 + 104k^3\lambda\mu a_2 - 24k\lambda a_0 a_2 - 36k\mu a_1 a_2 - 3k a_1 b_0^2,$$

$$- 6k\lambda a_2 b_0^2 + 21k^3\lambda^2 b_0 b_1 + 24k^3\mu b_0 b_1 - 12k a_0 b_0 b_1 - 18k\lambda a_1 b_0 b_1,$$

$$\left. - 24k\mu a_2 b_0 b_1 + 12k^3\lambda^3 b_1^2 + 78k^3\lambda\mu b_1^2 - 12k\lambda a_0 b_1^2 - 15k\mu a_1 b_1^2 \right) = 0,$$

$$\left(\frac{G'}{G}\right)^3 : -\frac{1}{2}\left(24k^3\lambda a_1 - 12k a_1^2 + 4c a_2 + 76k^3\lambda^2 a_2 + 80k^3\mu a_2 - 24k a_0 a_2, \right.$$

$$- 36k\lambda a_1 a_2 - 24k\mu a_2^2 - 6k a_2 b_0^2 + 36k^3\lambda b_0 b_1 - 18k a_1 b_0 b_1 - 24k\lambda a_2 b_0 b_1,$$

$$\left. + 57k^3\lambda^2 b_1^2 + 60k^3\mu b_1^2 - 12k a_0 b_1^2 + 15k\lambda a_1 b_1^2 - 18k\mu a_2 b_1^2 \right) = 0,$$

$$\left(\frac{G'}{G}\right)^4 : -\frac{1}{2}\left(12k^3 a_1 + 108k^3\lambda a_2 - 36k a_1 a_2 - 24k\lambda a_2^2 + 18k^3 b_0 b_1, \right.$$

$$\left. - 24k a_2 b_0 b_1 + 81k^3\lambda b_1^2 - 15k a_1 b_1^2 - 18k\lambda a_2 b_1^2 \right) = 0,$$

$$\left(\frac{G'}{G}\right)^5 : -\frac{1}{2}\left(48k^3 a_2 - 24k a_2^2 + 36k^3 b_1^2 - 18k a_2 b_0 b_1 \right) = 0. \tag{6.46}$$

Substituting eqs. (6.45) along with eq. (2.51) of Chapter 2 into eq. (6.43), then equating each coefficient of $(G'/G)(i = 0, 1, 2, ...)$ to zero, we can find another system of algebraic equations for a_i $(i = 0, 1, 2, ..., n)$, b_i $(i = 0, 1, 2, ..., n)$, λ, k, c, and μ as follows:

$$\left(\frac{G'}{G}\right)^0 : \frac{1}{2}\left(12k\mu a_1 b_0 - 2c\mu b_1 - 2k^3\lambda^2\mu b_1 - 4k^3\mu^2 b_1 + 12k\mu a_0 b_1 + 15k\mu b_0^2 b_1 \right) = 0,$$

$$\left(\frac{G'}{G}\right)^1 : -\frac{1}{2}\left(-12k\lambda a_1 b_0 - 24k\mu a_2 b_0 + 2c\lambda b_1 + 2k^3\lambda^3 b_1 + 16k^3\lambda\mu b_1, \right.$$

$$\left. -12k\lambda a_0 b_1 - 24k\mu a_1 b_1 - 15k\lambda b_0^2 b_1 - 30k\mu b_0 b_1^2 \right) = 0,$$

$$\left(\frac{G'}{G}\right)^2 : -\frac{1}{2}\left(-12ka_1b_0 - 24k\lambda a_2b_0 + 2cb_1 + 14k^3\lambda^2b_1 + 16k^3\mu b_1 - 12ka_0b_1 - 24k\lambda a_1b_1,\right.$$

$$\left.-36k\mu a_2b_1 - 15kb_0^2b_1 - 30k\lambda b_0b_1^2 - 15k\mu b_1^3\right) = 0,$$

$$\left(\frac{G'}{G}\right)^3 : -\frac{1}{2}\left(24ka_2b_0 + 24k^3\lambda b_1 - 24ka_1b_1 - 36k\lambda b_1a_2 - 30kb_0b_1^2 - 15k\lambda b_1^3\right) = 0,$$

$$\left(\frac{G'}{G}\right)^4 : -\frac{1}{2}\left(12k^3b_1 - 36ka_2b_1 - 15kb_1^3\right) = 0. \tag{6.47}$$

Solving the above algebraic eqs. (6.46) and (6.47), we have the following sets of coefficients for the solutions of eq. (6.45) as given below:

Case I:

$$c = c, \lambda = \lambda, k = k, a_0 = \frac{-2c + k^3\lambda^2 + 14k^3\mu + i\sqrt{3}k\lambda\sqrt{4ck + k^4\lambda^2 - 4k^4\mu}}{24k},$$

$$a_1 = \frac{1}{12}\left(9k^2\lambda + i\sqrt{3}\sqrt{4ck + k^4\lambda^2 - 4k^4\mu}\right), a_2 = \frac{3k^2}{4},$$

$$b_0 = -\frac{3ik^2\lambda - \sqrt{3}\sqrt{4ck + k^4\lambda^2 - 4k^4\mu}}{6k}, b_1 = -ik.$$

Case II:

$$c = c, \lambda = \lambda, k = k, a_0 = \frac{-2c + k^3\lambda^2 + 14k^3\mu - i\sqrt{3}k\lambda\sqrt{4ck + k^4\lambda^2 - 4k^4\mu}}{24k},$$

$$a_1 = \frac{1}{12}\left(9k^2\lambda - i\sqrt{3}\sqrt{4ck + k^4\lambda^2 - 4k^4\mu}\right), a_2 = \frac{3k^2}{4},$$

$$b_0 = -\frac{3ik^2\lambda + \sqrt{3}\sqrt{4ck + k^4\lambda^2 - 4k^4\mu}}{6k}, b_1 = -ik.$$

Case III:

$$c = c, \lambda = \lambda, k = k, a_0 = \frac{-2c + k^3\lambda^2 + 14k^3\mu - i\sqrt{3}k\lambda\sqrt{4ck + k^4\lambda^2 - 4k^4\mu}}{24k},$$

$$a_1 = \frac{1}{12}\left(9k^2\lambda - i\sqrt{3}\sqrt{4ck + k^4\lambda^2 - 4k^4\mu}\right), a_2 = \frac{3k^2}{4},$$

$$b_0 = -\frac{3ik^2\lambda - \sqrt{3}\sqrt{4ck + k^4\lambda^2 - 4k^4\mu}}{6k}, b_1 = ik.$$

Case IV:

$$c = c, \lambda = \lambda, k = k, a_0 = \frac{-2c + k^3\lambda^2 + 14k^3\mu + i\sqrt{3}k\lambda\sqrt{4ck + k^4\lambda^2 - 4k^4\mu}}{24k},$$

$$a_1 = \frac{1}{12}\left(9k^2\lambda + i\sqrt{3}\sqrt{4ck + k^4\lambda^2 - 4k^4\mu}\right), a_2 = \frac{3k^2}{4},$$

$$b_0 = -\frac{3ik^2\lambda + \sqrt{3}\sqrt{4ck + k^4\lambda^2 - 4k^4\mu}}{6k}, b_1 = ik.$$

Substituting the above obtained results into eq. (6.45) along with eqs. (2.52) of Chapter 2, we can find a series of exact solutions to eqs. (6.2) and (6.3).

From **Case I**, we obtain the following exact solutions:

1. When $\Delta = \lambda^2 - 4\mu > 0$, we obtain hyperbolic function solution as:

$$\Phi_{11} = \frac{-2c + k^3\lambda^2 + 14k^3\mu + i\sqrt{3}k\lambda\sqrt{4ck + k^4\lambda^2 - 4k^4\mu}}{24k},$$

$$+ \frac{1}{12}\left(9k^2\lambda + i\sqrt{3}\sqrt{4ck + k^4\lambda^2 - 4k^4\mu}\right)(N_1) + \frac{3k^2}{4}(N_1)^2 \tag{6.48}$$

$$\Psi_{11} = -\frac{3ik^2\lambda - \sqrt{3}\sqrt{4ck + k^4\lambda^2 - 4k^4\mu}}{6k} - ik(N_1). \tag{6.49}$$

$$\text{where } N_1 = \frac{\sqrt{\Delta}}{2}\left(\frac{C_1 \sinh\left(\frac{\sqrt{\Delta}}{2}\xi\right) + C_2 \cosh\left(\frac{\sqrt{\Delta}}{2}\xi\right)}{C_1 \cosh\left(\frac{\sqrt{\Delta}}{2}\xi\right) + C_2 \sinh\left(\frac{\sqrt{\Delta}}{2}\xi\right)}\right) - \frac{\lambda}{2}.$$

2. When $\Delta = \lambda^2 - 4\mu < 0$, we obtain trigonometric function solution as:

$$\Phi_{12} = \frac{-2c + k^3\lambda^2 + 14k^3\mu + i\sqrt{3}k\lambda\sqrt{4ck + k^4\lambda^2 - 4k^4\mu}}{24k},$$

$$+ \frac{1}{12}\left(9k^2\lambda + i\sqrt{3}\sqrt{4ck + k^4\lambda^2 - 4k^4\mu}\right)(N_2) + \frac{3k^2}{4}(N_2)^2 \tag{6.50}$$

$$\Psi_{12} = -\frac{3ik^2\lambda - \sqrt{3}\sqrt{4ck + k^4\lambda^2 - 4k^4\mu}}{6k} - ik(N_2). \tag{6.51}$$

$$\text{where } N_2 = \frac{\sqrt{-\Delta}}{2}\left(\frac{-C_1\sin\left(\frac{\sqrt{-\Delta}}{2}\xi\right) + C_2\cos\left(\frac{\sqrt{-\Delta}}{2}\xi\right)}{C_1\cos\left(\frac{\sqrt{-\Delta}}{2}\xi\right) + C_2\sin\left(\frac{\sqrt{-\Delta}}{2}\xi\right)}\right) - \frac{\lambda}{2}.$$

3. When $\Delta = \lambda^2 - 4\mu = 0$, we obtain the following solution as:

$$\Phi_{13} = \frac{-2c + k^3\lambda^2 + 14k^3\mu + i\sqrt{3}k\lambda\sqrt{4ck + k^4\lambda^2 - 4k^4\mu}}{24k}$$

$$+ \frac{1}{12}\left(9k^2\lambda + i\sqrt{3}\sqrt{4ck + k^4\lambda^2 - 4k^4\mu}\right)\left(\left(\frac{C_2}{C_1 + C_2\xi}\right) - \frac{\lambda}{2}\right) \tag{6.52}$$

$$+ \frac{3k^2}{4}\left(\left(\frac{C_2}{C_1 + C_2\xi}\right) - \frac{\lambda}{2}\right)^2,$$

$$\Psi_{13} = -\frac{3ik^2\lambda - \sqrt{3}\sqrt{4ck + k^4\lambda^2 - 4k^4\mu}}{6k} - ik\left(\left(\frac{C_2}{C_1 + C_2\xi}\right) - \frac{\lambda}{2}\right). \tag{6.53}$$

NOTE: Similarly, as the obtained solutions in **Case I**, we can formulate corresponding new exact solutions to eqs. (6.2) and (6.3) for **Cases II–IV**, which are omitted here.

6.5 Implementation of the Proposed Improved (G'/G)-Expansion Method for the Time-Fractional Kaup–Kupershmidt Equation

In this section, we apply the improved (G'/G)-expansion method to determine the exact solutions for the time-fractional Kaup–Kupershmidt equation (6.4).

By applying the fractional complex transform (2.54) of Chapter 2, eq. (6.4) can be reduced to the following non-linear ODE:

$$v\Phi'(\xi) + 45k\Phi^2(\xi)\Phi'(\xi) - \frac{75}{2}k^3\Phi'(\xi)\Phi''(\xi) - 15k^3\Phi(\xi)\Phi'''(\xi) + k^5\Phi^v(\xi) = 0. \tag{6.54}$$

By balancing the non-linear term and highest order derivative term in eq. (6.54), the value of n can be determined, which is $n = 2$ in this problem.

Therefore, we have the following ansatz:

$$\Phi(\xi) = a_0 + a_1F(\xi) + a_2F(\xi)^2, \tag{6.55}$$

where $F(\xi) = G'(\xi)/G(\xi)$ and $G(\xi)$ satisfies eq. (2.58) of Chapter 2.

New Exact Solutions of FDEs by (G'/G)-Expansion and Modified (G'/G)-Expansion Methods **165**

Substituting eq. (6.55) along with eq. (2.59) of Chapter 2 into eq. (6.54), and then equating each coefficient of $(G'/G)(i = 0,1,2,...)$ to zero, we can get a set of algebraic equations for a_i $(i = 0,1,2,..., n)$, A, B, C, k, and v as follows:

$$\left(\frac{G'}{G}\right)^0 : \frac{1}{2}\Big(90Aa_0^2 a_1 k - 150A^3 a_1 a_2 k^3 - 75A^2 a_1^2 Bk^3 - 180A^2 a_0 a_2 Bk^3 - 30Aa_0 a_1 B^2 k^3$$

$$- 60A^2 a_0 a_1(-1+C)k^3 + 2Aa_1 B^4 k^5 + 4A^2 B^2(15a_2 B + 11a_1(-1+C))k^5$$

$$+ 240A^3 a_2 B(-1+C)k^5 + 32A^3 a_1(-1+C)k^5 + 2Aa_1 v\Big) = 0$$

$$\left(\frac{G'}{G}\right)^1 : \frac{1}{2}\Big(180Aa_0 a_1^2 k + 180Aa_0^2 a_2 k + 90a_0^2 a_1 Bk - 300A^3 a_2^2 k^3 - 930A^2 a_1 a_2 Bk^3 - 180Aa_1^2 B^2 k^3,$$

$$- 180A^2 a_0 a_2 B^2 k^3 - 30a_0 a_1 B^3 k^3 - 60Aa_0 B(4a_2 B + 3a_1(-1+C))k^3$$

$$- 210A^2 a_1^2(-1+C)k^3 - 480A^2 a_0 a_2(-1+C)k^3 - 60Aa_0 a_1 B(-1+C)k^3$$

$$+ 2a_1 B^5 k^5 + 4AB^3(15a_2 B + 11a_1(-1+C))k^5 + 4AB^3(16a_2 B + 15a_1(-1+C))k^5$$

$$+ 240A^2 a_2 B^2(-1+C)k^5 + 16A^2 B(58a_2 B + 15a_1(-1+C))(-1+C)k^5$$

$$+ 544A^3 a_2(-1+C)k^5 + 32A^2 a_1 B(-1+C)^2 k^5 + 4Aa_2 v + 2a_1 Bv\Big) = 0,$$

$$(6.56)$$

and so on.

Solving the above algebraic equation (6.56), we have the set of coefficients for the solutions of eq. (6.55) as given below:

Case 1:

$$A = A, B = B, C = C, k = k, v = -\frac{1}{16}(-4A - B^2 + 4AC)^2 k^5,$$

$$(6.57)$$

$$a_0 = \frac{1}{12}(-8A + B^2 + 8AC)k^2, a_1 = B(-1+C)k^2, a_2 = (-1+C)^2 k^2.$$

Substituting the results obtained above into eq. (6.55) along with eqs. (2.60) to (2.63) of Chapter 2, we can obtain a series of exact solutions to eq. (6.4).

From **Case 1,** we obtain the following new exact solutions:

a. When $B \neq 0$ and $\Delta = B^2 + 4A - 4AC \geq 0$, we have the following exponential function solution:

$$\Phi_{11} = \frac{1}{12}(-8A + B^2 + 8AC)k^2 + B(-1+C)k^2 K_1 + (-1+C)^2 k^2 (K_1)^2,$$

$$(6.58)$$

where $K_1 = \dfrac{B}{2(1-C)} + \dfrac{B\sqrt{\Delta}}{2(1-C)} \left(\dfrac{C_1 \exp\left(\dfrac{\sqrt{\Delta}}{2}\xi\right) + C_2 \exp\left(\dfrac{-\sqrt{\Delta}}{2}\xi\right)}{C_1 \exp\left(\dfrac{\sqrt{\Delta}}{2}\xi\right) - C_2 \exp\left(\dfrac{-\sqrt{\Delta}}{2}\xi\right)} \right)$.

b. When $B \neq 0$ and $\Delta = B^2 + 4A - 4AC < 0$, we have the following trigonometric function solution:

$$\Phi_{12} = \frac{1}{12}(-8A + B^2 + 8AC)k^2 + B(-1+C)k^2 (K_2) + (-1+C)^2 k^2 (K_2)^2, \tag{6.59}$$

where $K_2 = \dfrac{B}{2(1-C)} + \dfrac{B\sqrt{-\Delta}}{2(1-C)} \left(\dfrac{iC_1 \cos\left(\dfrac{\sqrt{-\Delta}}{2}\xi\right) - C_2 \sin\left(\dfrac{\sqrt{-\Delta}}{2}\xi\right)}{iC_1 \sin\left(\dfrac{\sqrt{-\Delta}}{2}\xi\right) + C_2 \cos\left(\dfrac{\sqrt{-\Delta}}{2}\xi\right)} \right)$.

c. When $B = 0$ and $\Delta = A(1-C) \geq 0$, we have the following trigonometric function solution:

$$\Phi_{13} = \frac{1}{12}(-8A + B^2 + 8AC)k^2 + B(-1+C)k^2 (K_3) + (-1+C)^2 k^2 (K_3)^2, \tag{6.60}$$

where $K_3 = \dfrac{\sqrt{\Delta}}{(1-C)} \left(\dfrac{C_1 \cos\left(\sqrt{\Delta}\xi\right) + C_2 \sin\left(\sqrt{\Delta}\xi\right)}{C_1 \sin\left(\sqrt{\Delta}\xi\right) - C_2 \cos\left(\sqrt{\Delta}\xi\right)} \right)$.

d. When $B = 0$ and $\Delta = A(1-C) < 0$, we have the following hyperbolic function solution:

$$\Phi_{14} = \frac{1}{12}(-8A + B^2 + 8AC)k^2 + B(-1+C)k^2 (K_4) + (-1+C)^2 k^2 (K_4)^2, \tag{6.61}$$

where $\quad K_4 = \dfrac{\sqrt{-\Delta}}{(1-C)} \left(\dfrac{iC_1 \cosh\left(\sqrt{-\Delta}\xi\right) - C_2 \sinh\left(\sqrt{-\Delta}\xi\right)}{iC_1 \sinh\left(\sqrt{-\Delta}\xi\right) - C_2 \cosh\left(\sqrt{-\Delta}\xi\right)} \right) \quad$ and

$$\xi = kx - \frac{(-4A - B^2 + 4AC)^2 k^5 t^\alpha}{16\Gamma(\alpha + 1)}.$$

Case 2:

$$A = A, B = B, C = C, k = k, v = -11(-4A - B^2 + 4AC)^2 k^5,$$
$$a_0 = \frac{2}{3}(-8A + B^2 + 8AC)k^2, \, a_1 = 8B(-1+C)k^2, \, a_2 = 8(-1+C)^2 k^2$$

Substituting the results obtained above into eq. (6.55) along with eqs. (2.60) to (2.63) of Chapter 2, we can obtain a series of exact solutions to eq. (6.4).

New Exact Solutions of FDEs by (G'/G)-Expansion and Modified (G'/G)-Expansion Methods 167

From **Case 2**, we obtain the following new exact solutions:

a. When $B \neq 0$ and $\Delta = B^2 + 4A - 4AC \geq 0$, we have the following exponential function solution:

$$\Phi_{21} = \frac{2}{3}(-8A + B^2 + 8AC)k^2 + 8B(-1+C)k^2(K_1) + 8(-1+C)^2k^2(K_1)^2. \tag{6.62}$$

b. When $B \neq 0$ and $\Delta = B^2 + 4A - 4AC < 0$, we have the following trigonometric function solution:

$$\Phi_{22} = \frac{2}{3}(-8A + B^2 + 8AC)k^2 + 8B(-1+C)k^2(K_2) + 8(-1+C)^2k^2(K_2)^2. \tag{6.63}$$

c. When $B = 0$ and $\Delta = A(1-C) \geq 0$, we have the following trigonometric function solution:

$$\Phi_{23} = \frac{2}{3}(-8A + B^2 + 8AC)k^2 + 8B(-1+C)k^2(K_3) + 8(-1+C)^2k^2(K_3)^2. \tag{6.64}$$

d. When $B = 0$ and $\Delta = A(1-C) < 0$, we have the following hyperbolic function solution:

$$\Phi_{24} = \frac{2}{3}(-8A + B^2 + 8AC)k^2 + 8B(-1+C)k^2(K_4) + 8(-1+C)^2k^2(K_4)^2, \tag{6.65}$$

where $\xi = kx - \dfrac{11(-4A - B^2 + 4AC)^2 k^5 t^\alpha}{\Gamma(\alpha+1)}$.

6.6 Application of (G'/G)-Expansion Method and Improved (G'/G)-Expansion Method to the Time-Fractional Modified KdV Equation

This section emphasizes getting new exact solutions of the time-fractional modified KdV equation by using two novel analytical methods, viz., the (G'/G)-expansion method and improved (G'/G)-expansion method.

6.6.1 Application of (G'/G)-Expansion Method to the Time-Fractional Modified KdV Equation

In this section, we apply the (G'/G)-expansion method to determine the exact solutions for the time-fractional modified KdV equation (6.22).

By considering the complex transform (2.45) of Chapter 2, eq. (6.22) can be reduced to the following non-linear ODE:

$$v\Phi'(\xi) + pk\Phi^2(\xi)\Phi'(\xi) + qk^3\Phi'''(\xi) = 0. \tag{6.66}$$

By balancing the highest order derivative term and non-linear term in eq. (6.66), the value of n can be determined, which is $n = 1$ in this problem.

Therefore, we have the following ansatz:

$$\Phi(\xi) = a_0 + a_1 \left(\frac{G'}{G} \right), \tag{6.67}$$

where G satisfies eq. (2.50) of Chapter 2.

Substituting eq. (6.67) along with eq. (2.51) of Chapter 2, into eq. (6.66), and then equating each coefficient of $(i = 0, 1, 2, ...)$ to zero, we can obtain a set of algebraic equations for a_i $(i = 0, 1, 2, ..., n), \lambda, k, v$, and μ as follows:

$$\left(\frac{G'}{G} \right)^0 : -a_1 \left(a_0^2 k \mu p + \mu v + k^3 q \lambda^2 \mu + 2k^3 q \mu^2 \right) = 0.$$

$$\left(\frac{G'}{G} \right)^1 : -a_1 \left(a_0^2 k \lambda p + v \lambda + k^3 q \lambda^3 + 2a_0 a_1 k \mu p + 8k^3 q \lambda \mu \right) = 0.$$

$$\left(\frac{G'}{G} \right)^2 : -a_1 \left(a_0^2 k p + v + 2a_0 a_1 k p \lambda + 7k^3 q \lambda^2 + a_1^2 k \mu p + 8k^3 q \mu \right) = 0.$$

$$\left(\frac{G'}{G} \right)^3 : -a_1 \left(2a_0 a_1 k p + a_1^2 k \lambda p + 12k^3 q \lambda \right) = 0. \tag{6.68}$$

$$\left(\frac{G'}{G} \right)^4 : -a_1 \left(a_1^2 k p + 6k^3 q \right) = 0.$$

Solving the above algebraic equations (6.68), we have the set of coefficients for the solutions of eq. (6.67) as given below:

Case I:

$$\lambda = \lambda, k = k, \mu = \mu, v = \frac{1}{2} (\lambda^2 - 4\mu) k^3 q, a_0 = -\frac{ik\lambda \sqrt{\frac{3}{2} q}}{\sqrt{p}}, a_1 = -\frac{i\sqrt{6q}k}{\sqrt{p}}$$

Substituting the results obtained above into eq. (6.67) along with eqs. (2.52) of Chapter 2, we can obtain a series of exact solutions to eq. (6.22).

From **Case I**, we obtain the following exact solutions:

a. When $\Delta = \lambda^2 - 4\mu > 0$, we obtain hyperbolic function solution as:

$$\Phi_{11} = -\frac{ik\lambda \sqrt{\frac{3}{2} q}}{\sqrt{p}} - \frac{i\sqrt{6q}k}{\sqrt{p}} \left(\frac{\sqrt{\Delta}}{2} \left(\frac{C_1 \sinh\left(\frac{\sqrt{\Delta}}{2} \xi \right) + C_2 \cosh\left(\frac{\sqrt{\Delta}}{2} \xi \right)}{C_1 \cosh\left(\frac{\sqrt{\Delta}}{2} \xi \right) + C_2 \sinh\left(\frac{\sqrt{\Delta}}{2} \xi \right)} \right) - \frac{\lambda}{2} \right). \tag{6.69}$$

b. When $\Delta = \lambda^2 - 4\mu < 0$, we obtain trigonometric function solution as:

$$\Phi_{12} = -\frac{ik\lambda\sqrt{\frac{3}{2}}q}{\sqrt{p}} - \frac{i\sqrt{6q}k}{\sqrt{p}}\left(\frac{\sqrt{-\Delta}}{2}\left(\frac{-C_1\sin\left(\frac{\sqrt{-\Delta}}{2}\xi\right)+C_2\cos\left(\frac{\sqrt{-\Delta}}{2}\xi\right)}{C_1\cos\left(\frac{\sqrt{-\Delta}}{2}\xi\right)+C_2\sin\left(\frac{\sqrt{-\Delta}}{2}\xi\right)}\right) - \frac{\lambda}{2}\right). \tag{6.70}$$

c. When $\Delta = \lambda^2 - 4\mu = 0$, we obtain the following solution as:

$$\Phi_{13} = -\frac{ik\lambda\sqrt{\frac{3}{2}}q}{\sqrt{p}} - \frac{i\sqrt{6q}k}{\sqrt{p}}\left(\left(\frac{C_2}{C_1+C_2\xi}\right) - \frac{\lambda}{2}\right). \tag{6.71}$$

Case II:

$$\lambda = \lambda, k = k, \mu = \mu, v = \frac{1}{2}(\lambda^2 - 4\mu)k^3q, a_0 = \frac{ik\lambda\sqrt{\frac{3}{2}}q}{\sqrt{p}}, a_1 = \frac{i\sqrt{6q}k}{\sqrt{p}}.$$

Substituting the results obtained above into eq. (6.67) along with eqs. (2.52) of Chapter 2, we can obtain a series of exact solutions to eq. (6.22).

From **Case II**, we obtain the following exact solutions:

a. When $\Delta = \lambda^2 - 4\mu > 0$, we obtain hyperbolic function solution as:

$$\Phi_{21} = -\frac{ik\lambda\sqrt{\frac{3}{2}}q}{\sqrt{p}} + \frac{i\sqrt{6q}k}{\sqrt{p}}\left(\frac{\sqrt{\Delta}}{2}\left(\frac{C_1\sinh\left(\frac{\sqrt{\Delta}}{2}\xi\right)+C_2\cosh\left(\frac{\sqrt{\Delta}}{2}\xi\right)}{C_1\cosh\left(\frac{\sqrt{\Delta}}{2}\xi\right)+C_2\sinh\left(\frac{\sqrt{\Delta}}{2}\xi\right)}\right) - \frac{\lambda}{2}\right). \tag{6.72}$$

b. When $\Delta = \lambda^2 - 4\mu < 0$, we obtain trigonometric function solution as:

$$\Phi_{22} = \frac{ik\lambda\sqrt{\frac{3}{2}}q}{\sqrt{p}} + \frac{i\sqrt{6q}k}{\sqrt{p}}\left(\frac{\sqrt{-\Delta}}{2}\left(\frac{-C_1\sin\left(\frac{\sqrt{-\Delta}}{2}\xi\right)+C_2\cos\left(\frac{\sqrt{-\Delta}}{2}\xi\right)}{C_1\cos\left(\frac{\sqrt{-\Delta}}{2}\xi\right)+C_2\sin\left(\frac{\sqrt{-\Delta}}{2}\xi\right)}\right) - \frac{\lambda}{2}\right). \tag{6.73}$$

c. When $\Delta = \lambda^2 - 4\mu = 0$, we obtain the following solution as:

$$\Phi_{23} = \frac{ik\lambda\sqrt{\frac{3}{2}}q}{\sqrt{p}} + \frac{i\sqrt{6q}k}{\sqrt{p}}\left(\left(\frac{C_2}{C_1+C_2\xi}\right) - \frac{\lambda}{2}\right), \tag{6.74}$$

where $\xi = kx + \dfrac{vt^\alpha}{\Gamma(\alpha+1)}$.

From the proposed (G'/G)-expansion method, we obtained the six exact solutions for the time-fractional third order modified KdV equation.

6.6.1.1 The Numerical Simulation for Solution of Time-Fractional Third Order Modified KdV Equation Using Proposed (G'/G)-Expansion Method (Figures 6.5–6.13)

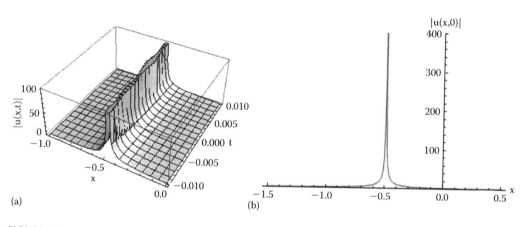

FIGURE 6.5
Case Ia: For $\alpha = 1$ (classical order). (a) The (G'/G)-expansion method solitary wave solution for $|u(x,t)|$ appears in eq. (6.69) of **Case I**, (b) corresponding solution for $|u(x,t)|$ when $t=0$, $C_1=1$, $C_2=2$, $p=6$, $q=1$, $\mu=1$, $\lambda=3$, and $\alpha=1$.

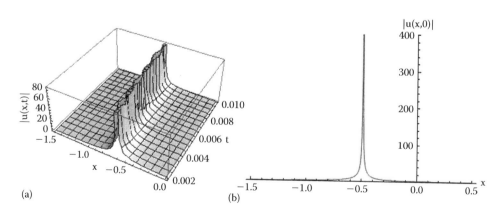

FIGURE 6.6
Case Ib: For $\alpha = 0.5$ (fractional order). (a) The (G'/G)-expansion method solitary wave solution for $|u(x,t)|$ appears in eq. (6.69) of **Case I**, (b) corresponding solution for $|u(x,t)|$ when $t=0$, $C_1=1$, $C_2=2$, $p=6$, $q=1$, $\mu=1$, $\lambda=3$, and $\alpha=0.5$.

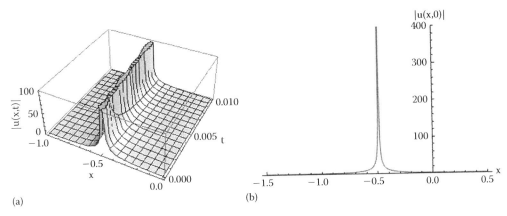

FIGURE 6.7
Case Ic: For $\alpha = 0.75$ (fractional order). (a) The (G'/G)-expansion method solitary wave solution for $|u(x,t)|$ appears in eq. (6.69) of **Case I**, (b) corresponding solution for $|u(x,t)|$ when $t = 0$, $C_1 = 1$, $C_2 = 2$, $p = 6$, $q = 1$, $\mu = 1$, $\lambda = 3$, and $\alpha = 0.75$.

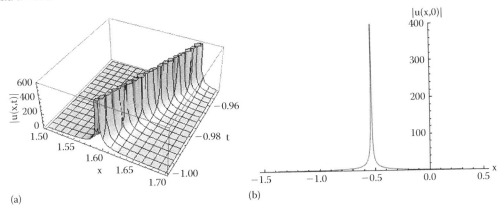

FIGURE 6.8
Case IIa: For $\alpha = 1$ (classical order). (a) The (G'/G)-expansion method solitary wave solution for $|u(x,t)|$ appears in eq. (6.69) of **Case I**, (b) corresponding solution for $|u(x,t)|$ when $t = 0$, $C_1 = 1$, $C_2 = 2$, $p = 6$, $q = 1$, $\mu = 1$, and $\alpha = 1$.

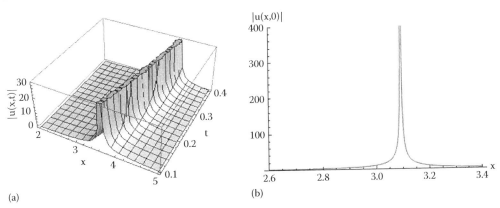

FIGURE 6.9
Case IIb: For $\alpha = 0.5$ (fractional order). (a) The (G'/G)-expansion method solitary wave solution for $|u(x,t)|$ appears in eq. (6.70) of **Case I**, (b) corresponding solution for $|u(x,t)|$ when $t = 0$, $C_1 = 1$, $C_2 = 2$, $p = 6$, $q = 1$, $\mu = 1$, $\lambda = 1$, and $\alpha = 0.5$.

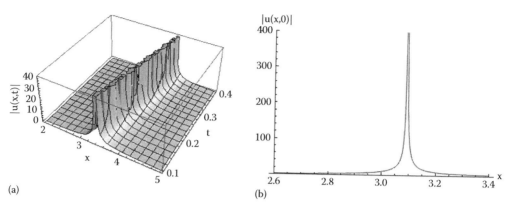

FIGURE 6.10
Case IIc: For $\alpha = 0.75$ (fractional order). (a) The (G'/G)-expansion method solitary wave solution for $|u(x,t)|$ appears in eq. (6.70) of **Case I**, (b) corresponding solution for $|u(x,t)|$ when $t = 0$, $C_1 = 1$, $C_2 = 2$, $p = 6$, $q = 1$, $\mu = 1$, $\lambda = 1$, and $\alpha = 0.75$.

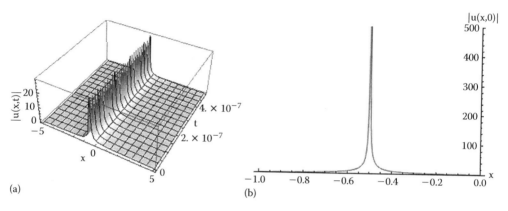

FIGURE 6.11
Case IIIa: For $\alpha = 1$ (classical order). (a) The (G'/G)-expansion method solitary wave solution for $|u(x,t)|$ appears in eq. (6.71) of **Case I**, (b) corresponding solution for $|u(x,t)|$ when $t = 0$, $C_1 = 1$, $C_2 = 2$, $p = 6$, $q = 1$, $\mu = 1$, and $\alpha = 1$.

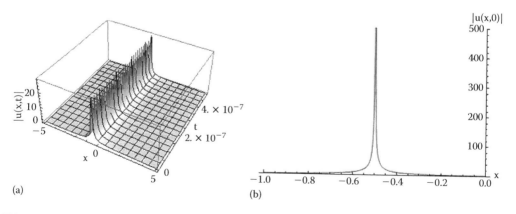

FIGURE 6.12
Case IIIb: For $\alpha = 0.5$ (fractional order). (a) The (G'/G)-expansion method solitary wave solution for $|u(x,t)|$ appears in eq. (6.71) of **Case I**, (b) corresponding solution for $|u(x,t)|$ when $t = 0$, $C_1 = 1$, $C_2 = 2$, $p = 6$, $q = 1$, $\mu = 1$, and $\alpha = 0.5$.

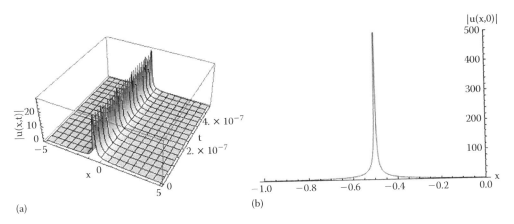

FIGURE 6.13
Case IIIc: For $\alpha = 0.75$ (fractional order). (a) The (G'/G)-expansion method solitary wave solution for $|u(x,t)|$ appears in eq. (6.71) of **Case I**, (b) corresponding solution for $|u(x,t)|$ when $t = 0$, $C_1 = 1$, $C_2 = 2$, $p = 6$, $q = 1$, $\mu = 1$, and $\alpha = 0.75$.

6.6.2 Implementation of Improved (G'/G)-Expansion Method for the Time-Fractional Modified KdV Equation

In this section, we apply the improved (G'/G)-expansion method to determine the exact solutions for the time-fractional modified KdV equation (6.22).

By applying fractional complex transform, eq. (2.54) of Chapter 2, eq. (6.22) can be reduced to the following non-linear ODE:

$$v\Phi'(\xi) + pk\Phi^2(\xi)\Phi'(\xi) + qk^3\Phi'''(\xi) = 0. \tag{6.75}$$

By balancing the highest order derivative term and non-linear term in eq. (6.75), the value of n can be determined, which is $n = 1$ in this problem.

Therefore, we have the following ansatz:

$$\Phi(\xi) = a_0 + a_1 F(\xi), \tag{6.76}$$

where $F(\xi) = G'(\xi)/G(\xi)$ and $G(\xi)$ satisfy eq. (2.58) of Chapter 2.

Substituting eq. (6.76) along with eq. (2.59) of Chapter 2 into eq. (6.75), and then equating each coefficient of $(G'/G)(i = 0, 1, 2, ...)$ to zero, we can obtain a set of algebraic equations for a_i ($i = 0, 1, 2, ..., n$), A, B, C, k, and v as follows:

$$\left(\frac{G'}{G}\right)^0 : a_1(Aa_0^2 kp + Ak^3 q(B^2 + 2A(-1+C)) + Av) = 0.$$

$$\left(\frac{G'}{G}\right)^1 : a_1(2Aa_0 a_1 kp + a_0^2 Bkp + b(B^2 + 2A(-1+C))k^3 q + 6AB(-1+C)k^3 q + Bv) = 0.$$

$$\left(\frac{G'}{G}\right)^2 : a_1\left(Aa_1^2kp + 2Ba_0a_1kp + a_0^2(-1+C)kp + 6B^2(-1+C)k^3q\right.$$

$$\left. + (B^2 + 2A(-1+C))(-1+C)^2k^3q + 6A(-1+C)^2k^3q + (-1+C)v\right) = 0,$$

$$\left(\frac{G'}{G}\right)^3 : a_1\left(Ba_1^2kp + 2(-1+C)a_0a_1k + 12B(-1+C)^2k^3q\right) = 0. \tag{6.77}$$

$$\left(\frac{G'}{G}\right)^4 : a_1\left(a_1^2(-1+C)kp + 6(-1+C)^3k^3q\right) = 0.$$

Solving the above algebraic equations (6.77), we have the set of coefficients for the solutions of eq. (6.76) as given below:

Case I:

$$A = A, B = B, C = C, k = k, v = -\frac{1}{2}(-4A - B^2 + 4AC)k^3q, a_0 = -\frac{ikB\sqrt{\frac{3}{2}}q}{\sqrt{p}}, a_1 = -\frac{i\sqrt{6q}(-1+C)k}{\sqrt{p}}.$$

Substituting the results obtained above into eq. (6.76) along with eqs. (2.60) to (2.63) of Chapter 2, we can obtain a series of exact solutions to eq. (6.22).

a. From **Case I**, we obtain the following exact solutions:

i. When $B \neq 0$ and $\Delta = B^2 + 4A - 4AC \geq 0$, we have the following exponential function solution:

$$\Phi_{11} = -\frac{ikB\sqrt{\frac{3}{2}}q}{\sqrt{p}} - \frac{i\sqrt{6q}(-1+C)k}{\sqrt{p}}\left(\frac{C_1\exp\left(\frac{\sqrt{\Delta}}{2}\xi\right) + C_2\exp\left(\frac{-\sqrt{\Delta}}{2}\xi\right)}{C_1\exp\left(\frac{\sqrt{\Delta}}{2}\xi\right) - C_2\exp\left(\frac{-\sqrt{\Delta}}{2}\xi\right)}\right). \tag{6.78}$$

ii. When $B \neq 0$ and $\Delta = B^2 + 4A - 4AC < 0$, we have the following trigonometric function solution:

$$\Phi_{12} = -\frac{ikB\sqrt{\frac{3}{2}}q}{\sqrt{p}} - \frac{i\sqrt{6q}(-1+C)k}{\sqrt{p}}\left(\frac{iC_1\cos\left(\frac{\sqrt{-\Delta}}{2}\xi\right) - C_2\sin\left(\frac{\sqrt{-\Delta}}{2}\xi\right)}{iC_1\sin\left(\frac{\sqrt{-\Delta}}{2}\xi\right) + C_2\cos\left(\frac{\sqrt{-\Delta}}{2}\xi\right)}\right). \tag{6.79}$$

iii. When $B = 0$ and $\Delta = A(1-C) \geq 0$, we have the following trigonometric function solution:

$$\Phi_{13} = -\frac{ikB\sqrt{\frac{3}{2}q}}{\sqrt{p}} - \frac{i\sqrt{6q}(-1+C)k}{\sqrt{p}}\left(\frac{C_1\cos\left(\sqrt{\Delta}\xi\right) + C_2\sin\left(\sqrt{\Delta}\xi\right)}{C_1\sin\left(\sqrt{\Delta}\xi\right) - C_2\cos\left(\sqrt{\Delta}\xi\right)}\right). \tag{6.80}$$

iv. When $B = 0$ and $\Delta = A(1-C) < 0$, we have the following hyperbolic function solution:

$$\Phi_{14} = -\frac{ikB\sqrt{\frac{3}{2}q}}{\sqrt{p}} - \frac{i\sqrt{6q}(-1+C)k}{\sqrt{p}}\left(\frac{iC_1\cosh\left(\sqrt{-\Delta}\xi\right) - C_2\sinh\left(\sqrt{-\Delta}\xi\right)}{iC_1\sinh\left(\sqrt{-\Delta}\xi\right) - C_2\cosh\left(\sqrt{-\Delta}\xi\right)}\right). \tag{6.81}$$

Case II:

$$A = A, B = B, C = C, k = k, v = -\frac{1}{2}(-4A - B^2 + 4AC)k^3q, a_0 = \frac{ikB\sqrt{\frac{3}{2}q}}{\sqrt{p}}, a_1 = \frac{i\sqrt{6q}(-1+C)k}{\sqrt{p}}.$$

Substituting the results obtained above into eq. (6.76) along with eqs. (2.60) to (2.63) of Chapter 2, we can obtain a series of exact solutions to eq. (6.22).

b. From **Case II**, we obtain the following exact solutions:

i. When $B \neq 0$ and $\Delta = B^2 + 4A - 4AC \geq 0$, we have the following exponential function solution:

$$\Phi_{21} = \frac{ikB\sqrt{\frac{3}{2}q}}{\sqrt{p}} + \frac{i\sqrt{6q}(-1+C)k}{\sqrt{p}}\left(\frac{C_1\exp\left(\frac{\sqrt{\Delta}}{2}\xi\right) + C_2\exp\left(\frac{-\sqrt{\Delta}}{2}\xi\right)}{C_1\exp\left(\frac{\sqrt{\Delta}}{2}\xi\right) - C_2\exp\left(\frac{-\sqrt{\Delta}}{2}\xi\right)}\right). \tag{6.82}$$

ii. When $B \neq 0$ and $\Delta = B^2 + 4A - 4AC < 0$, we have the following trigonometric function solution:

$$\Phi_{22} = \frac{ikB\sqrt{\frac{3}{2}q}}{\sqrt{p}} + \frac{i\sqrt{6q}(-1+C)k}{\sqrt{p}}\left(\frac{iC_1\cos\left(\frac{\sqrt{-\Delta}}{2}\xi\right) - C_2\sin\left(\frac{\sqrt{-\Delta}}{2}\xi\right)}{iC_1\sin\left(\frac{\sqrt{-\Delta}}{2}\xi\right) + C_2\cos\left(\frac{\sqrt{-\Delta}}{2}\xi\right)}\right). \tag{6.83}$$

iii. When $B = 0$ and $\Delta = A(1-C) \geq 0$, we have the following trigonometric function solution:

$$\Phi_{23} = -\frac{ikB\sqrt{\frac{3}{2}q}}{\sqrt{p}} + \frac{i\sqrt{6q}(-1+C)k}{\sqrt{p}}\left(\frac{C_1\cos\left(\sqrt{\Delta}\xi\right) + C_2\sin\left(\sqrt{\Delta}\xi\right)}{C_1\sin\left(\sqrt{\Delta}\xi\right) - C_2\cos\left(\sqrt{\Delta}\xi\right)}\right). \tag{6.84}$$

iv. When $B = 0$ and $\Delta = A(1-C) < 0$, we have the following hyperbolic function solution:

$$\Phi_{24} = -\frac{ikB\sqrt{\frac{3}{2}q}}{\sqrt{p}} + \frac{i\sqrt{6q}(-1+C)k}{\sqrt{p}}\left(\frac{iC_1\cosh\left(\sqrt{-\Delta}\xi\right) - C_2\sinh\left(\sqrt{-\Delta}\xi\right)}{iC_1\sinh\left(\sqrt{-\Delta}\xi\right) - C_2\cosh\left(\sqrt{-\Delta}\xi\right)}\right), \tag{6.85}$$

where $\xi = kx + \dfrac{vt^{\alpha}}{\Gamma(\alpha+1)}$.

From the proposed improved (G'/G)-expansion method, we obtained the eight exact solutions for the time-fractional third order modified KdV equation.

6.6.2.1 The Numerical Simulation for Solution of Time-Fractional Third Order Modified KdV Equation Using Proposed Improved (G′/G)-Expansion Method (Figures 6.14–6.25)

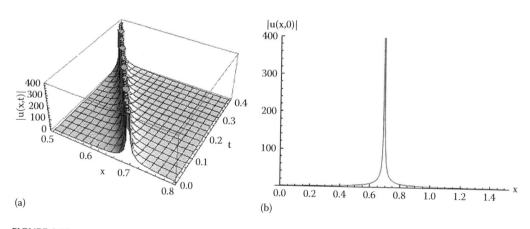

FIGURE 6.14
Case Ia: For $\alpha = 1$ (classical order). (a) The improved (G'/G)-expansion method solitary wave solution for $|u(x,t)|$ appears in eq. (6.78) of **Case I**, (b) corresponding solution for $|u(x,t)|$ when $t = 0$, $C_1 = 1$, $C_2 = 2$, $p = 6$, $q = 1$, $A = 0$, $B = 1$, $C = -1$, and $\alpha = 1$.

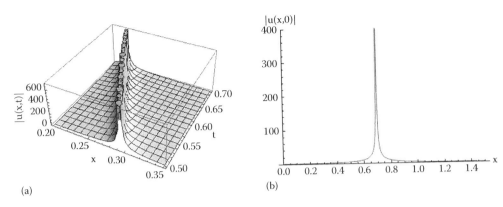

FIGURE 6.15
Case Ib: For $\alpha = 0.5$ (fractional order). (a) The improved (G'/G)-expansion method solitary wave solution for $|u(x,t)|$ appears in eq. (6.78) of **Case I**, (b) corresponding solution for $|u(x,t)|$ when $t = 0$, $C_1 = 1, C_2 = 2$, $p = 6$, $q = 1$, $A = 0$, $B = 1$, $C = -1$, and $\alpha = 0.5$.

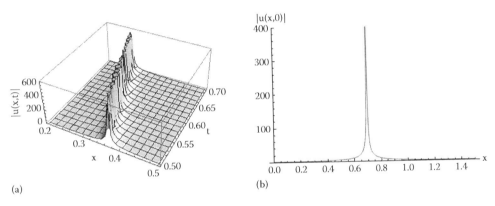

FIGURE 6.16
Case Ic: For $\alpha = 0.75$ (fractional order). (a) The improved (G'/G)-expansion method solitary wave solution for $|u(x,t)|$ appears in eq. (6.78) of **Case I**, (b) corresponding solution for $|u(x,t)|$ when $t = 0$, $C_1 = 1$, $C_2 = 2$, $p = 6$, $q = 1$, $A = 0$, $B = 1$, $C = -1$, and $\alpha = 0.75$.

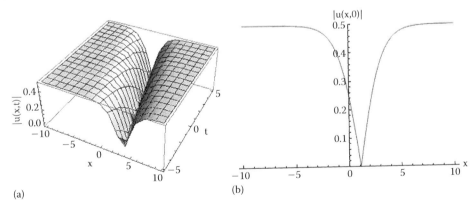

FIGURE 6.17
Case IIa: For $\alpha = 1$ (classical order). (a) The improved (G'/G)-expansion method solitary wave solution for $|u(x,t)|$ appears in eq. (6.79) of **Case I**, (b) corresponding solution for $|u(x,t)|$ when $t = 0$, $C_1 = 1$, $C_2 = 2$, $p = 6$, $q = 1$, $A = 0$, $B = -1$, $C = -1$, and $\alpha = 1$.

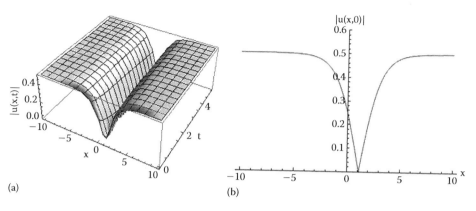

FIGURE 6.18
Case IIb: For $\alpha = 0.5$ (fractional order). (a) The improved (G'/G)-expansion method solitary wave solution for $|u(x,t)|$ appears in eq. (6.79) of **Case I**, (b) corresponding solution for $|u(x,t)|$ when $t = 0$, $C_1 = 1$, $C_2 = 2$, $p = 6$, $q = 1$, $A = 0$, $B = -1$, $C = -1$, and $\alpha = 0.5$.

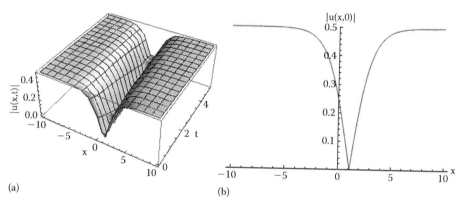

FIGURE 6.19
Case IIc: For $\alpha = 0.75$ (fractional order). (a) The improved (G'/G)-expansion method solitary wave solution for $|u(x,t)|$ appears in eq. (6.79) of **Case I**, (b) corresponding solution for $|u(x,t)|$ when $t = 0$, $C_1 = 1$, $C_2 = 2$, $p = 6$, $q = 1$, $A = 0$, $B = -1$, $C = -1$, and $\alpha = 0.75$.

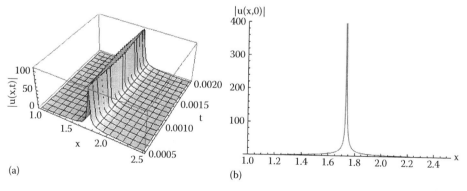

FIGURE 6.20
Case IIIa: For $\alpha = 1$ (classical order). (a) The improved (G'/G)-expansion method solitary wave solution for $|u(x,t)|$ appears in eq. (6.80) of **Case I**, (b) corresponding solution for $|u(x,t)|$ when $t = 0$, $C_1 = 1$, $C_2 = 2$, $p = 6$, $q = 1$, $A = 2$, $B = 0$, $C = 4$, and $\alpha = 1$.

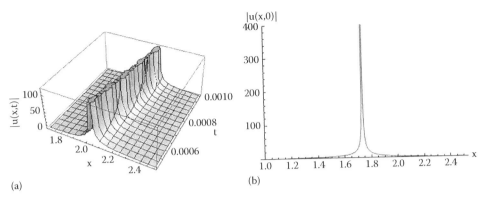

FIGURE 6.21
Case IIIb: For $\alpha = 0.5$ (fractional order). (a) The improved (G'/G)-expansion method solitary wave solution for $|u(x,t)|$ appears in eq. (6.80) of **Case I**, (b) corresponding solution for $|u(x,t)|$ when $t = 0$, $C_1 = 1$, $C_2 = 2$, $p = 6$, $q = 1$, $A = 2$, $B = 0$, $C = 4$, and $\alpha = 0.5$.

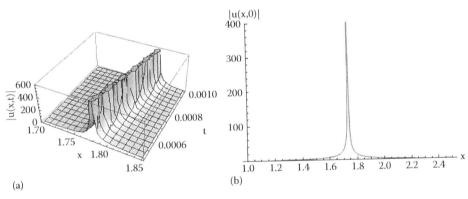

FIGURE 6.22
Case IIIc: For $\alpha = 0.75$ (fractional order). (a) The improved (G'/G)-expansion method solitary wave solution for $|u(x,t)|$ appears in eq. (6.80) of **Case I**, (b) corresponding solution for $|u(x,t)|$ when $t = 0$, $C_1 = 1$, $C_2 = 2$, $p = 6$, $q = 1$, $A = 2$, $B = 0$, $C = 4$, and $\alpha = 0.75$.

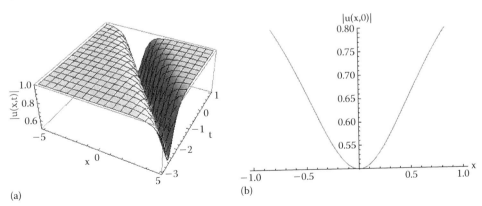

FIGURE 6.23
Case IVa: For $\alpha = 1$ (classical order). (a) The improved (G'/G)-expansion method solitary wave solution for $|u(x,t)|$ appears in eq. (6.81) of **Case I**, (b) corresponding solution for $|u(x,t)|$ when $t = 0$, $C_1 = 1$, $C_2 = 2$, $p = 6$, $q = 1$, $A = -1$, $B = 0$, $C = -2$, and $\alpha = 1$.

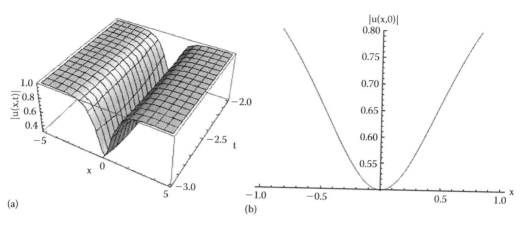

FIGURE 6.24
Case IVb: For $\alpha = 0.5$ (fractional order). (a) The improved (G'/G)-expansion method solitary wave solution for $|u(x,t)|$ appears in eq. (6.81) of **Case I**, (b) corresponding solution for $|u(x,t)|$ when $t = 0$, $C_1 = 1$, $C_2 = 2$, $p = 6$, $q = 1$, $A = -1$, $B = 0$, $C = -2$, and $\alpha = 0.5$.

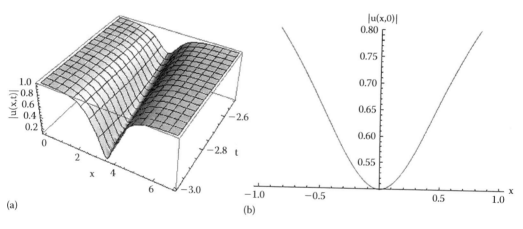

FIGURE 6.25
Case IVc: For $\alpha = 0.75$ (fractional order). (a) The improved (G'/G)-expansion method solitary wave solution for $|u(x,t)|$ appears in eq. (6.81) of **Case I**, (b) corresponding solution for $|u(x,t)|$ when $t = 0$, $C_1 = 1$, $C_2 = 2$, $p = 6$, $q = 1$, $A = -1$, $B = 0$, $C = -2$, and $\alpha = 0.75$.

In the present numerical simulation, three exact solutions of eq. (6.22) have been used to draw the graphs as shown in Figures 6.5 to 6.25 for different values of α. We have drawn the solitary wave three-dimensional and the corresponding two-dimensional solutions graphs of the obtained exact solution of eq. (6.22). We concluded that in the above, all cases for the solution graphs of the time-fractional third order mKdV equation represent the solitary wave solutions with regard to both fractional order and classical order solutions. These exact solutions obtained from the proposed methods are the new solutions for the modified KdV equation, which will find the outcomes of non-linearity, inhomogeneity, and dispersive nature of the waves.

6.7 Conclusion

In this chapter, we determine the solitary wave exact solutions of non-linear evolution equations, namely, the time-fractional modified Kawahara equation, fractional coupled JM equation, time-fractional modified KdV equation, and the time-fractional Kaup–Kupershmidt equation by the (G'/G)-expansion method and improved (G'/G)-expansion method. By taking the view of local fractional derivative defined on fractals, we have considered fractional complex transform, which can easily convert a fractional differential equation into its equivalent ordinary differential form. So, fractional complex transform is extremely effective for solving fractional differential equations. The focused methods have many advantages: they are straightforward and concise. Furthermore, this study shows that the proposed method is quite efficient. The performance of this method is reliable and effective and gives the exact solitary wave solutions which have been presented by three-dimensional and two-dimensional solutions graphs. With the successful implementation of the proposed methods, we have found many new exact solutions, which may be useful for describing certain non-linear physical phenomena in fluid. Since the (G'/G)-expansion method yields many parameters and arbitrary constants, so, in this case, the degree of freedom of solution is high for which it is difficult to handle the solutions in practical purpose.

Moreover, the improved (G'/G)-expansion method provides new and more general type traveling wave solutions than the (G'/G)-expansion methods which are significant to reveal the pertinent features of the physical phenomenon. These methods clearly avoid linearization, unrealistic assumptions, and, hence, it provides effective exact solutions. The proposed methods are very efficient and powerful techniques in finding the solutions of time-fractional non-linear partial differential equations. The suggested methods are very effective, powerful, standard, direct, and easily computerizable techniques providing new exact solutions of partial differential equations. Moreover, the processes can be relevant to identical constitution of non-linear wave equations to gain the knowledge of solitary waves and the outcome of non-linearity, inhomogeneity, and dispersive nature of the medium.

7

New Exact Solutions of Fractional Differential Equations by Proposed Tanh and Modified Kudryashov Methods

7.1 Introduction

Many important science and engineering problems involved to study on non-linear evolution equations. Thus, it is very important to find exact solutions for analyzing the physical nature of the solutions. In this chapter, by using the tanh method and modified Kudryashov method, the exact solutions for the fractional non-linear differential equations have been analyzed.

Based on the proposed tanh method, a series of new exact solutions of the fractional non-linear differential equations (FNDEs) have been obtained. It is shown that the proposed tanh method is straightforward and concise, and its applications are promising. The modified Kudryashov method has been successfully used to seek exact traveling wave solutions of the FNDEs. As a result, a series of new exact solutions are obtained, which have not been reported yet. The solution procedure is very simple and the traveling wave solutions are expressed by the hyperbolic functions, the trigonometric functions, and the rational functions. It has been shown that both proposed methods provide a very effective and powerful mathematical tool for solving non-linear evolution equations.

7.2 Outline of Present Study

The present chapter is devoted to construct the new exact solutions for the time-fractional fifth-order Sawada–Kotera equation, time-fractional fifth-order modified Sawada–Kotera equation, time-fractional fifth–order Kuramoto–Sivashinsky (K–S) equation, and time-fractional coupled Jaulent–Miodek equation using a relatively new technique, namely the

183

tanh method. Also the time-fractional fifth–order modified Sawada–Kotera equation has been solved by the modified Kudryashov method.

7.2.1 Time-Fractional Fifth-Order S–K Equation

Let us consider the time-fractional fifth-order Sawada–Kotera (S–K) eqs. [66,70,234]:

$$_0D_t^\alpha u + 5u^2u_x + 5u_xu_{xx} + 5uu_{xxx} + u_{xxxxx} = 0, \tag{7.1}$$

where $0 < \alpha \le 1$.

It is well known that the Sawada–Kotera equation is a remarkable unidirectional non-linear evolution equation belonging to the completely integrable hierarchy of higher order Korteweg–de Vries (KdV) equations and has many sets of conservation laws [235]. Numerous properties of the S–K equation, like multisoliton solutions, conserved quantities, Bäcklund transformation, Darboux transformation, and so on [201,234–238] have been researched in the past. Furthermore, the S–K equation is frequently preferable in many physical situations as an equation which allows us to model waves that propagate in opposite directions.

The present chapter is devoted to construct the exact solutions for the time-fractional fifth-order Sawada–Kotera equation using a relatively new technique, namely, the tanh method [67–70] via fractional complex transform [60–64]. To the best of the information of the author, no previous research work has been done using the proposed technique for solving fractional partial differential equations (FPDEs).

7.2.2 Time-Fractional Fifth-Order Modified S–K Equation

Here, we have considered the time-fractional fifth-order modified Sawada–Kotera (mS–K) equation [239,240] in following form:

$$_0D_t^\alpha u - u_{xxxxx} + \left(5u_xu_{xx} + 5uu_x^2 + 5u^2u_{xx} - u^5\right)_x = 0. \tag{7.2}$$

where $0 < \alpha \le 1$.

The eq. (7.2) is directly derived from the time-fractional fifth-order S–K equation [59,185,241]:

$$_0D_t^\alpha u + 5u^2u_x + 5u_xu_{xx} + 5uu_{xxx} + u_{xxxxx} = 0, \tag{7.3}$$

and the time-fractional fifth-order Kaup–Kupershmidt (K–K) equation [68,242,243]:

$$_0D_t^\alpha u + 45u^2u_x - 15pu_xu_{xx} - 15uu_{xxx} + u_{xxxxx} = 0, \tag{7.4}$$

by applying the Miura transformations. The time-fractional fifth-order modified Sawada–Kotera equation has the combined physical nature of both fractional S–K and K–K equations.

The present chapter is dedicated to build the exact solitary solutions for time-fractional fifth-order mS–K equation utilizing fairly new techniques, in particular, the tanh method [67–70] and modified Kudryashov method [71,72]. To the best of the knowledge of author, no prior exploration work has been carried out using the proposed methods for solving the time-fractional fifth-order modified Sawada–Kotera equation.

7.2.3 Time-Fractional K–S Equation

Let us consider the time-fractional K–S equation [66, 244–248]:

$$_0D_t^\alpha u + auu_x + bu_{xx} + ku_{xxxx} = 0, \tag{7.5}$$

where $0 < \alpha \le 1, a, b, k$ are arbitrary constants.

The K–S equation (7.5) represents the variations of the position of a flame front, the motion of a fluid going down a vertical wall, or a spatially uniform oscillating chemical reaction in a homogeneous medium [247]. This equation was examined as a prototypical example of spatiotemporal chaos in one space dimension [248]. Moreover, this equation was originally derived in the context of plasma instabilities, flame front propagation, and phase turbulence in a reaction-diffusion system [248].

The present chapter is devoted to construct the exact solutions for the time-fractional Kuramoto–Sivashinsky equation using a relatively new technique, namely, the tanh method [67–70], with the help of fractional complex transform [60–64]. To the best of the information of the author, no previous research work has been done using the proposed technique for solving FPDEs.

NOTE: We have also presented the exact solutions of time-fractional coupled Jaulent–Miodek equations presented in eqs. (6.2) and (6.3) of Chapter 6 by using the tanh method.

7.3 Implementation of Proposed Tanh Method for the Exact Solutions of Time-Fractional Fifth-Order S–K Equation

In this section, we apply the proposed tanh method to determine the exact solutions for the time-fractional fifth-order Sawada–Kotera equation (7.1).

By applying the fractional complex transform (2.65) of Chapter 2, eq. (7.1) can be reduced to the following non-linear ordinary differential equation (ODE):

$$-cv\Phi'(\xi) + 5c\Phi^2(\xi)\Phi'(\xi) + 5c^3\Phi'(\xi)\Phi''(\xi) + 5c^3\Phi(\xi)\Phi'''(\xi) + c^5\Phi^V(\xi) = 0. \tag{7.6}$$

By reducing eq. (7.6), we get:

$$-v\Phi'(\xi) + 5\Phi^2(\xi)\Phi'(\xi) + 5c^2\Phi'(\xi)\Phi''(\xi) + 5c^2\Phi(\xi)\Phi'''(\xi) + c^4\Phi^V(\xi) = 0. \tag{7.7}$$

Further, by integrating eq. (7.7) with respect to ξ, we get:

$$-v\Phi(\xi) + \frac{5}{3}\Phi^3(\xi) + 5c^2\Phi(\xi)\Phi''(\xi) + c^4\Phi^{IV}(\xi) = 0. \tag{7.8}$$

By balancing the highest order derivative term and non-linear term in eq. (7.8), the value of n can be determined, which is $n = 2$ in this problem.

Therefore, by using eq. (2.67) of Chapter 2, we have the following ansatz:

$$\Phi(\xi) = a_0 + a_1 Y + a_2 Y^2, \tag{7.9}$$

where $Y = \tanh(\xi)$.

According to eq. (2.68) of Chapter 2, we have:

$$\frac{d\Phi}{d\xi} \rightarrow \left(1 - Y^2\right)\left(a_1 + 2a_2 Y\right),$$

$$\frac{d^2\Phi}{d\xi^2} \rightarrow \left(1 - Y^2\right)\left(-2Y\left(a_1 + 2a_2 Y\right) + 2a_2\left(1 - Y^2\right)\right),$$

$$\frac{d^4\Phi}{d\xi^4} \rightarrow -8Y\left(1 - Y^2\right)\left(3Y^2 - 2\right)\left(a_1 + 2a_2 Y\right) + 8a_2\left(1 - Y^2\right)^2\left(9Y^2 - 2\right). \tag{7.10}$$

Putting these values of eq. (7.10) along with eq. (7.9) into eq. (7.8), then after collecting all terms with the same degree of Y^i ($i = 0, 1, 2, ...$), we can obtain a set of algebraic equations for a_i ($i = 0, 1, 2, ..., n$), v and c as follows:

Coefficient of Y^0 : $-va_0 + \dfrac{5a_0^3}{3} - 16c^4 a_2 + 10c^2 a_0 a_2 = 0.$

Coefficient of Y^1 : $16c^4 a_1 - va_1 - 10c^2 a_0 a_1 + 5a_0^2 a_1 + 10c^2 a_1 a_2 = 0.$

Coefficient of Y^2 : $-10c^2 a_1^2 + 5a_0 a_1^2 + 136c^4 a_2 - va_2 - 40c^2 a_0 a_2 + 5a_2 a_0^2 + 10c^2 a_2^2 = 0.$

Coefficient of Y^3 : $-40c^4 a_1 + 10c^2 a_0 a_1 + \dfrac{5a_1^3}{3} - 50c^2 a_1 a_2 + 10a_0 a_1 a_2 = 0.$

Coefficient of Y^4 : $10c^2 a_1^2 - 240c^4 a_2 + 30c^2 a_0 a_2 + 5a_1^2 a_2 - 40c^2 a_2^2 + 5a_0 a_2^2 = 0.$

Coefficient of Y^5 : $24c^4 a_1 + 40c^2 a_1 a_2 + 5a_1 a_2^2 = 0.$

Coefficient of Y^6 : $120c^4 a_2 + 30c^2 a_2^2 + \dfrac{5a_2^3}{3} = 0. \tag{7.11}$

Solving the above algebraic equations (7.11), we have the following sets of coefficients for the solutions of eq. (7.9) as given below:

Case I:

$$c = c, v = 16c^4, a_0 = 6c^2, a_1 = 0, a_2 = -6c^2.$$

For **Case I**, we have the following solution:

$$\Phi_{11} = 6c^2 \operatorname{sech}^2(\xi), \tag{7.12}$$

where $\xi = c\left(x - \dfrac{16c^4 t^\alpha}{\Gamma(\alpha + 1)}\right).$

Case II:

$$c = c, v = \frac{8\left(-15c^4 - \sqrt{105}c^4\right)}{-15 + \sqrt{105}}, a_0 = \frac{15c^2 - \sqrt{105}c^2}{5}, a_1 = 0, a_2 = -6c^2.$$

For **Case II**, we have the following solution:

$$\Phi_{21} = \frac{15c^2 - \sqrt{105}c^2}{5} - 6c^2 \tanh^2 \xi, \tag{7.13}$$

where $\xi = c\left(x - \dfrac{8\left(-15c^4 - \sqrt{105}c^4\right)t^\alpha}{-15 + \sqrt{105}\,\Gamma(\alpha + 1)} \right).$

Case III:

$$c = c, v = \frac{8\left(15c^4 - \sqrt{105}c^4\right)}{15 + \sqrt{105}}, a_0 = \frac{15c^2 + \sqrt{105}c^2}{5}, a_1 = 0, a_2 = -6c^2.$$

For **Case III**, we have the following solution:

$$\Phi_{31} = \frac{15c^2 + \sqrt{105}c^2}{5} - 6c^2 \tanh^2 \xi, \tag{7.14}$$

where $\xi = c\left(x - \dfrac{8\left(15c^4 - \sqrt{105}c^4\right)t^\alpha}{15 + \sqrt{105}\,\Gamma(\alpha + 1)} \right).$

7.3.1 The Numerical Simulations for Solutions of Time-Fractional Fifth-Order S–K Equation Using Proposed Tanh Method

In this present numerical experiment, three exact solutions of eq. (7.1) have been used to draw the graphs as shown in Figures 7.1 to 7.6 for different fractional order values of α.

7.3.1.1 Numerical Results and Discussion (Figures 7.1–7.6)

Case 1:
   a. For $\alpha = 0.75$ (fractional order)
   b. For $\alpha = 0.25$ (fractional order)

Case 2:
   a. For $\alpha = 0.75$ (fractional order)
   b. For $\alpha = 0.25$ (fractional order)

Case 3:
   a. For $\alpha = 0.75$ (fractional order)
   b. For $\alpha = 0.25$ (fractional order)

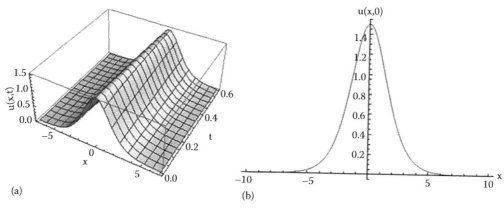

FIGURE 7.1
Case 1a: For $\alpha = 0.75$ (fractional order). (a) The tanh method solitary wave solution for $u(x,t)$ appears in eq. (7.12) of **Case I**, (b) corresponding solution for $u(x,t)$ when $t = 0$, $c = 0.5$, and $\alpha = 0.75$.

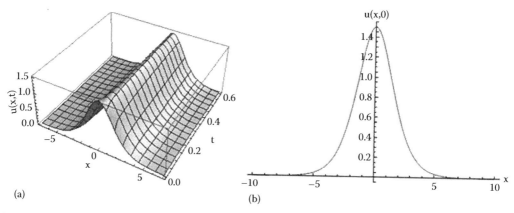

FIGURE 7.2
Case 2a: For $\alpha = 0.75$ (fractional order). (a) The tanh method solitary wave solution for $u(x,t)$ appears in eq. (7.12) of **Case I**, (b) corresponding solution for $u(x,t)$ when $t = 0$, $c = 0.5$, and $\alpha = 0.25$.

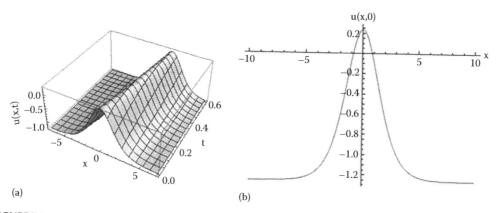

FIGURE 7.3
Case 1b: For $\alpha = 0.25$ (fractional order). (a) The tanh method solitary wave solution for $u(x,t)$ appears in eq. (7.13) of **Case II**, (b) corresponding solution for $u(x,t)$ when $t = 0$, $c = 0.5$, and $\alpha = 0.75$.

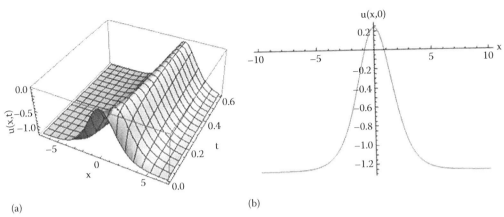

FIGURE 7.4
Case 2b: For $\alpha = 0.25$ (fractional order). (a) The tanh method solitary wave solution for $u(x,t)$ appears in eq. (7.13) of **Case II**, (b) corresponding solution for $u(x,t)$ when $t = 0$, $c = 0.5$, and $\alpha = 0.25$.

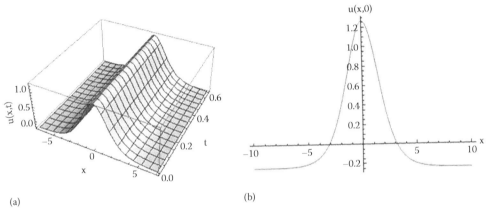

FIGURE 7.5
Case 3a: For $\alpha = 0.75$ (fractional order). (a) The tanh method solitary wave solution for $u(x,t)$ appears in eq. (7.14) of **Case III**, (b) corresponding solution for $u(x,t)$ when $t = 0$, $c = 0.5$, and $\alpha = 0.75$.

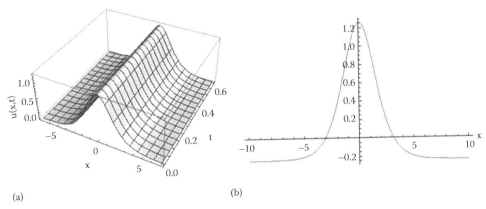

FIGURE 7.6
Case 3b: For $\alpha = 0.25$ (fractional order). (a) The tanh method solitary wave solution for $u(x,t)$ appears in eq. (7.14) of **Case III**, (b) corresponding solution for $u(x,t)$ when $t = 0$, $c = 0.5$, and $\alpha = 0.25$.

In the present numerical simulation, we have drawn the solitary wave three dimensional and corresponding two-dimensional solutions graphs of the obtained exact solutions of eq. (7.1) for fractional order. Figures 7.1 to 7.6 give a good impression about the change of amplitude and width of the soliton due to the variation of the fractional order. Remarkably, both two-dimensional and three-dimensional graphs describe the behavior of $u(x,t)$ in space x at time t corresponding to different values of the fractional order. The behavior shows that increase of the fractional parameter changes the nature of the solitary wave solution. This means that the fractional parameter can be used to modify the shape of solitary wave without the change of the non-linearity and dispersion effects in the medium.

7.4 Application of Proposed Tanh and Modified Kudryashov Methods for the Exact Solitary Wave Solutions of Time-Fractional Fifth-Order mS–K Equation

In the present analysis, we have applied proposed the tanh method and modified Kudryashov method for getting the exact traveling wave solutions for the time-fractional fifth-order mS–K equation.

7.4.1 Application of Proposed Tanh Method for the Exact Solitary Wave Solutions of Time-Fractional Fifth-Order mS–K Equation

This section represents the implementation of the proposed tanh method for finding the exact solutions of eq. (7.2).

With the aid of the fractional complex transform (2.65) of Chapter 2, eq. (7.2) may also be written as the following ODE form:

$$-v\Phi'(\xi) - c^4\Phi^V(\xi) + 5c^3\left(\Phi''(\xi)^2 + \Phi'(\xi)\Phi'''(\xi)\right) + 5c^3\left(\Phi'(\xi)^3 + 2\Phi(\xi)\Phi'(\xi)\Phi''(\xi)\right)$$
$$+ 5c^2\left(\Phi(\xi)^2\Phi'''(\xi) + 2\Phi(\xi)\Phi'(\xi)\Phi''(\xi)\right) - 5\Phi'(\xi)\Phi^4(\xi) = 0. \tag{7.15}$$

Again by integrating eq. (7.15) once with respect to ξ, we have:

$$-v\Phi(\xi) - c^4\Phi^{IV}(\xi) + 5c^3\Phi'(\xi)\Phi''(\xi) + 5c^2\Phi(\xi)\Phi'(\xi)^2 + 5c^2\Phi(\xi)^2\Phi''(\xi) - \Phi^5(\xi) = 0. \tag{7.16}$$

Then, from eq. (7.16), the integer value of n can be determined by balancing the highest order derivative that is $\Phi^{IV}(\xi)$ and non-linear term that is $\Phi(\xi)^2\Phi''(\xi)$. Here, for this problem, $n = 1$.

Now, by using eq. (2.67) of Chapter 2, we have the following ansatz:

$$\Phi(\xi) = a_0 + a_1 Y, \tag{7.17}$$

where $Y = \tanh(\xi)$.

New Exact Solutions of FDEs by Proposed Tanh and Modified Kudryashov Methods   191

From eq. (2.68) of Chapter 2, we have:

$$\frac{d\Phi}{d\xi} \to a_1\left(1-Y^2\right),$$

$$\frac{d^2\Phi}{d\xi^2} \to -2a_1 Y\left(1-Y^2\right), \tag{7.18}$$

$$\frac{d^4\Phi}{d\xi^4} \to -8a_1 Y\left(1-Y^2\right)\left(3Y^2-2\right).$$

Then, by substituting these values of eq. (7.18) together with eq. (7.17) into eq. (7.16), and then collecting all terms with the same degree of Y^i $(i=0,1,2,...)$ together, we will get a set of algebraic equations for a_i $(i=0,1,2,...,n)$, v and c:

Coefficient of $Y^0 : -v a_0 - a_0^5 + 5c^2 a_0 a_1^2 = 0.$

Coefficient of $Y^1 : -16c^4 a_1 - v a_1 - 10c^2 a_0^2 a_1 + 5a_0^4 a_1 - 10c^3 a_1^2 + 5c^2 a_1^3 = 0.$

Coefficient of $Y^2 : -30c^2 a_0 a_1^2 - 10a_0^3 a_1^2 = 0.$

Coefficient of $Y^3 : 40c^4 a_1 + 10c^2 a_0^2 a_1 + 20c^3 a_1^2 - 20c^2 a_1^3 - 10a_0^2 a_1^3 = 0. \tag{7.19}$

Coefficient of $Y^4 : 25c^2 a_0 a_1^2 - 5a_0 a_1^4 = 0.$

Coefficient of $Y^5 : -24c^4 a_1 - 10c^3 a_1^2 + 15c^2 a_1^3 - a_1^5 = 0.$

Solving the obtained algebraic systems obtained in eq. (7.19), we have the following sets of solutions of eq. (7.17) presented as below:

Case I:

$$c = c, v = -c^4, a_0 = 0, a_1 = -c.$$

We have the following exact solution for **Case I:**

$$\Phi_{11} = -c \tanh(\xi), \tag{7.20}$$

where $\xi = c\left(x + \dfrac{c^4 t^\alpha}{\Gamma(\alpha+1)}\right).$

Case II:

$$c = c, v = -16c^4, a_0 = 0, a_1 = 2c.$$

We have the following exact solution for **Case II**:

$$\Phi_{21} = 2c\tanh(\xi), \tag{7.21}$$

where $\xi = c\left(x - \dfrac{16c^4 t^\alpha}{\Gamma(\alpha+1)}\right)$.

7.4.1.1 The Numerical Simulations for Presenting the Nature of the Solutions of Fractional Fifth-Order mS–K Equations by Utilizing Proposed Tanh Method

In this section, we have drawn the three-dimensional solitary wave solutions graphs which explain the physical meaning of each solution of the governing mS–K equation by utilizing eqs. (7.20) and (7.21).

7.4.1.1.1 Numerical Results and Discussion (Figures 7.7–7.10)

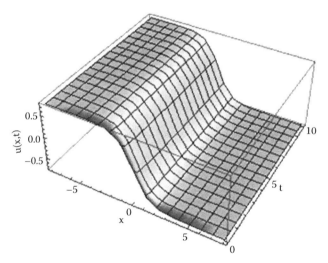

FIGURE 7.7
Case 1a: For $\alpha = 0.25$ (fractional order). The three-dimensional solitary wave solution graph obtained by using the tanh method for $u(x,t)$ appears in eq. (7.20) of **Case I**, when $c = 0.75$ and $\alpha = 0.25$.

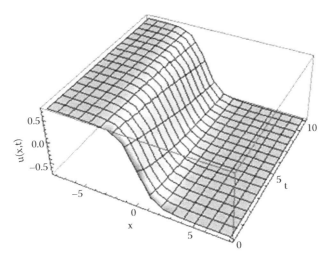

FIGURE 7.8
Case 1b: For $\alpha = 0.75$ (fractional order). The three-dimensional solitary wave solution graph obtained by using the tanh method for $u(x,t)$ appears in eq. (7.20) of **Case I**, when $c = 0.75$ and $\alpha = 0.75$.

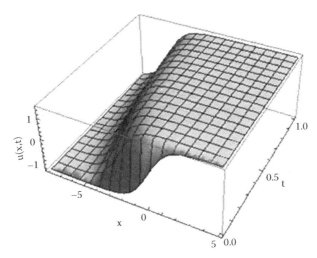

FIGURE 7.9
Case 2a: For $\alpha = 0.25$ (fractional order). The three-dimensional solitary wave solution graph obtained by using the tanh method for $u(x,t)$ appears in eq. (7.21) of **Case II**, when $c = 0.5$ and $\alpha = 0.25$.

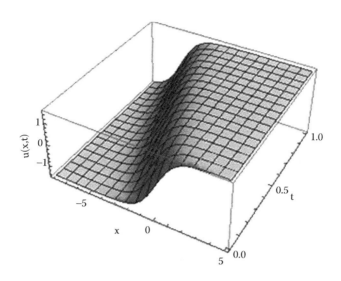

FIGURE 7.10
Case 2b: For $\alpha = 0.75$ (fractional order). The three-dimensional solitary wave solution graph obtained by using the tanh method for $u(x,t)$ appears in eq. (7.21) of **Case II**, when $c = 0.5$ and $\alpha = 0.75$.

7.4.2 Application of Proposed Modified Kudryashov Method for the Exact Solitary Wave Solutions of Time-Fractional Fifth-Order mS–K Equation

In the present analysis, we have applied the modified Kudryashov method for getting the exact traveling wave solutions for the time-fractional fifth-order mS–K equation.

Through applying the fractional complex transform (7.12), eq. (7.2) may also be written as the following ODE form:

$$\gamma \Psi'(\zeta) + l^5 \Psi^V(\zeta) + 5l^4 \left(\Psi''(\zeta)^2 + \Psi'(\zeta)\Psi'''(\zeta) \right) + 5l^3 \left(\Psi'(\zeta)^3 + 2\Psi(\zeta)\Psi'(\zeta)\Psi''(\zeta) \right)$$
$$+ 5l^3 \left(\Psi(\zeta)^2 \Psi'''(\zeta) + 2\Psi(\zeta)\Psi'(\zeta)\Psi''(\zeta) \right) - 5l\Psi'(\zeta)\Psi^4(\zeta) = 0. \quad (7.22)$$

By integrating eq. (7.22) once with respect to ζ, we have:

$$\gamma \Psi(\zeta) - l^4 \Psi^{IV}(\zeta) + 5l^3 \Psi'(\zeta)\Psi''(\zeta) + 5l^2 \Psi(\zeta)\Psi'(\zeta)^2 + 5l^2 \Psi(\zeta)^2 \Psi''(\zeta) - \Psi^5(\zeta) = 0. \quad (7.23)$$

Then, from eq. (7.23), the integer value of m can be determined by balancing the highest order derivative that is $\Psi^{IV}(\zeta)$ and non-linear term that is $\Psi(\zeta)^2 \Psi''(\zeta)$. Here, for this problem, $m = 1$.

So by using eq. (7.14), we have the following ansatz:

$$\Psi(\zeta) = b_0 + b_1 Q(\zeta), \quad (7.24)$$

where $Q(\zeta) = \dfrac{1}{1 \pm a^\zeta}$ and b_0, b_1 should be determined later.

Then, by substituting the derivatives of function $Q(\zeta)$ with respect to ζ into eq. (7.23) together with eq. (7.24), we obtain a system of algebraic equations, which are given as:

New Exact Solutions of FDEs by Proposed Tanh and Modified Kudryashov Methods 195

Coefficient of $Q^0 : \gamma b_0 + l b_0^5 = 0.$

Coefficient of $Q^1 : 5l b_0^4 b + \gamma a_1 + 5l^3 b_0^2 b_1 \log^2 a - l^5 b_1 \log^4 a = 0.$

Coefficient of $Q^2 : 10 l b_0^3 b_1^2 - 15 l^3 b_0^2 b_1 \log^2 a + 15 l^3 b_1^2 b_0 \log^2 a$
$$- 5l^4 b_1^2 \log^3 a + 15 l^5 b_1 \log^4 a = 0.$$

Coefficient of $Q^3 : -10 l b_0^2 b_1^3 + 10 l^3 b_0^2 b_1 \log^2 a - 40 l^3 b_1^2 b_0 \log^2 a + 10 l^3 b_1^3 \log^2 a \qquad (7.25)$
$$+ 20 l^4 b_1^2 \log^3 a - 50 l^5 b_1 \log^4 a = 0.$$

Coefficient of $Q^4 : -5 l b_0 b_1^4 + 25 l^3 b_1^2 b_0 \log^2 a - 25 l^3 b_1^3 \log^2 a,$
$$- 25 l^4 b_1^2 \log^3 a + 60 l^5 b_1 \log^4 a = 0.$$

Coefficient of $Q^5 : -l^4 b_1^5 + 15 l^3 b_1^3 \log^2 a + 10 l^4 b_1^2 \log^3 a - 24 l^5 a_1 \log^4 a = 0.$

Solving the above system, we obtain the following two families of solutions, which is described as follows:

Case I:

$$\gamma = \frac{1}{16} l^5 \log^4 a, a_0 = -\frac{1}{2} l \log a, a_1 = l \log a.$$

We have the following exact solution for **Case I:**

$$\Psi_{11} = l \log a \left(-1 + \frac{1 - \tan Fs(\zeta/2)}{2} \right), \qquad (7.26)$$

$$\Psi_{12} = l \log a \left(-1 + \frac{1 - \cot Fs(\zeta/2)}{2} \right), \qquad (7.27)$$

where $\zeta = lx + \dfrac{l^5 \log^4 a t^\alpha}{16 \Gamma(\alpha + 1)}.$

Case II:

$$\gamma = \frac{1}{16} l^5 \log^4 a, a_0 = l \log a, a_1 = -2 l \log a.$$

We have the following exact solution for **Case II:**

$$\Psi_{21} = -l \log a \tan Fs\left(\frac{\zeta}{2} \right), \qquad (7.28)$$

$$\Psi_{22} = -l \log a \cot Fs\left(\frac{\zeta}{2}\right), \tag{7.29}$$

where $\zeta = lx + \dfrac{l^5 \log^4 a t^\alpha}{16\Gamma(\alpha+1)}$.

7.4.2.1 The Numerical Simulations for Solutions of Time-Fractional Fifth-Order mS–K Equations by Using Modified Kudryashov Method

In this section, we have drawn the three-dimensional solitary wave solutions graphs which explain the physical meaning of each solution of the governing mS–K equation by utilizing eqs. (7.26) and (7.28).

7.4.2.1.1 Numerical Results and Discussion (Figures 7.11–7.14)

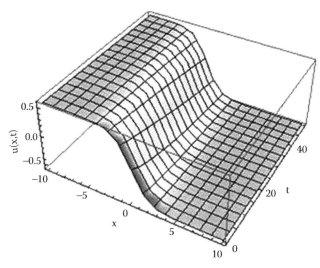

FIGURE 7.11
Case 1a: For $\alpha = 0.75$ (fractional order). The three-dimensional solitary wave solution graph obtained by using the modified Kudryashov method for $u(x,t)$ appears in eq. (7.26) of **Case I**, when $l = 0.5$, $a = 10$, and $\alpha = 0.75$.

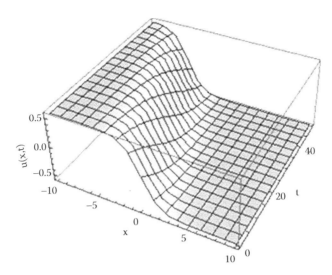

FIGURE 7.12
Case 1b: For $\alpha = 1$ (classical order). The three-dimensional solitary wave solution graph obtained by using the modified Kudryashov method for $u(x,t)$ appears in eq. (7.26) of **Case I**, when $l = 0.5$, $a = 10$, and $\alpha = 1$.

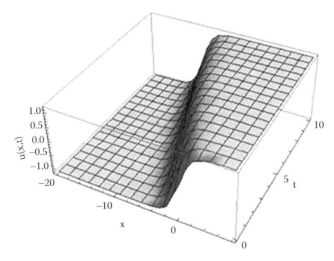

FIGURE 7.13
Case 2a: For $\alpha = 0.75$ (fractional order). The three-dimensional solitary wave solution graph obtained by using the modified Kudryashov method for $u(x,t)$ appears in eq. (7.28) of **Case II**, when $l = 0.5$, $a = 10$, and $\alpha = 0.75$.

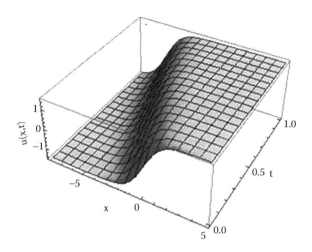

FIGURE 7.14
Case 2b: For $\alpha = 1$ (classical order). The three-dimensional solitary wave solution graph obtained by using the modified Kudryashov method for $u(x,t)$ appears in eq. (7.28) of **Case II**, when $l = 0.5$, $a = 10$, and $\alpha = 1$.

In the present section, some solutions graphs for fractional mS–K have been presented for visualizing the underlying mechanism of the governing equation. Figures 7.7 to 7.14 demonstrate a great impact concerning the exchange of amplitude and nature of the soliton because of the variant of the fractional order. The present three-dimensional graphs describe the behavior of $u(x, t)$ in space x and time t, in the case of the fractional order system. The behavior represents that increase of the fractional parameter that changes the wave nature of the solitary wave solution.

7.5 Implementation of Proposed Tanh Method for the Exact Solutions of Time-Fractional K–S Equation

In this section, we apply the proposed tanh method to determine the exact solutions for the time-fractional Kuramoto–Sivashinsky eq. (7.5).

By applying the fractional complex transform (2.65) of Chapter 2, eq. (7.5) can be reduced to the following non-linear ODE:

$$-cv\Phi'(\xi) + ac\Phi(\xi)\Phi'(\xi) + bc^2\Phi''(\xi) + kc^4\Phi^{IV}(\xi) = 0. \tag{7.30}$$

By reducing eq. (7.30), we get:

$$-v\Phi'(\xi) + a\Phi(\xi)\Phi'(\xi) + bc\Phi''(\xi) + kc^3\Phi^{IV}(\xi) = 0. \tag{7.31}$$

Further, by integrating eq. (7.31) with respect to ξ, we get:

$$-v\Phi(\xi) + \frac{a}{2}\Phi^2(\xi) + bc\Phi'(\xi) + kc^3\Phi'''(\xi) = 0. \tag{7.32}$$

New Exact Solutions of FDEs by Proposed Tanh and Modified Kudryashov Methods **199**

By balancing the highest order derivative term and non-linear term in eq. (7.32), the value of n can be determined, which is $n = 3$ in this problem.

Therefore, by using eq. (2.67) of Chapter 2, we have the following ansatz:

$$\Phi(\xi) = a_0 + a_1 Y + a_2 Y^2 + a_3 Y^3, \tag{7.33}$$

where $Y = \tanh(\xi)$.

According to eq. (2.68) of Chapter 2, we have:

$$\frac{d\Phi}{d\xi} \to \left(1 - Y^2\right)\left(a_1 + 2a_2 Y + 3a_3 Y^2\right),$$

$$\frac{d^2\Phi}{d\xi^2} \to \left(1 - Y^2\right)\left(-2Y\left(a_1 + 2a_2 Y + 3a_3 Y^2\right) + \left(2a_2 + 6a_3 Y\right)\left(1 - Y^2\right)\right), \tag{7.34}$$

$$\frac{d^3\Phi}{d\xi^3} \to 2\left(1 - Y^2\right)\left(3Y^2 - 1\right)\left(a_1 + 2a_2 + 3a_3 Y^2\right) - 6Y(2a_2 + 6a_3 Y)\left(1 - Y^2\right)^2 + 6a_3\left(1 - Y^2\right)^3.$$

Putting these values of eq. (7.34) along with eq. (7.33) into eq. (7.32), then after collecting all terms of the same degree of Y^i ($i = 0, 1, 2, \ldots$) and simultaneously equating to zero, we can obtain a set of algebraic equations for a_i ($i = 0, 1, 2, \ldots, n$), v, and c as follows:

Coefficient of Y^0 : $-va_0 + \dfrac{aa_0^2}{2} - 2c^3 ka_1 + cba_1 + 6c^3 ka_3 = 0.$

Coefficient of Y^1 : $-va_1 + aa_0 a_1 - 16c^3 ka_2 - 2cba_2 = 0.$

Coefficient of Y^2 : $8c^3 ka_1 - cba_1 + \dfrac{aa_1^2}{2} - va_2 + aa_0 a_2 - 60c^3 ka_3 + 3cba_3 = 0.$

Coefficient of Y^3 : $40c^3 ka_2 - 2cba_2 + aa_1 a_2 - va_3 + aa_0 a_3 = 0.$ $\qquad$ (7.35)

Coefficient of Y^4 : $-6c^3 ka_1 + \dfrac{aa_2^2}{2} + 114c^3 ka_3 - 3cba_3 + \alpha a_1 a_3 = 0.$

Coefficient of Y^5 : $-18c^3 ka_2 + aa_2 a_3 = 0.$

Coefficient of Y^6 : $-51c^3 ka_3 + \dfrac{aa_3^2}{2} = 0.$

Solving the above algebraic equations (7.35), we have the following sets of coefficients for the solutions generated from eq. (7.33) as given below:

Case I:

$$c = -\frac{1}{2}\sqrt{\frac{11b}{19k}}, v = -\frac{30}{19}\sqrt{\frac{11}{19k}}b^{\frac{3}{2}}, a_0 = -\frac{30}{19a}\sqrt{\frac{11}{19k}}b^{\frac{3}{2}}, a_1 = \frac{135}{19a}\sqrt{\frac{11}{19k}}b^{\frac{3}{2}}, a_2 = 0, a_3 = -\frac{165}{19a}\sqrt{\frac{11}{19k}}b^{\frac{3}{2}}.$$

For **Case I**, we have the following solution:

$$\Phi_{11} = -\frac{15}{19a}\sqrt{\frac{11}{19k}}b^{\frac{3}{2}}\left(2 - 9\tanh(\xi) + 11\tanh^3(\xi)\right),\tag{7.36}$$

where $\xi = -\dfrac{1}{2}\sqrt{\dfrac{11b}{19k}}\left(x + \sqrt{\dfrac{11}{19k}}\dfrac{30b^{\frac{3}{2}}t^\alpha}{19\Gamma(\alpha+1)}\right)$.

Case II:

$$c = -\frac{1}{2}\sqrt{\frac{11b}{19k}}, v = \frac{30}{19}\sqrt{\frac{11}{19k}}b^{\frac{3}{2}}, a_0 = \frac{30}{19a}\sqrt{\frac{11}{19k}}b^{\frac{3}{2}}, a_1 = \frac{135}{19a}\sqrt{\frac{11}{19k}}b^{\frac{3}{2}}, a_2 = 0, a_3 = -\frac{165}{19a}\sqrt{\frac{11}{19k}}b^{\frac{3}{2}}.$$

For **Case II**, we have the following solution:

$$\Phi_{21} = \frac{15}{19a}\sqrt{\frac{11}{19k}}b^{\frac{3}{2}}\left(2 + 9\tanh(\xi) - 11\tanh^3(\xi)\right),\tag{7.37}$$

where $\xi = -\dfrac{1}{2}\sqrt{\dfrac{11b}{19k}}\left(x - \sqrt{\dfrac{11}{19k}}\dfrac{30b^{\frac{3}{2}}t^\alpha}{19\Gamma(\alpha+1)}\right)$.

Case III:

$$c = \frac{1}{2}\sqrt{\frac{11b}{19k}}, v = -\frac{30}{19}\sqrt{\frac{11}{19k}}b^{\frac{3}{2}}, a_0 = -\frac{30}{19a}\sqrt{\frac{11}{19k}}b^{\frac{3}{2}}, a_1 = -\frac{135}{19a}\sqrt{\frac{11}{19k}}b^{\frac{3}{2}}, a_2 = 0, a_3 = \frac{165}{19a}\sqrt{\frac{11}{19k}}b^{\frac{3}{2}}.$$

For **Case III**, we have the following solution:

$$\Phi_{31} = -\frac{15}{19a}\sqrt{\frac{11}{19k}}b^{\frac{3}{2}}\left(2 + 9\tanh(\xi) + 11\tanh^3(\xi)\right),\tag{7.38}$$

where $\xi = \dfrac{1}{2}\sqrt{\dfrac{11b}{19k}}\left(x + \sqrt{\dfrac{11}{19k}}\dfrac{30b^{\frac{3}{2}}t^\alpha}{19\Gamma(\alpha+1)}\right)$.

Case IV:

$$c = \frac{1}{2}\sqrt{\frac{11b}{19k}}, v = \frac{30}{19}\sqrt{\frac{11}{19k}}b^{\frac{3}{2}}, a_0 = \frac{30}{19a}\sqrt{\frac{11}{19k}}b^{\frac{3}{2}}, a_1 = -\frac{135}{19a}\sqrt{\frac{11}{19k}}b^{\frac{3}{2}}, a_2 = 0, a_3 = \frac{165}{19a}\sqrt{\frac{11}{19k}}b^{\frac{3}{2}}.$$

For **Case IV**, we have the following solution:

$$\Phi_{41} = \frac{15}{19a}\sqrt{\frac{11}{19k}}b^{\frac{3}{2}}\left(2 - 9\tanh(\xi) + 11\tanh^3(\xi)\right),\tag{7.39}$$

where $\xi = \dfrac{1}{2}\sqrt{\dfrac{11b}{19k}}\left(x - \sqrt{\dfrac{11}{19k}}\,\dfrac{30b^{\frac{3}{2}}t^{\alpha}}{19\Gamma(\alpha+1)}\right).$

Case V:

$$c = -\frac{1}{2}\sqrt{\frac{-b}{19k}}\,, v = -\frac{30}{19}\sqrt{\frac{-1}{19k}}\,b^{\frac{3}{2}}, a_0 = -\frac{30}{19a}\sqrt{\frac{-1}{19k}}\,b^{\frac{3}{2}}, a_1 = -\frac{45}{19a}\sqrt{\frac{-1}{19k}}\,b^{\frac{3}{2}}, a_2 = 0, a_3 = \frac{15}{19a}\sqrt{\frac{-1}{19k}}\,b^{\frac{3}{2}}.$$

For **Case V**, we have the following solution:

$$\Phi_{51} = -\frac{15}{19a}\sqrt{\frac{-1}{19k}}\,b^{\frac{3}{2}}\left(2 + 3\tanh(\xi) - \tanh^3(\xi)\right), \tag{7.40}$$

where $\xi = -\dfrac{1}{2}\sqrt{\dfrac{-b}{19k}}\left(x + \sqrt{\dfrac{-1}{19k}}\,\dfrac{30b^{\frac{3}{2}}t^{\alpha}}{19\Gamma(\alpha+1)}\right).$

Case VI:

$$c = -\frac{1}{2}\sqrt{\frac{-b}{19k}}\,, v = \frac{30}{19}\sqrt{\frac{-1}{19k}}\,b^{\frac{3}{2}}, a_0 = \frac{30}{19a}\sqrt{\frac{-1}{19k}}\,b^{\frac{3}{2}}, a_1 = -\frac{45}{19a}\sqrt{\frac{-1}{19k}}\,b^{\frac{3}{2}}, a_2 = 0, a_3 = \frac{15}{19a}\sqrt{\frac{-1}{19k}}\,b^{\frac{3}{2}}.$$

For **Case VI**, we have the following solution:

$$\Phi_{61} = \frac{15}{19a}\sqrt{\frac{-1}{19k}}\,b^{\frac{3}{2}}\left(2 - 3\tanh(\xi) + \tanh^3(\xi)\right), \tag{7.41}$$

where $\xi = -\dfrac{1}{2}\sqrt{\dfrac{-b}{19k}}\left(x - \sqrt{\dfrac{-1}{19k}}\,\dfrac{30b^{\frac{3}{2}}t^{\alpha}}{19\Gamma(\alpha+1)}\right).$

Case VII:

$$c = \frac{1}{2}\sqrt{\frac{-b}{19k}}\,, v = -\frac{30}{19}\sqrt{\frac{-1}{19k}}\,b^{\frac{3}{2}}, a_0 = -\frac{30}{19a}\sqrt{\frac{-1}{19k}}\,b^{\frac{3}{2}}, a_1 = \frac{45}{19a}\sqrt{\frac{-1}{19k}}\,b^{\frac{3}{2}}, a_2 = 0, a_3 = -\frac{15}{19a}\sqrt{\frac{-1}{19k}}\,b^{\frac{3}{2}}.$$

For **Case VII**, we have the following solution:

$$\Phi_{71} = -\frac{15}{19a}\sqrt{\frac{-1}{19k}}\,b^{\frac{3}{2}}\left(2 - 3\tanh(\xi) + \tanh^3(\xi)\right), \tag{7.42}$$

where $\xi = \dfrac{1}{2}\sqrt{\dfrac{-b}{19k}}\left(x+\sqrt{\dfrac{-1}{19k}}\dfrac{30b^{\frac{3}{2}}t^{\alpha}}{19\Gamma(\alpha+1)}\right).$

Case VIII:

$$c=\dfrac{1}{2}\sqrt{\dfrac{-b}{19k}}, v=\dfrac{30}{19}\sqrt{\dfrac{-1}{19k}}b^{\frac{3}{2}}, a_0=\dfrac{30}{19a}\sqrt{\dfrac{-1}{19k}}b^{\frac{3}{2}}, a_1=\dfrac{45}{19b}\sqrt{\dfrac{-1}{19k}}b^{\frac{3}{2}}, a_2=0, a_3=-\dfrac{15}{19a}\sqrt{\dfrac{-1}{19k}}b^{\frac{3}{2}}.$$

For **Case VIII**, we have the following solution:

$$\Phi_{81} = \dfrac{15}{19a}\sqrt{\dfrac{-1}{19k}}b^{\frac{3}{2}}\left(2+3\tanh(\xi)-\tanh^{3}(\xi)\right), \tag{7.43}$$

where $\xi = \dfrac{1}{2}\sqrt{\dfrac{-b}{19k}}\left(x-\sqrt{\dfrac{-1}{19k}}\dfrac{30b^{\frac{3}{2}}t^{\alpha}}{19\Gamma(\alpha+1)}\right).$

7.5.1 The Numerical Simulations for Time-Fractional K–S Equation Obtained by New Proposed Tanh Method

In this present numerical experiment, two exact solutions of eq. (7.5) have been used to draw the graphs as shown in Figures 7.15 and 7.16 for different fractional order values of α.

In the present numerical simulation, we have drawn the traveling wave three-dimensional solutions surfaces and corresponding two-dimensional solution graphs for the obtained exact solutions of eq. (7.5) in the case of fractional order time derivative.

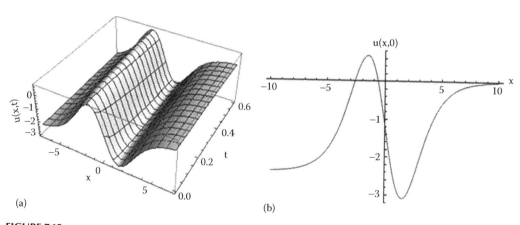

FIGURE 7.15
Case 1: For $\alpha = 0.5$ (fractional order). (a) The traveling wave solution for $u(x,t)$ appears in eq. (7.36) of **Case I**, (b) corresponding solution for $u(x,t)$, when $t=0, a=1, b=1, k=1,$ and $\alpha=0.5$.

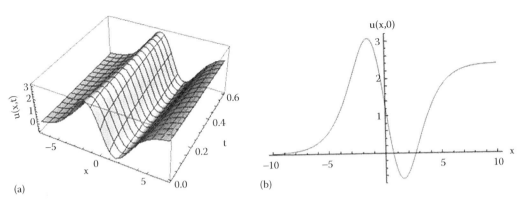

FIGURE 7.16
Case 2: For $\alpha = 0.5$ (fractional order). (a) The traveling wave solution for $u(x,t)$ appears in eq. (7.37) of **Case II**, (b) corresponding solution for $u(x,t)$, when $t = 0$, $a = 1$, $b = 1$, $k = 1$, and $\alpha = 0.5$.

7.6 Implementation of Proposed Tanh Method for the Exact Solutions of Time-Fractional Coupled JM Equations

In the present analysis, we implement the proposed tanh method to determine the new exact solutions for time-fractional coupled Jaulent–Miodek equations (6.2) and (6.3) of Chapter 6.

By applying the following fractional complex transform:

$$u(x,t) = \Phi(\xi), v(x,t) = \Psi(\xi), \xi = x - \frac{v t^\alpha}{\Gamma(\alpha+1)},$$

the eqs.(6.2) and (6.3) of Chapter 6 can be reduced to the following non-linear ODEs:

$$-v\Phi'(\xi) + \Phi'''(\xi) + \frac{3}{2}\Psi(\xi)\Psi'''(\xi) + \frac{9}{2}\Psi'(\xi)\Psi''(\xi), \tag{7.44}$$

$$-6\Phi(\xi)\Phi'(\xi) - 6\Phi(\xi)\Psi(\xi)\Psi'(\xi) - \frac{3}{2}\Phi'(\xi)\Psi^2(\xi) = 0.$$

and

$$-v\Psi'(\xi) + \Psi'''(\xi) - 6\Phi'(\xi)\Psi(\xi) - 6\Phi(\xi)\Psi'(\xi) - \frac{15}{2}\Psi'(\xi)\Psi^2(\xi) = 0. \tag{7.45}$$

Let $\Phi(\xi) = a_0 + \sum_{i=1}^{n_1} a_i Y^i$ and $\Psi(\xi) = b_0 + \sum_{i=1}^{n_2} b_i Y^i,$ \qquad (7.46)

where $Y = \tanh(\xi)$.

By balancing the highest order derivative term and non-linear term in eqs. (7.44) and (7.45), the values of n_1 and n_2 can be determined, which are $n_1 = 2$ and $n_2 = 1$ in this problem.

Therefore, by eq. (7.46), we have the following ansatz:

$$\Phi(\xi) = a_0 + a_1 Y + a_2 Y^2,$$

$$\Psi(\xi) = b_0 + b_1 Y. \tag{7.47}$$

According to eq. (2.68) of Chapter 2, we have:

$$\frac{d\Phi}{d\xi} \to (1 - Y^2)(a_1 + 2a_2 Y),$$

$$\frac{d^2\Phi}{d\xi^2} \to (1 - Y^2)(-2Y(a_1 + 2a_2 Y) + 2a_2(1 - Y^2)),$$

$$\frac{d^3\Phi}{d\xi^3} \to -12Y(1 - Y^2)^2 a_2 + 2Y(1 - Y^2)(3Y^2 - 1)(a_1 + 2a_2 Y),$$

$$\frac{d\Psi}{d\xi} \to b_1(1 - Y^2),$$

$$\frac{d^2\Psi}{d\xi^2} \to -2Y b_1(1 - Y^2),$$

$$\frac{d^3\Psi}{d\xi^3} \to 2Y b_1(1 - Y^2)(3Y^2 - 1). \tag{7.48}$$

Putting these values of eqs. (7.48) together with eq. (7.47) into eq. (7.44), then collecting all the like terms with the same degree of Y^i $(i = 0, 1, 2, ...)$, we can get a system of algebraic equations for a_i $(i = 0, 1, 2, ..., n)$, b_i $(i = 0, 1, 2, ..., n)$, and v as follows:

Coefficient of Y^0: $-4a_1 - 2va_1 - 12a_0 a_1 - 3a_1 b_0^2 - 6b_0 b_1 - 12a_0 b_0 b_1 = 0$.

Coefficient of Y^1: $-12a_1^2 - 32a_2 - 4va_2 - 24a_0 a_2 - 6a_2 b_0^2 - 18a_1 b_0 b_1 - 24b_1^2 - 12a_0 b_1^2 = 0$.

Coefficient of Y^2: $16a_1 + 2va_1 + 12a_0 a_1 - 36a_1 a_2 + 3a_1 b_0^2 + 24b_0 b_1$

$$+ 12a_0 b_0 b_1 - 24a_2 b_0 b_1 - 15a_1 b_1^2 = 0.$$

Coefficient of Y^3: $12a_1^2 + 80a_2 + 4va_2 + 24a_0 a_2 - 24a_2^2 + 6a_2 b_0^2$

$$+ 18a_1 b_0 b_1 + 60b_1^2 + 12a_0 b_1^2 - 18a_2 b_1^2 = 0. \tag{7.49}$$

$$\text{Coefficient of } Y^4 : 12a_1 - 36a_1a_2 + 18b_0b_1 - 24a_2b_0b_1 - 15a_1b_1^2 = 0.$$

$$\text{Coefficient of } Y^5 : -48a_2 + 24a_2^2 - 36b_1^2 + 18a_2b_1^2 = 0.$$

Putting these values of eqs. (7.48) together with eq. (7.47) into eq. (7.45), then collecting all the like terms with the same degree of Y^i $(i = 0,1,2,...)$, we can obtain another system of algebraic equations for a_i $(i = 0,1,2,...,n)$, b_i $(i = 0,1,2,...,n)$, and v as follows:

$$\text{Coefficient of } Y^0 : -12a_1b_0 - 4b_1 - 2vb_1 - 12a_0b_1 - 15b_1b_0^2 = 0.$$

$$\text{Coefficient of } Y^1 : -24a_2b_0 - 24a_1b_1 - 30b_0b_1^2 = 0.$$

$$\text{Coefficient of } Y^2 : 12a_1b_0 + 16b_1 + 2vb_1 + 12a_0b_1 - 36a_2b_1 + 15b_0^2b_1 - 15b_1^3 = 0. \qquad (7.50)$$

$$\text{Coefficient of } Y^3 : 24a_2b_0 + 24a_1b_1 + 30b_0b_1^2 = 0.$$

$$\text{Coefficient of } Y^4 : -12b_1 + 36a_2b_1 + 15b_1^3 = 0.$$

Solving the above algebraic eqs. (7.49) and (7.50), we have the following sets of coefficients for the solutions of eq. (7.47) as given below:

Case I:

$$v = v, a_0 = \frac{1}{12}(-7 + v), a_1 = -\frac{i\sqrt{1-v}}{2\sqrt{3}}, a_2 = \frac{3}{4}, b_0 = -\frac{\sqrt{1-v}}{\sqrt{3}}, b_1 = -i.$$

For **Case I**, we have the following solution:

$$\Phi_{11} = \frac{1}{12}(-7 + v) - \frac{i\sqrt{1-v}}{2\sqrt{3}}\tanh(\xi) + \frac{3}{4}\tanh^2(\xi), \qquad (7.51)$$

$$\Psi_{11} = -\frac{\sqrt{1-v}}{\sqrt{3}} - i\tanh(\xi),$$

where $\xi = x - \dfrac{v\, t^\alpha}{\Gamma(\alpha + 1)}$.

Case II:

$$v = v, a_0 = \frac{1}{12}(-7 + v), a_1 = \frac{i\sqrt{1-v}}{2\sqrt{3}}, a_2 = \frac{3}{4}, b_0 = \frac{\sqrt{1-v}}{\sqrt{3}}, b_1 = -i.$$

For **Case II**, we have the following solution:

$$\Phi_{21} = \frac{1}{12}(-7 + v) + \frac{i\sqrt{1-v}}{2\sqrt{3}}\tanh(\xi) + \frac{3}{4}\tanh^2(\xi), \qquad (7.52)$$

$$\Psi_{21} = \frac{\sqrt{1-v}}{\sqrt{3}} - i\tanh(\xi),$$

where $\xi = x - \dfrac{v\,t^\alpha}{\Gamma(\alpha+1)}$.

Case III:

$$v = v, a_0 = \frac{1}{12}(-7+v), a_1 = \frac{i\sqrt{1-v}}{2\sqrt{3}}, a_2 = \frac{3}{4}, b_0 = -\frac{\sqrt{1-v}}{\sqrt{3}}, b_1 = i.$$

For **Case III**, we have the following solution:

$$\Phi_{31} = \frac{1}{12}(-7+v) + \frac{i\sqrt{1-v}}{2\sqrt{3}}\tanh(\xi) + \frac{3}{4}\tanh^2(\xi), \tag{7.53}$$

$$\Psi_{31} = -\frac{\sqrt{1-v}}{\sqrt{3}} + i\tanh(\xi),$$

where $\xi = x - \dfrac{v\,t^\alpha}{\Gamma(\alpha+1)}$.

Case IV:

$$v = v, a_0 = \frac{1}{12}(-7+v), a_1 = -\frac{i\sqrt{1-v}}{2\sqrt{3}}, a_2 = \frac{3}{4}, b_0 = \frac{\sqrt{1-v}}{\sqrt{3}}, b_1 = i.$$

For **Case IV**, we have the following solution:

$$\Phi_{41} = \frac{1}{12}(-7+v) - \frac{i\sqrt{1-v}}{2\sqrt{3}}\tanh(\xi) + \frac{3}{4}\tanh^2(\xi), \tag{7.54}$$

$$\Psi_{41} = \frac{\sqrt{1-v}}{\sqrt{3}} + i\tanh(\xi),$$

where $\xi = x - \dfrac{v\,t^\alpha}{\Gamma(\alpha+1)}$.

7.6.1 The Numerical Simulations for Solutions of Time-Fractional Coupled JM Equations Using Tanh Method

In this present numerical experiment, the exact solutions for eqs. (6.2) and (6.3) of Chapter 6 obtained by using the proposed tanh method presented in eq. (7.51) have been used to draw the three-dimensional and two-dimensional solution graphs as shown in Figures 7.17 and 7.18.

In the present numerical simulation, we have presented the three-dimensional and two-dimensional solution graphs for coupled JM equations by the proposed tanh method.

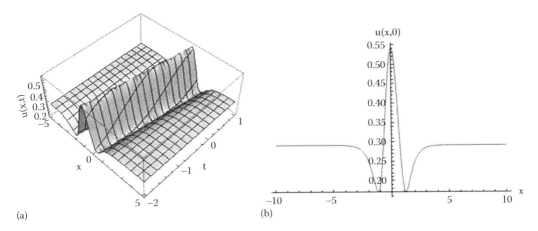

FIGURE 7.17
(a) The tanh method three-dimensional solitary wave solution for $u(x,t)$ appears in eq. (7.51) as Φ_{11} of **Case I**, (b) corresponding two-dimensional solution graph for $u(x,t)$ when $t=0$, $v=0.5$, and $\alpha=1$.

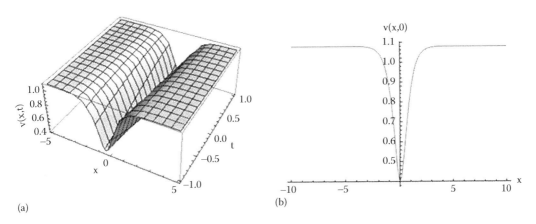

FIGURE 7.18
(a) The tanh method three dimensional solitary wave solution for $v(x,t)$ appears in eq. (7.51) as Ψ_{11} of **Case I**, (b) corresponding two dimensional solution graph for $v(x,t)$ when $t=0$, $v=0.5$, and $\alpha=1$.

We have concluded that the nature of the solution graph is solitary, which can be used to describe many physical phenomenon.

7.7 Conclusion

In this chapter, we determine the solitary wave exact solutions of non-linear evolution equations, namely, time-fractional fifth-order Sawada–Kotera equation, time-fractional fifth-order modified Sawada–Kotera equation and time-fractional fifth-order Kuramoto–Sivashinsky (K–S) equation, and time-fractional coupled Jaulent–Miodek equations by proposed tanh method via fractional complex transform. We have also applied the modified Kudryashov method to solve time-fractional fifth-order mS–K equation. By applying the proposed tanh method, we obtain solitary wave solutions. Both methods provide new and more general type solitary wave solutions which are significant to reveal the pertinent features of the physical phenomenon. The fractional complex transform can easily convert a fractional differential equation into its equivalent ordinary differential equation form. So, fractional complex transform has been efficiently used for solving fractional differential equation.

Also, in the proposed tanh method, the solution contains only one parameter. Hence, the degree of freedom of the solution obtained by the tanh method is much less than other solutions that are obtained by other methods. Three-dimensional plots of some of the investigated solutions by the tanh method have been also drawn to visualize the underlying dynamics of such results. The tanh method has many advantages: it is straightforward and concise. Furthermore, this survey indicates that the proposed method is rather effective for solving time-fractional fifth-order Sawada–Kotera equation. The execution of this method is reliable and efficient and provides the exact solitary wave solutions. This method clearly avoids linearization, unrealistic assumptions, and therefore it provides efficient exact solutions. Therefore, the proposed method is very effective and a powerful technique in determining the exact solutions of time-fractional non-linear partial differential equations.

Here, we have utilized the proposed tanh method and modified Kudryashov method to solve the fractional non-linear evolution equation, viz., the time-fractional fifth-order mS–K equation conveniently. As a consequence, two new hyperbolic solutions by using the tanh method and four new generalized Fibonacci hyperbolic solutions by using the modified Kudryashov method for the time-fractional fifth-order mS–K equation have been effectively found. In order to make convenient implementation of the newly proposed tanh and modified Kudryashov methods, the fractional complex transform has been effectively used here for converting the time-fractional mS–K equations to ordinary differential equations. By using the obtained solutions, we have drawn the three-dimensional graphs, which give the nature of the solutions as solitary wave. The executions of the proposed approaches are reliable and provide the certain new solitary wave solutions. Although both methods are clearly avoiding unrealistic assumptions, linearization, but the modified Kudryashov method gives more exact solutions of the discussed non-linear mS–K equation, and, for this reason, it presents effective approach for finding more exact wave solutions for the fractional order mS–K equation.

Also, in this chapter, several traveling wave exact solutions of the time-fractional Kuramoto–Sivashinsky equation have been successfully obtained by a proposed tanh method with the help of fractional complex transform. Furthermore, the present analysis indicates that the proposed method is effective and efficient for solving the time-fractional Kuramoto–Sivashinsky equation.

By using the obtained solutions, we have drawn the three-dimensional graphs, which give the nature of the solutions as solitary wave. The execution of the proposed approach is reliable and provides the certain new solitary wave solutions. Furthermore, this survey indicates that the proposed methods are rather effective for solving time-fractional partial differential equations.

The proposed tanh method renders solutions with physical phenomena. Also, the other suggested method is very effective, powerful, direct, and easily computerizable technique providing new exact solutions of fractional differential equations.

8

New Exact Solutions of Fractional Differential Equations by Proposed Novel Method

8.1 Introduction

In recent years, non-linear fractional evolution equations in mathematical physics play an important role in various fields, such as fluid mechanics, plasma physics, optical fibers, and so on. It is well known that the traveling wave solutions of non-linear fractional evolution equations play an important role in the study of non-linear wave phenomena. Such wave phenomena are observed in fluid dynamics, plasma, elastic media, optical fibers, etc.

The objective of this chapter is to employ a novel analytical method for getting the new exact solutions for fractional coupled non-linear equations like the time-fractional Korteweg–de Vries (KdV)–Burgers equation, time-fractional KdV-modified (mKdV) equation, time-fractional coupled Schrödinger–KdV (SK) equations, and time-fractional coupled Schrödinger–Boussinesq (SB) equations in plasma physics.

8.2 Outline of Present Study

In the past few years, a great deal of attention has been intended by the researchers on the study of non-linear evolution equations that appeared in mathematical physics. Especially KdV type equations [249,250] have been paid of more attention due to their traveling wave nature along the water surface and various applications in solid-state physics, plasma physics, and quantum field theory, and Schrödinger types of equations [251,252] are known to describe the quantum mechanical behavior [251,252]. The development of instability associated with the envelope modulation of a high frequency wave packet coupled to a low frequency wave field is presented by coupled non-linear equations like coupled Schrödinger–KdV and coupled Schrödinger–Boussinesq equations in plasma physics [253].

8.2.1 Time-Fractional KdV–Burgers Equation

The KdV equation was first discovered by Boussinesq [254] and mathematically introduced by Korteweg and de Vries [255]. For studying the problems of the drift of liquids containing fuel bubble, the fluid flow in elastic tubes, the control KdV equation will also be lowered to the KdV–Burgers (KdVB) equation, which is the same as the KdV equation if a viscous dissipation term is introduced and it represents mathematical model of waves on shallow water surfaces. The Burger equation is the special case of the KdVB equation which has been found to describe many types of physical phenomenon like turbulence [256] and approximate theory of flow through a shock wave traveling in viscous fluid [257].

Here, we considered the time-fractional KdV–Burgers equation [258–260] as:

$$_0D_t^\alpha u + \varepsilon u u_x + \eta u_{xx} + v u_{xxx} = 0, \tag{8.1}$$

where ε, η, v are constants, α denotes the order of fractional derivative whose range is $0 < \alpha \le 1$.

There are many numerical and analytical methods proposed in the recent past for the solving of the KdVB equation. Zaki [261] has used the B-spline collocation method for finding the numerical approximate solution for the KdVB equation. The approximate solution of the KdVB equation has been studied by Kaya [262] and Wang et al. [263]. The exact traveling solutions of the KdVB equation have been established by Feng [264], Jeffrey and Xu [265], Bikbaev [266], Bona and Schonbek [267], and Feng and Knobel [268].

8.2.2 Time-Fractional Combined KdV–mKdV Equation

The KdV–mKdV equation mainly describes the propagation of bounded particle of the atmosphere dust-acoustic solitary waves, internal solitary waves in shallow seas, and ion acoustic waves in plasmas with negative ions [269–271].

Here, the governing equation is the time-fractional combined KdV–mKdV equation [269], which is given as:

$$_0D_t^\alpha u + a u u_x + b u^2 u_x + \gamma u_{xxx} = 0, \tag{8.2}$$

where a, b, γ are the constants, α is the fractional order whose range is $0 < \alpha \le 1$.

Up until now, many researchers have investigated the solutions for the KdV–mKdV equation. The notable works of Naher and Abdullah [272], Bekir [273], Triki et al. [274], Gómez Sierra et al. [275], Zhang and Tian [276], Krishnan and Peng [277], Yang and Tang [278], and Lu and Shi [279] are involving the exact traveling wave solutions for the KdV–mKdV equation.

The fractional differential equations can be described best in discontinuous media and the fractional order is equivalent to their fractional dimensions. Fractal media which are complex, appear in different fields of engineering and physics. In this context, the local fractional calculus theory is very important for modeling problems for fractal mathematics and engineering on Cantorian space in fractal media.

8.2.3 Time-Fractional Coupled Schrödinger–KdV Equations

Let us consider the time-fractional coupled SK equations [280, 281]:

$$i \, _0D_t^\alpha u - u_{xx} - uv = 0,$$

$$_0D_t^\alpha v + 6vv_x + v_{xxx} + \left(|u|^2 \right)_x = 0, \tag{8.3}$$

where the α symbolizes the order of fractional derivative, whose range is $0 < \alpha \le 1$.

New Exact Solutions of Fractional Differential Equations by Proposed Novel Method   211

The Schrödinger–KdV equation describes various processes, such as dust-acoustic, Langmuir, and electromagnetic waves in dusty plasma [282–284]. Various methods like unified algebraic method [285], hybrid of Fourier transform method [286], and variational iteration method [287] have been used for finding the solutions of the Schrödinger–KdV equation.

8.2.4 Time-Fractional Coupled SB Equations

Let us consider time-fractional coupled SB equations [288, 289]:

$$i\varepsilon\,{}_0D_t^\alpha u + \frac{3}{2}u_{xx} - \frac{1}{2}uv = 0,$$

$$\,{}_0D_t^{2\alpha}v - v_{xx} - v_{xxxx} - v_{xx}^2 - \frac{1}{4}\left(\left|u\right|^2\right)_{xx} = 0, \tag{8.4}$$

where u is the complex valued function which represents the short wave amplitude of media, and v is the real-valued function which represents the long wave amplitude of media. The $\varepsilon > 0$ denotes the ratio between the electron number with respect to ion number, and the α is the fractional order whose range is $0 < \alpha \le 1$.

The coupled Schrödinger–Boussinesq equations are originated from non-linear magnetosonic and upper-hybrid waves in magnetized plasma [290]. It also describes the diatomic lattice system [291], the dynamics of Langmuir soliton formation, and the interaction in plasma [292–295]. Various methods like multi-symplectic scheme [288], Fourier spectral method [289], conservative difference scheme [296], (G'/G)-expansion method [297], and extended simplest equation method [298] have been used for finding solution for the coupled Schrödinger–Boussinesq equation.

Our main objective here is to find new exact solutions of the time-fractional KdV–Burgers equation, time-fractional combined KdV–mKdV equation, time-fractional coupled SK equations, and time-fractional coupled SB equations by applying a reliable and relatively new analytical method.

8.3 Exact Solutions for Time-Fractional KdV–Burgers Equation

In this section, the newly proposed method has been applied here for finding the exact solutions for eq. (8.1).

By using the following fractional complex transform:

$$u(x,t) = \Phi(\xi),\ \xi = cx + \frac{kt^\alpha}{\Gamma(\alpha+1)},$$

in eq. (8.1), we have the following non-linear ordinary differential equation (ODE):

$$k\Phi'(\xi) + \varepsilon c\Phi(\xi)\Phi'(\xi) + \eta c^2\Phi''(\xi) + vc^3\Phi'''(\xi) = 0. \tag{8.5}$$

Let us consider the ansatz:

$$\Phi(\xi) = a_0 + \sum_{i=1}^{n} a_i \phi^i. \tag{8.6}$$

The dominant terms with highest order singularity of eq. (8.5) are $\varepsilon c \Phi(\xi) \Phi'(\xi)$ and $v c^3 \Phi'''(\xi)$. The maximum value of pole is 2 which yields $n = 2$.

Therefore, by eq. (8.6), we have the following ansatz:

$$\Phi(\xi) = a_0 + a_1 \phi + a_2 \phi^2, \tag{8.7}$$

where ϕ satisfies eq. (2.80) of Chapter 2.

Substituting eq. (8.7) along with eqs. (2.81) of Chapter 2 into eq. (8.5), and then equating each coefficient of $\phi^i (i = 0, 1, 2, ...)$ to zero, we can find a system of algebraic equations for a_0, a_1, a_2, c, and k as follows:

$$\phi^1 : a_1 k + \varepsilon c a_0 a_1 + \eta a_1 c^2 + v c^3 a_1 = 0.$$

$$\phi^2 : -a_1 k + 2 k a_2 - \varepsilon c a_1 a_0 + \varepsilon c a_1^2 + 2\varepsilon c a_0 a_2 - 3\eta c^2 a_1 + 4\eta c^2 a_2 - 7 c^3 v a_1 + 8 v c^3 a_2 = 0.$$

$$\phi^3 : -2 k a_2 - \varepsilon c a_1^2 - 2\varepsilon c a_2 a_0 + 3\varepsilon c a_1 a_2 + 2\eta c^2 a_1 - 10\eta c^2 a_2 + 12 c^3 v a_1 - 38 v c^3 a_2 = 0. \tag{8.8}$$

$$\phi^4 : -3\varepsilon c a_1 a_2 + 2\varepsilon c a_2^2 + 6\eta c^2 a_2 - 12 c^3 v a_1 + 54 v c^3 a_2 = 0.$$

$$\phi^5 : -2\varepsilon c a_2^2 - 24 v c^3 a_2 = 0.$$

Solving the above algebraic eqs. (8.8), we have the following sets of coefficients for the solutions of eq. (8.7) as given below:

Case I:

$$c = -\frac{\eta}{5v}, \, k = k, \, a_0 = \frac{6\eta^3 + 125 k v^2}{25\varepsilon \eta v}, \, a_1 = 0, \, a_2 = -\frac{12\eta^2}{25\varepsilon v}.$$

For **Case I**, we have the following solution:

$$\Phi_{11} = \frac{250 k v^2 - 6\eta^3 \sec h^2 \left(\dfrac{\xi}{2} \right)(-1 + \sinh(\xi))}{50\varepsilon \eta v}, \tag{8.9}$$

where $\xi = -\dfrac{\eta x}{5v} + \dfrac{k t^\alpha}{\Gamma(\alpha + 1)}.$

Case II:

$$c = \frac{\eta}{5v}, k = k, a_0 = -\frac{6\eta^3 + 125kv^2}{25\varepsilon\eta v}, a_1 = \frac{24\eta^2}{25\varepsilon v}, a_2 = -\frac{12\eta^2}{25\varepsilon v}.$$

For **Case II**, we have the following solution:

$$\Phi_{21} = \frac{-250kv^2 + 6\eta^3 \operatorname{sech}^2\left(\frac{\xi}{2}\right)(1+\sinh(\xi))}{50\varepsilon\eta v}, \tag{8.10}$$

where $\xi = \dfrac{\eta x}{5v} + \dfrac{kt^\alpha}{\Gamma(\alpha+1)}$.

8.3.1 Numerical Simulations for Time-Fractional KdV–Burgers Equation

In this section, the eq. (8.9) has been used for drawing the solution graphs for the time-fractional KdV–Burgers equation in the case of both classical and fractional orders.

The present section contains the numerical simulations for the time-fractional KdV–Burgers equation. We have presented the solution graphs for the time-fractional KdV–Burgers equation for both classical and fractional order. Figures 8.1 to 8.3 show the evolution of the antikink-solitary wave solutions for eq. (8.1). From the above study, we conclude that the solitary waves maintain their shape when traveling down, and main evolution characteristics of the solitary waves do not change during the propagation.

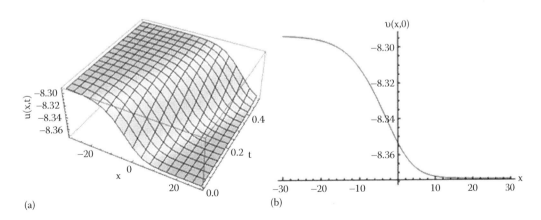

FIGURE 8.1
Case I: For $\alpha = 1$ (classical order). (a) The three-dimensional antikink-solitary wave solution graph for $u(x,t)$ appears in eq. (8.9) as Φ_{11} in **Case I**, when $\varepsilon = -6$, $v = 1$, $\eta = 1$, $k = 10$, and $\alpha = 1$, (b) the corresponding two-dimensional solution graph for $u(x,t)$ when $t = 0$.

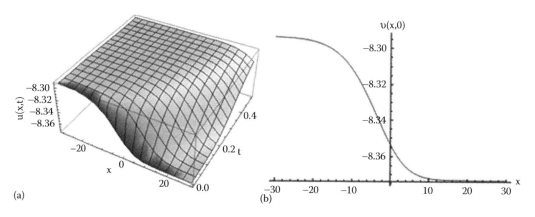

FIGURE 8.2
Case II: For $\alpha = 0.5$ (fractional order). (a) The three-dimensional antikink-solitary wave solution graph for $u(x,t)$ appears in eq. (8.9) as Φ_{11} in **Case I**, when $\varepsilon = -6$, $v = 1$, $\eta = 1$, $k = 10$, and $\alpha = 0.5$, (b) the corresponding two-dimensional solution graph for $u(x,t)$ when $t = 0$.

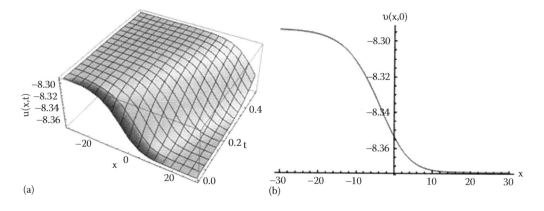

FIGURE 8.3
Case III: For $\alpha = 0.75$ (fractional order). (a) The three-dimensional antikink-solitary wave solution graph for $u(x,t)$ appears in eq. (8.9) as Φ_{11} in **Case I**, when $\varepsilon = -6$, $v = 1$, $\eta = 1$, $k = 10$, and $\alpha = 0.75$, (b) the corresponding two-dimensional solution graph for $u(x,t)$ when $t = 0$.

8.4 Exact Solutions for Time-Fractional KdV–mKdV Equation

In this section, the proposed method has been applied here for finding the exact solutions for eq. (8.2). By using the following fractional complex transform:

$$u(x,t) = \Phi(\xi), \; \xi = cx + \frac{kt^\alpha}{\Gamma(\alpha+1)},$$

New Exact Solutions of Fractional Differential Equations by Proposed Novel Method   215

in eq. (8.2), we have the following non-linear ODE:

$$k\Phi'(\xi) + ac\Phi(\xi)\Phi'(\xi) + bc^3\Phi^2(\xi)\Phi'(\xi) + \gamma c^3\Phi'''(\xi) = 0. \tag{8.11}$$

Let us consider the ansatz:

$$\Phi(\xi) = a_0 + \sum_{i=1}^{n} a_i\phi^i. \tag{8.12}$$

The dominant terms with highest order singularity of eq. (8.11) are $bc^3\Phi^2(\xi)\Phi'(\xi)$ and $\gamma c^3\Phi'''(\xi)$. The maximum value of pole is 1 which yields $n = 1$.

Therefore, by eq. (8.12), we have the following ansatz:

$$\Phi(\xi) = a_0 + a_1\phi, \tag{8.13}$$

where ϕ satisfies eq. (2.80) of Chapter 2.

Substituting eq. (8.13) along with eqs. (2.81) of Chapter 2 into eq. (8.11), and then equating each coefficient of $\phi^i (i = 0, 1, 2, ...)$ to zero, we can find a system of algebraic equations for a_0, a_1, c, and k as follows:

$$\phi^1 : aca_0a_1 + bc^3a_0^2a_1 + a_1k + \gamma c^3a_1 = 0.$$

$$\phi^2 : -aca_1a_0 + aca_1^2 - bc^3a_0^2a_1 + 2bc^3a_0a_1^2 - a_1k + 7\gamma c^3a_1ka_2 = 0. \tag{8.14}$$

$$\phi^3 : -aca_1^2 - 2bc^3a_0a_1^2 + bc^3a_1^3 + 12\gamma c^3a_1 = 0.$$

$$\phi^4 : -bc^3a_1^3 - 6\gamma c^3a_1 = 0.$$

Solving the above algebraic eqs. (8.14), we have the following sets of coefficients for the solutions of eq. (8.11) as given below:

Case I:

$$c = c, \quad k = \frac{a^2 + 2b\gamma c^4}{4bc}, \quad a_0 = -\frac{(a + i\sqrt{6b\gamma}c^2)}{2bc^2}, \quad a_1 = \frac{i\sqrt{6\gamma}}{\sqrt{b}}.$$

For **Case I**, we have the following solution:

$$\Phi_{11} = -\frac{\dfrac{a}{c^2} - i\sqrt{6b\gamma}c^2\tanh\left(\dfrac{\xi}{2}\right)}{2b}, \tag{8.15}$$

where $\xi = cx + \dfrac{(a^2 + 2b\gamma c^4)t^\alpha}{4bc\Gamma(\alpha + 1)}$.

Case II:

$$c = c, \quad k = \frac{a^2 + 2b\gamma c^4}{4bc}, \quad a_0 = -\frac{(a - i\sqrt{6b\gamma}\, c^2)}{2bc^2}, \quad a_1 = -\frac{i\sqrt{6\gamma}}{\sqrt{b}}.$$

For **Case II**, we have the following solution:

$$\Phi_{21} = -\frac{\dfrac{a}{c^2} + i\sqrt{6b\gamma}\, c^2 \tanh\left(\dfrac{\xi}{2}\right)}{2b}, \tag{8.16}$$

where $\xi = cx + \dfrac{(a^2 + 2b\gamma c^4) t^\alpha}{4bc\Gamma(\alpha + 1)}$.

8.4.1 Numerical Simulations for Time-Fractional KdV–mKdV Equation

In this section, the eq. (8.15) has been used for drawing the solution graphs for the time-fractional KdV–mKdV equation in the case of both classical and fractional orders. As the solution obtained in eq. (8.15) is complex in nature, we have taken here the absolute value of present solutions for obtaining the three-dimensional and the corresponding two-dimensional graphs.

The present section contains the numerical simulations for the time-fractional KdV–mKdV equation. We have presented the solution graphs for the time-fractional KdV–mKdV equation for both classical and fractional order. Figures 8.4 to 8.6 show the evolution of the kink-solitary wave solutions for eq. (8.2). From the above study, we conclude that the solitary waves maintain their shape when traveling down, and main evolution characteristics of the solitary waves do not change during the propagation.

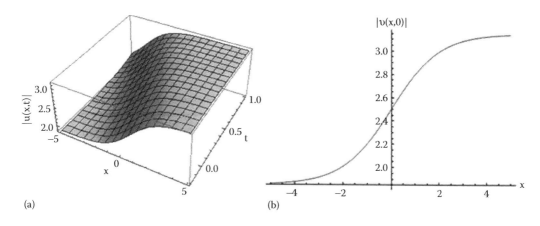

FIGURE 8.4

Case I: For $\alpha = 1$ (classical order). (a) The three-dimensional kink-solitary wave solution graph for $|u(x,t)|$ appears in eq. (8.15) as Φ_{11} in **Case I**, when $a = 3$, $\gamma = -1/6$, $b = 0.6$, $c = 1$, and $\alpha = 1$, (b) the corresponding two-dimensional solution graph for $|u(x,t)|$ when $t = 0$.

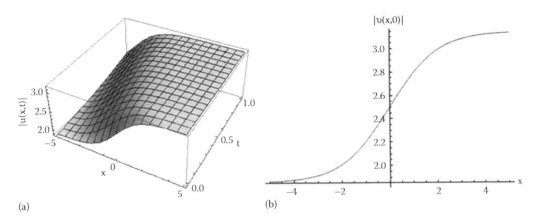

FIGURE 8.5
Case II: For $\alpha = 0.5$ (fractional order). (a) The three-dimensional kink-solitary wave solution graph for $|u(x,t)|$ appears in eq. (8.15) as Φ_{11} in **Case I**, when $a = 3$, $\gamma = -1/6$, $b = 0.6$, $c = 1$, and $\alpha = 0.5$, (b) the corresponding two-dimensional solution graph for $|u(x,t)|$ when $t = 0$.

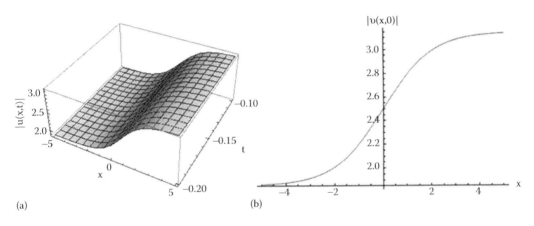

FIGURE 8.6
Case III: For $\alpha = 0.75$ (fractional order). (a) The three-dimensional kink-solitary wave solution graph for $|u(x,t)|$ appears in eq. (8.15) as Φ_{11} in **Case I**, when $a = 3$, $\gamma = -1/6$, $b = 0.6$, $c = 1$, and $\alpha = 0.75$, (b) the corresponding two-dimensional solution graph for $|u(x,t)|$ when $t = 0$.

8.5 Exact Solutions for Time-Fractional Coupled Schrödinger–KdV Equations

In this section, the newly proposed method has been applied here for finding the exact solutions for eqs. (8.3).

By using the fractional complex transform eq. (2.77) of Chapter 2 in eqs. (8.3), we have the following non-linear ODE:

$$\Phi(\zeta)\Psi(\zeta) + (r - k^2)\Phi(\zeta) + c^2\Phi''(\zeta)^2 + i(-\gamma + 2kc)\Phi'(\zeta) = 0, \qquad (8.17)$$

$$\gamma\Psi'(\zeta) + 6c\Psi(\zeta)\Psi'(\zeta) + c^3\Psi'''(\zeta) - 2c\Phi(\zeta)\Phi'(\zeta) = 0, \qquad (8.18)$$

Again, the eq. (8.17) can be written as:

$$\Phi(\zeta)\Psi(\zeta)+(r-k^2)\Phi(\zeta)+c^2\Phi''(\zeta)^2=0, \tag{8.19}$$

where $\gamma = 2kc$.

By integrating eq. (8.18) once with respect to ζ and putting $\gamma = 2kc$, we have:

$$2k\Psi(\zeta)+3\Psi^2(\zeta)+c^2\Psi''(\zeta)-\Phi^2(\zeta)=0. \tag{8.20}$$

Let the ansatz be:

$$\Phi(\zeta)=a_0+\sum_{i=1}^{n}a_i\phi^i \text{ and } \Psi(\zeta)=b_0+\sum_{i=1}^{m}b_i\phi^i. \tag{8.21}$$

The dominant terms with highest order singularity of eq. (8.19) are $\Phi(\zeta)\Psi(\zeta)$ and $c^2\Phi''(\zeta)^2$. The maximum value of pole is 2, that means here $n = 2$. Similarly, the dominant terms with highest order singularity of eq. (8.20) are $3\Psi^2(\zeta)$ and $c^2\Psi''(\zeta)$. The maximum value of pole is 2, that means here $m = 2$.

Therefore, by eq. (8.21), we have the following ansatz:

$$\Phi(\zeta)=a_0+a_1\phi+a_2\phi^2 \text{ and } \Psi(\zeta)=b_0+b_1\phi+b_2\phi^2, \tag{8.22}$$

where ϕ satisfies eq. (2.80) of Chapter 2.

Substituting eq. (8.22) along with eqs. (2.81) of Chapter 2 into eqs. (8.19) and (8.20), then equating each coefficient of φ^i ($i = 0, 1, 2, ...$) to zero, we can find a system of algebraic equations for $a_0, a_1, a_2, b_0, b_1, b_2, c, k,$ and r as follows:

$$\phi^0: a_0b_0+a_0(-k^2+r)=0, -a_0^2+3b_0^2+2b_0k=0.$$

$$\phi^1: a_1b_0+a_0b_1+a_1c^2+a_1(-k^2+r)=0, -2a_0a_1+6b_0b_1+c^2b_1+2b_1k=0.$$

$$\phi^2: a_2b_0+a_1b_1+a_0b_2-3a_1c^2+4a_2c^2+a_2(-k^2+r)=0,$$

$$-a_1^2-2a_0a_2+3b_1^2+6b_0b_2-3c^2b_1+4c^2b_2+2b_2k=0. \tag{8.23}$$

$$\phi^3: a_2b_1+a_1b_2+2a_1c^2-10a_2c^2=0, -2a_1a_2+6b_1b_2+2c^2b_1-10c^2b_2=0.$$

$$\phi^4: a_2b_2+6a_2c^2=0, -a_2^2+3b_2^2++6c^2b_2=0.$$

Solving the above algebraic eqs. (8.23), we have the following sets of coefficients for the solutions of eqs. (8.19) and (8.20) as given below:

Case I:

$$c=c, k=-\frac{c^2}{2}, r=\frac{c^2(-4+c^2)}{4}, a_0=0, a_1=-6\sqrt{2}c^2, a_2=6\sqrt{2}c^2,$$

$$b_0=0, b_1=6c^2, b_2=-6c^2.$$

For **Case I**, we have the following solution:

$$\Phi_{11} = -\frac{6\sqrt{2}c^2(\cosh(\zeta)+\sinh(\zeta))}{1+\cosh(\zeta)+\sinh(\zeta)} + \frac{6\sqrt{2}c^2(\cosh(2\zeta)+\sinh(2\zeta))}{(1+\cosh(\zeta)+\sinh(\zeta))^2},$$

$$\Psi_{11} = \frac{6c^2(\cosh(\zeta)+\sinh(\zeta))}{1+\cosh(\zeta)+\sinh(\zeta)} - \frac{6c^2(\cosh(2\zeta)+\sinh(2\zeta))}{(1+\cosh(\zeta)+\sinh(\zeta))^2},$$

$$(8.24)$$

where $\zeta = cx + \dfrac{\gamma t^\alpha}{\Gamma(\alpha+1)}$.

Case II:

$$c = c,\ k = -\frac{c^2}{2},\ r = \frac{c^2(8+3c^2)}{12},\ a_0 = \sqrt{2}c^2,\ a_1 = -6\sqrt{2}c^2,\ a_2 = 6\sqrt{2}c^2,$$

$$b_0 = -\frac{2c^2}{3},\ b_1 = 6c^2,\ b_2 = -6c^2.$$

For **Case II**, we have the following solution:

$$\Phi_{21} = \sqrt{2}c^2 - \frac{6\sqrt{2}c^2(\cosh(\zeta)+\sinh(\zeta))}{1+\cosh(\zeta)+\sinh(\zeta)} + \frac{6\sqrt{2}c^2(\cosh(2\zeta)+\sinh(2\zeta))}{(1+\cosh(\zeta)+\sinh(\zeta))^2},$$

$$\Psi_{21} = -\frac{2c^2}{3} + \frac{6c^2(\cosh(\zeta)+\sinh(\zeta))}{1+\cosh(\zeta)+\sinh(\zeta)} - \frac{6c^2(\cosh(2\zeta)+\sinh(2\zeta))}{(1+\cosh(\zeta)+\sinh(\zeta))^2},$$

$$(8.25)$$

where $\zeta = cx + \dfrac{\gamma t^\alpha}{\Gamma(\alpha+1)}$.

Case III:

$$c = c,\ k = \frac{c^2}{2},\ r = \frac{c^2(4+c^2)}{4},\ a_0 = \sqrt{2}c^2,\ a_1 = -6\sqrt{2}c^2,\ a_2 = 6\sqrt{2}c^2,\ b_0 = -c^2,$$

$$b_1 = 6c^2,\ b_2 = -6c^2.$$

For **Case III**, we have the following solution:

$$\Phi_{31} = \sqrt{2}c^2 - \frac{6\sqrt{2}c^2(\cosh(\zeta)+\sinh(\zeta))}{1+\cosh(\zeta)+\sinh(\zeta)} + \frac{6\sqrt{2}c^2(\cosh(2\zeta)+\sinh(2\zeta))}{(1+\cosh(\zeta)+\sinh(\zeta))^2},$$

$$\Psi_{31} = -c^2 + \frac{6c^2(\cosh(\zeta)+\sinh(\zeta))}{1+\cosh(\zeta)+\sinh(\zeta)} - \frac{6c^2(\cosh(2\zeta)+\sinh(2\zeta))}{(1+\cosh(\zeta)+\sinh(\zeta))^2},$$

$$(8.26)$$

where $\zeta = cx + \dfrac{\gamma t^\alpha}{\Gamma(\alpha+1)}$.

Case IV:

$$c = c, k = \frac{c^2}{2}, r = \frac{c^2(-8+3c^2)}{12}, a_0 = 0, a_1 = -6\sqrt{2}c^2, a_2 = 6\sqrt{2}c^2, b_0 = -\frac{c^2}{3},$$

$$b_1 = 6c^2, b_2 = -6c^2$$

For **Case IV**, we have the following solution:

$$\Phi_{41} = -\frac{6\sqrt{2}c^2(\cosh(\zeta)+\sinh(\zeta))}{1+\cosh(\zeta)+\sinh(\zeta)} + \frac{6\sqrt{2}c^2(\cosh(2\zeta)+\sinh(2\zeta))}{(1+\cosh(\zeta)+\sinh(\zeta))^2},$$

$$\Psi_{41} = -\frac{c^2}{3} + \frac{6c^2(\cosh(\zeta)+\sinh(\zeta))}{1+\cosh(\zeta)+\sinh(\zeta)} - \frac{6c^2(\cosh(2\zeta)+\sinh(2\zeta))}{(1+\cosh(\zeta)+\sinh(\zeta))^2}, \tag{8.27}$$

where $\zeta = cx + \dfrac{\gamma t^\alpha}{\Gamma(\alpha+1)}.$

Case V:

$$c = c, k = -\frac{c^2}{2}, r = \frac{c^2(-4+c^2)}{4}, a_0 = 0, a_1 = 6\sqrt{2}c^2, a_2 = -6\sqrt{2}c^2, b_0 = 0,$$

$$b_1 = 6c^2, b_2 = -6c^2.$$

For **Case V**, we have the following solution:

$$\Phi_{51} = \frac{6\sqrt{2}c^2(\cosh(\zeta)+\sinh(\zeta))}{1+\cosh(\zeta)+\sinh(\zeta)} - \frac{6\sqrt{2}c^2(\cosh(2\zeta)+\sinh(2\zeta))}{(1+\cosh(\zeta)+\sinh(\zeta))^2},$$

$$\Psi_{51} = \frac{6c^2(\cosh(\zeta)+\sinh(\zeta))}{1+\cosh(\zeta)+\sinh(\zeta)} - \frac{6c^2(\cosh(2\zeta)+\sinh(2\zeta))}{(1+\cosh(\zeta)+\sinh(\zeta))^2}, \tag{8.28}$$

where $\zeta = cx + \dfrac{\gamma t^\alpha}{\Gamma(\alpha+1)}.$

Case VI:

$$c = c, k = -\frac{c^2}{2}, r = \frac{c^2(8+3c^2)}{12}, a_0 = -\sqrt{2}c^2, a_1 = 6\sqrt{2}c^2, a_2 = -6\sqrt{2}c^2, b_0 = -\frac{2c^2}{3},$$

$$b_1 = 6c^2, b_2 = -6c^2.$$

For **Case VI**, we have the following solution:

$$\Phi_{61} = -\sqrt{2}c^2 + \frac{6\sqrt{2}c^2(\cosh(\zeta)+\sinh(\zeta))}{1+\cosh(\zeta)+\sinh(\zeta)} - \frac{6\sqrt{2}c^2(\cosh(2\zeta)+\sinh(2\zeta))}{(1+\cosh(\zeta)+\sinh(\zeta))^2},$$

$$\Psi_{61} = -\frac{2c^2}{3} + \frac{6c^2(\cosh(\zeta)+\sinh(\zeta))}{1+\cosh(\zeta)+\sinh(\zeta)} - \frac{6c^2(\cosh(2\zeta)+\sinh(2\zeta))}{(1+\cosh(\zeta)+\sinh(\zeta))^2}, \tag{8.29}$$

where $\zeta = cx + \dfrac{\gamma t^\alpha}{\Gamma(\alpha+1)}.$

New Exact Solutions of Fractional Differential Equations by Proposed Novel Method **221**

Case VII:

$$c=c, k=\frac{c^2}{2}, r=\frac{c^2(4+c^2)}{4}, a_0=-\sqrt{2}c^2, a_1=6\sqrt{2}c^2, a_2=-6\sqrt{2}c^2, b_0=-c^2,$$

$$b_1=6c^2, b_2=-6c^2.$$

For **Case VII**, we have the following solution:

$$\Phi_{71}=-\sqrt{2}c^2+\frac{6\sqrt{2}c^2(\cosh(\zeta)+\sinh(\zeta))}{1+\cosh(\zeta)+\sinh(\zeta)}-\frac{6\sqrt{2}c^2(\cosh(2\zeta)+\sinh(2\zeta))}{(1+\cosh(\zeta)+\sinh(\zeta))^2},$$

$$\Psi_{71}=-c^2+\frac{6c^2(\cosh(\zeta)+\sinh(\zeta))}{1+\cosh(\zeta)+\sinh(\zeta)}-\frac{6c^2(\cosh(2\zeta)+\sinh(2\zeta))}{(1+\cosh(\zeta)+\sinh(\zeta))^2}, \tag{8.30}$$

where $\zeta=cx+\dfrac{\gamma t^\alpha}{\Gamma(\alpha+1)}$.

Case VIII:

$$c=c, k=\frac{c^2}{2}, r=\frac{c^2(-8+3c^2)}{12}, a_0=0, a_1=6\sqrt{2}c^2, a_2=-6\sqrt{2}c^2, b_0=-\frac{c^3}{3},$$

$$b_1=6c^2, b_2=-6c^2.$$

For **Case VIII**, we have the following solution:

$$\Phi_{81}=\frac{6\sqrt{2}c^2(\cosh(\zeta)+\sinh(\zeta))}{1+\cosh(\zeta)+\sinh(\zeta)}-\frac{6\sqrt{2}c^2(\cosh(2\zeta)+\sinh(2\zeta))}{(1+\cosh(\zeta)+\sinh(\zeta))^2},$$

$$\Psi_{81}=-\frac{c^2}{3}+\frac{6c^2(\cosh(\zeta)+\sinh(\zeta))}{1+\cosh(\zeta)+\sinh(\zeta)}-\frac{6c^2(\cosh(2\zeta)+\sinh(2\zeta))}{(1+\cosh(\zeta)+\sinh(\zeta))^2}, \tag{8.31}$$

where $\zeta=cx+\dfrac{\gamma t^\alpha}{\Gamma(\alpha+1)}$.

8.5.1 Numerical Simulations for Time-Fractional Coupled SK Equations

In this section, the eq. (8.24) has been used for presenting the solution graphs for time-fractional SK equations in the case of both classical and fractional orders.

The present section contains the numerical simulations for time-fractional coupled SK equations. We have presented the solution graphs for time-fractional coupled SK equations for both classical and fractional order. Figures 8.7 to 8.10 show the evolution of the solitary wave solutions for eqs. (8.3).

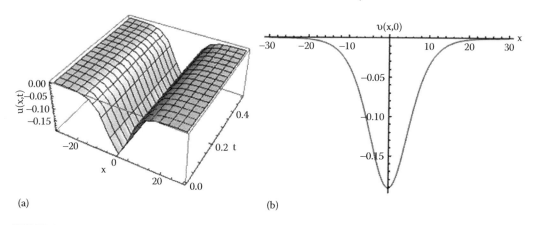

FIGURE 8.7
Case I: For $\alpha = 1$ (classical order). (a) The three-dimensional solitary wave graph for $u(x,t)$ appears in eq. (8.24) as Φ_{11} in **Case 1**, when $c = 0.3$ and $\alpha = 1$, (b) the corresponding two-dimensional graph for $u(x,t)$ when $t = 0$.

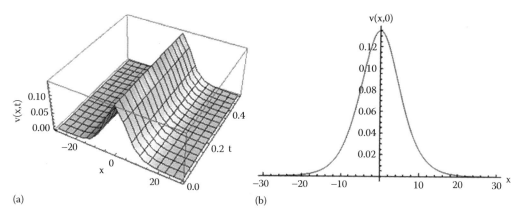

FIGURE 8.8
Case I: For $\alpha = 1$ (classical order). (a) The three-dimensional solitary wave graph for $v(x,t)$ appears in eq. (8.24) as Ψ_{11} in **Case I**, when and $\alpha = 1$, (b) the corresponding two-dimensional graph for $v(x,t)$ when $t = 0$.

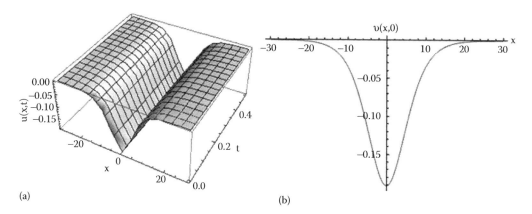

FIGURE 8.9
Case II: For $\alpha = 0.5$ (fractional order). (a) The three-dimensional solitary wave graph for $u(x,t)$ appears in eq. (8.24) as Φ_{11} in **Case I**, when $c = 0.3$ and $\alpha = 0.5$, (b) the corresponding two-dimensional graph for $u(x,t)$ when $t = 0$.

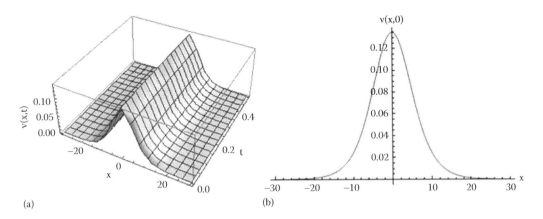

FIGURE 8.10
Case II: For $\alpha = 0.5$ (fractional order). (a) The three-dimensional solitary wave graph for $v(x,t)$ appears in eq. (8.24) as Ψ_{11} in **Case I**, when and $\alpha = 0.5$, (b) the corresponding two-dimensional graph for $v(x,t)$ when $t = 0$.

8.6 Exact Solutions for Time-Fractional Coupled SB Equations

In this section, the newly proposed method has been used here for getting the exact solutions for eqs. (8.4).

By using the fractional complex transform eq. (2.77) of Chapter 2 into eqs. (8.4), we have the following non-linear ODEs:

$$-\left(r\varepsilon + \frac{3}{2}k^2\right)\Phi(\zeta) + \frac{3}{2}c^2\Phi''(\zeta) - \frac{1}{2}\Phi(\zeta)\Psi(\zeta) + i(r\varepsilon + 3kc)\Phi'(\zeta) = 0, \tag{8.32}$$

$$(\gamma^2 - c^2)\Psi''(\zeta) - c^4\Psi^{IV}(\zeta) - c^2(\Psi^2(\zeta))'' - \frac{1}{4}c^2(\Phi^2(\zeta))'' = 0. \tag{8.33}$$

Again, the eq. (8.32) can be written as:

$$-\left(2r\varepsilon + 3k^2\right)\Phi(\zeta) + 3c^2\Phi''(\zeta) - \Phi(\zeta)\Psi(\zeta) = 0, \tag{8.34}$$

where $\gamma = -\dfrac{3kc}{\varepsilon}$.

By integrating eq. (8.33) twice with respect to ζ and putting $\gamma = -\dfrac{3kc}{\varepsilon}$, we have:

$$\left(\left(-\frac{3kc}{\varepsilon}\right)^2 - c^2\right)\Psi(\zeta) - c^4\Psi''(\zeta) - c^2\Psi^2(\zeta) - \frac{1}{4}c^2\Phi^2(\zeta) = 0. \tag{8.35}$$

Let the ansatz be:

$$\Phi(\zeta) = a_0 + \sum_{i=1}^{n} a_i\varphi^i \text{ and } \Psi(\zeta) = b_0 + \sum_{i=1}^{m} b_i\varphi^i. \tag{8.36}$$

The dominant terms with highest order singularity of eq. (8.34) are $\Phi(\zeta)'\Psi(\zeta)$ and $3c^2\Phi''(\zeta)$. The maximum value of pole is 2, that means here $n = 2$. Similarly, the dominant terms with highest order singularity in eq. (8.35) are $c^4\Psi''(\zeta)$ and $c^2\Psi^2(\zeta)$. The maximum value of pole is 2, that means here $m = 2$.

Therefore, by eq. (8.36), we have the following ansatz:

$$\Phi(\zeta) = a_0 + a_1\phi + a_2\phi^2 \text{ and } \Psi(\zeta) = b_0 + b_1\phi + b_2\phi^2, \tag{8.37}$$

where ϕ satisfies eq. (2.80) of Chapter 2.

Substituting eq. (8.37) along with eqs. (2.81) of Chapter 2 into eqs. (8.34) and (8.35), then equating each coefficient of $\varphi^i (i = 0, 1, 2, ...)$ to zero, we can find a system of algebraic equations for $a_0, a_1, a_2, b_0, b_1, b_2, c, k,$ and r as follows:

$$\phi^0 : -a_0b_0 + a_0(3k^2 + 2\varepsilon r) = 0, \; -\frac{1}{4}a_0^2c^2 - b_0^2c^2 + b_0\left(-c^2 + \frac{9c^2k^2}{\varepsilon^2}\right) = 0.$$

$$\phi^1 : -a_1b_0 - a_0b_1 - 3a_1c^2 + a_1(3k^2 + 2\varepsilon r) = 0, \; -\frac{1}{2}a_0a_1c^2 - 2b_0b_1c^2 + c^4b_1 + b_1\left(-c^2 + \frac{9c^2k^2}{\varepsilon^2}\right) = 0.$$

$$\phi^2 : -a_2b_0 - a_1b_1 - a_0b_2 - 9a_1c^2 + 12a_2c^2 + a_2(3k^2 + 2r\varepsilon) = 0,$$

$$-\frac{1}{4}a_1^2c^2 - \frac{1}{2}a_0a_2c^2 - b_1^2c^2 - 2b_0b_2c^2 + 3c^4b_1 - 4c^4b_2 + b_2\left(-c^2 + \frac{9c^2k^2}{\varepsilon^2}\right) = 0. \tag{8.38}$$

$$\phi^3 : -a_2b_1 - a_1b_2 + 6a_1c^2 - 30a_2c^2 = 0, \; -\frac{1}{2}a_1a_2c^2 - 2b_1b_2c^2 - 2c^4b_1 + 10c^4b_2 = 0.$$

$$\phi^4 : -a_2b_2 + 18a_2c^2 = 0, \; -\frac{1}{4}a_2^2c^2 - b_2^2c^2 - 6c^4b_2 = 0.$$

Solving the above algebraic eqs. (8.38), we have the following sets of coefficients for the solutions of eqs. (8.34) and (8.35) as given below:

Case I:

$$c = c, k = -\frac{1}{3}\sqrt{1+c^2}\,\varepsilon, \; r = \frac{-9c^2 - \varepsilon^2 - c^2\varepsilon^2}{6\varepsilon}, \; a_0 = 0, \; a_1 = -24i\sqrt{3}c^2, \; a_2 = 24i\sqrt{3}c^2,$$

$$b_0 = 0, \; b_1 = -18c^2, \; b_2 = 18c^2.$$

For **Case I**, we have the following solutions:

$$\Phi_{11} = -\frac{24i\sqrt{3}c^2(\cosh(\zeta) + \sinh(\zeta))}{1 + \cosh(\zeta) + \sinh(\zeta)} + \frac{24i\sqrt{3}c^2(\cosh(2\zeta) + \sinh(2\zeta))}{(1 + \cosh(\zeta) + \sinh(\zeta))^2},$$

$$\Psi_{11} = -\frac{18c^2(\cosh(\zeta) + \sinh(\zeta))}{1 + \cosh(\zeta) + \sinh(\zeta)} + \frac{18c^2(\cosh(2\zeta) + \sinh(2\zeta))}{(1 + \cosh(\zeta) + \sinh(\zeta))^2}, \tag{8.39}$$

where $\zeta = cx + \dfrac{\gamma t^\alpha}{\Gamma(\alpha + 1)}$.

Case II:

$$c = c, k = -\frac{1}{3}\sqrt{1+c^2}\,\varepsilon, r = \frac{-12c^2 - \varepsilon^2 + c^2\varepsilon^2}{6\varepsilon}, a_0 = 0, a_1 = -24i\sqrt{3}c^2, a_2 = 24i\sqrt{3}c^2,$$

$$b_0 = -c^2, b_1 = -18c^2, b_2 = 18c^2.$$

For **Case II**, we have the following solutions:

$$\Phi_{21} = -\frac{24i\sqrt{3}c^2(\cosh(\zeta) + \sinh(\zeta))}{1 + \cosh(\zeta) + \sinh(\zeta)} + \frac{24i\sqrt{3}c^2(\cosh(2\zeta) + \sinh(2\zeta))}{(1 + \cosh(\zeta) + \sinh(\zeta))^2},$$

$$\Psi_{21} = -c^2 - \frac{18c^2(\cosh(\zeta) + \sinh(\zeta))}{1 + \cosh(\zeta) + \sinh(\zeta)} + \frac{18c^2(\cosh(2\zeta) + \sinh(2\zeta))}{(1 + \cosh(\zeta) + \sinh(\zeta))^2}, \tag{8.40}$$

where $\zeta = cx + \dfrac{\gamma t^\alpha}{\Gamma(\alpha + 1)}$.

Case III:

$$c = c, k = -\frac{1}{3}\sqrt{1+c^2}\,\varepsilon, r = \frac{-9c^2 - \varepsilon^2 - c^2\varepsilon^2}{6\varepsilon}, a_0 = 0, a_1 = 24i\sqrt{3}c^2, a_2 = -24i\sqrt{3}c^2,$$

$$b_0 = 0, b_1 = -18c^2, b_2 = 18c^2.$$

For **Case III**, we have the following solutions:

$$\Phi_{31} = \frac{24i\sqrt{3}c^2(\cosh(\zeta) + \sinh(\zeta))}{1 + \cosh(\zeta) + \sinh(\zeta)} - \frac{24i\sqrt{3}c^2(\cosh(2\zeta) + \sinh(2\zeta))}{(1 + \cosh(\zeta) + \sinh(\zeta))^2},$$

$$\Psi_{31} = -\frac{18c^2(\cosh(\zeta) + \sinh(\zeta))}{1 + \cosh(\zeta) + \sinh(\zeta)} + \frac{18c^2(\cosh(2\zeta) + \sinh(2\zeta))}{(1 + \cosh(\zeta) + \sinh(\zeta))^2}, \tag{8.41}$$

where $\zeta = cx + \dfrac{\gamma t^\alpha}{\Gamma(\alpha + 1)}$.

Case IV:

$$c = c, k = -\frac{1}{3}\sqrt{1-c^2}\,\varepsilon, r = \frac{-12c^2 - \varepsilon^2 + c^2\varepsilon^2}{6\varepsilon}, a_0 = 0, a_1 = 24i\sqrt{3}c^2, a_2 = -24i\sqrt{3}c^2,$$

$$b_0 = -c^2, b_1 = -18c^2, b_2 = 18c^2.$$

For **Case IV**, we have the following solutions:

$$\Phi_{41} = \frac{24i\sqrt{3}c^2(\cosh(\zeta)+\sinh(\zeta))}{1+\cosh(\zeta)+\sinh(\zeta)} - \frac{24i\sqrt{3}c^2(\cosh(2\zeta)+\sinh(2\zeta))}{(1+\cosh(\zeta)+\sinh(\zeta))^2},$$

$$\Psi_{41} = -c^2 - \frac{18c^2(\cosh(\zeta)+\sinh(\zeta))}{1+\cosh(\zeta)+\sinh(\zeta)} + \frac{18c^2(\cosh(2\zeta)+\sinh(2\zeta))}{(1+\cosh(\zeta)+\sinh(\zeta))^2}, \qquad (8.42)$$

where $\zeta = cx + \dfrac{\gamma t^\alpha}{\Gamma(\alpha+1)}$.

Case V:

$$c = c, k = -\frac{1}{3}\sqrt{1+c^2}\,\varepsilon, r = \frac{12c^2 - \varepsilon^2 - c^2\varepsilon^2}{6\varepsilon}, a_0 = 4i\sqrt{3}c^2, a_1 = -24i\sqrt{3}c^2, a_2 = 24i\sqrt{3}c^2,$$

$$b_0 = 4c^2, b_1 = -18c^2, b_2 = 18c^2.$$

For **Case V**, we have the following solutions:

$$\Phi_{51} = 4i\sqrt{3}c^2 - \frac{24i\sqrt{3}c^2(\cosh(\zeta)+\sinh(\zeta))}{1+\cosh(\zeta)+\sinh(\zeta)} + \frac{24i\sqrt{3}c^2(\cosh(2\zeta)+\sinh(2\zeta))}{(1+\cosh(\zeta)+\sinh(\zeta))^2},$$

$$\Psi_{51} = 4c^2 - \frac{18c^2(\cosh(\zeta)+\sinh(\zeta))}{1+\cosh(\zeta)+\sinh(\zeta)} + \frac{18c^2(\cosh(2\zeta)+\sinh(2\zeta))}{(1+\cosh(\zeta)+\sinh(\zeta))^2}, \qquad (8.43)$$

where $\zeta = cx + \dfrac{\gamma t^\alpha}{\Gamma(\alpha+1)}$.

Case VI:

$$c = c, k = -\frac{1}{3}\sqrt{1-c^2}\,\varepsilon, r = \frac{9c^2 - \varepsilon^2 - c^2\varepsilon^2}{6\varepsilon},$$

$$a_0 = 4i\sqrt{3}c^2, a_1 = -24i\sqrt{3}c^2, a_2 = 24i\sqrt{3}c^2,$$

$$b_0 = 3c^2, b_1 = -18c^2, b_2 = 18c^2.$$

For **Case VI**, we have the following solutions:

$$\Phi_{61} = 4i\sqrt{3}c^2 - \frac{24i\sqrt{3}c^2(\cosh(\zeta)+\sinh(\zeta))}{1+\cosh(\zeta)+\sinh(\zeta)} + \frac{24i\sqrt{3}c^2(\cosh(2\zeta)+\sinh(2\zeta))}{(1+\cosh(\zeta)+\sinh(\zeta))^2},$$

$$\Psi_{61} = 3c^2 - \frac{18c^2(\cosh(\zeta)+\sinh(\zeta))}{1+\cosh(\zeta)+\sinh(\zeta)} + \frac{18c^2(\cosh(2\zeta)+\sinh(2\zeta))}{(1+\cosh(\zeta)+\sinh(\zeta))^2}, \qquad (8.44)$$

where $\zeta = cx + \dfrac{\gamma t^\alpha}{\Gamma(\alpha+1)}$.

New Exact Solutions of Fractional Differential Equations by Proposed Novel Method   **227**

Case VII:

$$c = c, k = -\frac{1}{3}\sqrt{1+c^2}\varepsilon, r = \frac{12c^2 - \varepsilon^2 - c^2\varepsilon^2}{6\varepsilon}, a_0 = -4i\sqrt{3}c^2, a_1 = 24i\sqrt{3}c^2, a_2 = -24i\sqrt{3}c^2,$$

$$b_0 = 4c^2, b_1 = -18c^2, b_2 = 18c^2.$$

For **Case VII**, we have the following solutions:

$$\Phi_{71} = -4i\sqrt{3}c^2 + \frac{24i\sqrt{3}c^2(\cosh(\zeta)+\sinh(\zeta))}{1+\cosh(\zeta)+\sinh(\zeta)} - \frac{24i\sqrt{3}c^2(\cosh(2\zeta)+\sinh(2\zeta))}{(1+\cosh(\zeta)+\sinh(\zeta))^2},$$

$$\Psi_{71} = 4c^2 - \frac{18c^2(\cosh(\zeta)+\sinh(\zeta))}{1+\cosh(\zeta)+\sinh(\zeta)} + \frac{18c^2(\cosh(2\zeta)+\sinh(2\zeta))}{(1+\cosh(\zeta)+\sinh(\zeta))^2}, \tag{8.45}$$

where $\zeta = cx + \dfrac{\gamma t^\alpha}{\Gamma(\alpha+1)}$.

Case VIII:

$$c = c, k = -\frac{1}{3}\sqrt{1-c^2}\varepsilon, r = \frac{9c^2 - \varepsilon^2 + c^2\varepsilon^2}{6\varepsilon}, a_0 = -4i\sqrt{3}c^2, a_1 = 24i\sqrt{3}c^2,$$

$$a_2 = -24i\sqrt{3}c^2, b_0 = 3c^2, b_1 = -18c^2, b_2 = 18c^2$$

For **Case VIII**, we have the following solutions:

$$\Phi_{81} = -4i\sqrt{3}c^2 + \frac{24i\sqrt{3}c^2(\cosh(\zeta)+\sinh(\zeta))}{1+\cosh(\zeta)+\sinh(\zeta)} - \frac{24i\sqrt{3}c^2(\cosh(2\zeta)+\sinh(2\zeta))}{(1+\cosh(\zeta)+\sinh(\zeta))^2},$$

$$\Psi_{81} = 3c^2 - \frac{18c^2(\cosh(\zeta)+\sinh(\zeta))}{1+\cosh(\zeta)+\sinh(\zeta)} + \frac{18c^2(\cosh(2\zeta)+\sinh(2\zeta))}{(1+\cosh(\zeta)+\sinh(\zeta))^2}, \tag{8.46}$$

where $\zeta = cx + \dfrac{\gamma t^\alpha}{\Gamma(\alpha+1)}$.

8.6.1 Numerical Simulations for Time-Fractional Coupled SB Equations

In this section, the eq. (8.39) has been used for presenting the solution graphs for time-fractional coupled SB equations in the case of both classical and fractional orders. As the solution obtained in eq. (8.39) for $u(x,t)$ is complex in nature, we have taken here the absolute value of $u(x,t)$ for obtaining the three-dimensional and the corresponding two-dimensional graphs.

The present section contains the numerical simulations for both time-fractional coupled SB equations. We have presented the solution graphs for time-fractional coupled SB equations for both classical and fractional order. Figures 8.11 to 8.14 show the evolution of the solitary wave solutions for eqs. (8.4).

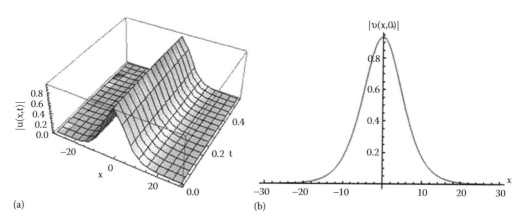

FIGURE 8.11
Case I: For $\alpha = 1$ (classical order). (a) The three-dimensional solitary wave graph for $|u(x,t)|$ appears in eq. (8.39) as Φ_{11} in **Case I**, when $\varepsilon = 0.5$, $c = 0.3$, and $\alpha = 1$, (b) the corresponding two-dimensional graph for $|u(x,t)|$ when $t = 0$.

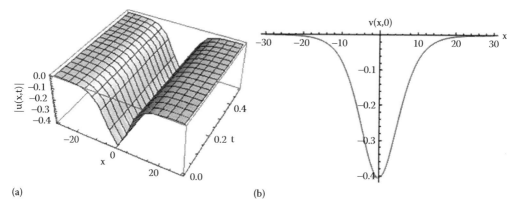

FIGURE 8.12
Case I: For $\alpha = 1$ (classical order). (a) The three-dimensional solitary wave graph for $v(x,t)$ appears in eq. (8.39) as Ψ_{11} in **Case I**, when $c = 0.3$ and $\alpha = 1$, (b) the corresponding two-dimensional solution graph for $v(x,t)$ when $t = 0$.

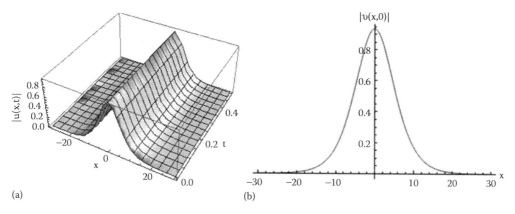

FIGURE 8.13
Case II: For $\alpha = 0.5$ (fractional order). (a) The three-dimensional solitary wave graph for $|u(x,t)|$ appears in eq. (8.39) as Φ_{11} in **Case I**, when $\varepsilon = 0.5$, $c = 0.3$, and $\alpha = 0.5$, (b) the corresponding two-dimensional solution graph for $|u(x,t)|$ when $t = 0$.

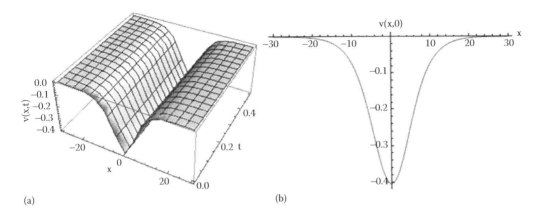

FIGURE 8.14
Case II: For $\alpha = 0.5$ (fractional order). (a) The three-dimensional solitary wave solution graph for $v(x,t)$ appears in eq. (8.39) as Ψ_{11} in **Case I**, when $c = 0.3$ and $\alpha = 0.5$, (b) the corresponding two-dimensional solution graph for $v(x,t)$ when $t = 0$.

8.7 Conclusion

In this chapter, we have presented the new method for exact solutions of the time-fractional KdV–Burgers equation, time-fractional KdV–mKdV equation, time-fractional coupled SK equations, and time-fractional coupled SB equations. We have used fractional complex transform accompanied by the properties of local fractional calculus for transferring the fractional partial differential equations to ordinary differential equations. The most essential interest of the proposed new method is that it takes less computation for obtaining the exact solutions. The obtained exact solutions have also been used here for performing the numerical simulations. From the numerical simulations, we have analyzed the nature of solutions in physical form as solitary waves. From the above analysis, we have concluded that the proposed new method is an effective and powerful technique for the computation of exact traveling wave solutions.

9

New Exact Solutions of Fractional Coupled Differential Equations by Jacobi Elliptic Function Method

9.1 Introduction

Numerous phenomena in physics and applied science have been described effectively by developing the models utilizing fractional calculus, which is also known as the theory of derivatives and integrals of non-integer order. In this way, appearances of fractional order derivatives make the study more fascinating and challenging. Thus, fractional differential equations have gained so much attention for researchers for describing the non-linear phenomena in mathematical physics.

In the past decades, a lot of methods have been developed to construct traveling wave solutions to non-linear fractional evolution equations; out of them, the solutions obtained by the Jacobi elliptic function method have got importance due to its double periodic wave nature.

9.2 Outline of Present Study

In this chapter, the new exact solutions of time-fractional coupled Drinfeld–Sokolov–Wilson equations have been derived by using a new reliable analytical method. We applied a comparatively new method, viz., the Jacobi elliptic function method for finding some new exact solutions of time-fractional coupled Drinfeld–Sokolov–Wilson equations. The fractional complex transform have been used here along with the property of local fractional calculus for reduction of fractional partial differential equations (FPDEs) to ordinary differential equations (ODEs).

Due to vital applications in engineering and physics, the author is influenced to solve the following coupled Drinfeld–Sokolov–Wilson (DSW) equations with fractional order time derivatives, which are given as [299,300]:

$$_0D_t^\alpha u + avv_x = 0, \tag{9.1}$$

$$_0D_t^\alpha v + bv_{xxx} + \gamma uv_x + \varepsilon vu_x = 0, \tag{9.2}$$

where $a, b, \gamma, \varepsilon$ are the non-zero parameters, and the α symbolizes the order of fractional derivative with $0 < \alpha \leq 1$. When $\alpha = 1$, the eqs. (9.1) and (9.2) are reduced to classical

DSW equations, which were firstly introduced by Drinfeld and Sokolov [301,302] and studied by Wilson [303].

The DSW equations arise in shallow water flow models [301–303], when special assumptions are used to simplify the shallow water equations. Many algebraic and numerical solutions of DSW equations have been studied by researchers due to their wide application for describing as a model of water waves. Mainly, Santillana and Dawson [304] have used the Galerkin method to solve DSW equations numerically. He et al. [305] and Xue-Qin and Hong-Yan [306] have used the *F*-expansion method to solve classical DSW equations. The modified simple equation (MSE) method has been used by Khan et al. [307] for getting traveling wave solutions for DSW equations. Conservation laws of DSW equations have been studied by Matjila et al. [308].

Analytical methods [185,249,309–312] are devised to uniformly construct a series of exact solutions for general integrable and non-integrable non-linear evolution equations. Compared with most existing analytical methods, the Jacobi elliptic function method [79–85] not only gives new and more general solutions, but also supplies a tenet to classify the various types of exact solutions with the values of some parameters. Our main objective here is to find new double periodic solutions of time-fractional coupled DSW equations by applying a reliable and relatively new analytical method, viz., the Jacobi elliptic function method by means of complex transform.

9.3 Implementation of Proposed Method for Exact Solutions for Time-Fractional Coupled Drinfeld–Sokolov–Wilson Equations

In this section, the Jacobi elliptic method has been applied for obtaining the exact solutions for time-fractional coupled Drinfeld–Sokolov–Wilson equations.

By using the fractional complex transform eq. (2.83) of Chapter 2 in eqs. (9.1) and (9.2), we have the following non-linear ODEs:

$$c\Phi'(\zeta) + ak\Psi(\zeta)\Psi'(\zeta) = 0, \tag{9.3}$$

$$c\Psi'(\zeta) + bk^3\Psi'''(\zeta) + \gamma k\Phi(\zeta)\Psi'(\zeta) + \varepsilon k\Psi(\zeta)\Phi'(\zeta) = 0. \tag{9.4}$$

By integrating, the eq. (9.3) can be written as:

$$\Phi(\zeta) = -\frac{ak\Psi^2(\zeta)}{2c}. \tag{9.5}$$

By putting eq. (9.5) in eq. (9.4) and integrating once with respect to ζ, we have:

$$2c^2\Psi(\zeta) + 2cbk^3\Psi''(\zeta) - \frac{ak^2(\gamma + 2\varepsilon)\Psi^3(\zeta)}{3} = 0. \tag{9.6}$$

Let

$$\Psi(\zeta) = A_0 + \sum_{i=1}^{N} \left(A_i \phi^i(\zeta) + B_i \phi^{-i}(\zeta) \right). \tag{9.7}$$

New Exact Solutions of FCDEs by Jacobi Elliptic Function Method   233

The dominant terms with highest order singularity of eq. (9.6) are $\Psi''(\zeta)$ and $\Psi^3(\zeta)$. The maximum value of pole is 1, that means here, $N = 1$.

Therefore, by eq. (9.7), we have the following ansatz:

$$\Psi(\zeta) = A_0 + A_1\phi(\zeta) + B_1\phi^{-1}(\zeta), \tag{9.8}$$

where $\phi(\zeta)$ satisfies eq. (2.86) of Chapter 2.

Substituting eq. (9.8) along with eqs. (2.86) of Chapter 2 into eq. (9.6), then equating each coefficient of $\phi^i(i = 0,1,2,\dots)$ to zero, we can find a system of algebraic equations for A_0, A_1, B_1, c and k as follows:

$$\phi^0 : B_1k^2(-12bckr + aB_1^2(\gamma + 2\varepsilon)) = 0,$$

$$\phi^1 : 3aA_0B_1^2k^2(\gamma + 2\varepsilon) = 0,$$

$$\phi^2 : 3B_1(-2c^2 - 2bck^3p + ak^2(A_0^2 + A_1B_1)(\gamma + 2\varepsilon)) = 0, \tag{9.9}$$

$$\phi^3 : A_0(-6c^2 + ak^2(A_0^2 + 6A_1B_1)(\gamma + 2\varepsilon)) = 0, \tag{9.9}$$

$$\phi^4 : 3A_1(-2c^2 + 2bck^3p + ak^2(A_0^2 + 6A_1B_1)(\gamma + 2\varepsilon)) = 0,$$

$$\phi^5 : 3aA_0A_1^2k^2(\gamma + 2\varepsilon) = 0.$$

$$\phi^6 : A_1k^2(-12bckq + aA_1^2(\gamma + 2\varepsilon)) = 0.$$

Solving the above algebraic eqs. (9.9), we have the following sets of coefficients:

Set I:

$$c = -bk^3p,\ k = k,\ A_0 = 0,\ A_1 = -\frac{2\sqrt{3pq}bk^2}{\sqrt{-a(\gamma + 2\varepsilon)}},\ B_1 = 0. \tag{9.10}$$

Set II:

$$c = -bk^3p,\ k = k,\ A_0 = 0,\ A_1 = \frac{2\sqrt{3pq}bk^2}{\sqrt{-a(\gamma + 2\varepsilon)}},\ B_1 = 0. \tag{9.11}$$

Set III:

$$c = -bk^3p,\ k = k,\ A_0 = 0,\ A_1 = 0,\ B_1 = -\frac{2\sqrt{3pr}bk^2}{\sqrt{-a(\gamma + 2\varepsilon)}}. \tag{9.12}$$

Set IV:

$$c = -bk^3p,\ k = k,\ A_0 = 0,\ A_1 = 0,\ B_1 = \frac{2\sqrt{3pr}bk^2}{\sqrt{-a(\gamma + 2\varepsilon)}}. \tag{9.13}$$

Set V:

$$c = -bk^3p - \frac{6M_1}{abkq(\gamma + 2\varepsilon)},\ k = k,\ A_0 = 0,\ A_1 = -\frac{2\sqrt{-3ab^2k^4pq(\gamma + 2\varepsilon) + 6M_1}}{a(\gamma + 2\varepsilon)},$$

$$B_1 = -\frac{2ab^2k^4qr(\gamma+2\varepsilon)\sqrt{-3ab^2k^4pq(\gamma+2\varepsilon)-18M_1}}{a^2b^2k^4q(\gamma+2\varepsilon)^2\sqrt{qr}},$$

where

$$M_1 = \sqrt{a^2b^4k^8q^3r(\gamma+2\varepsilon)^2}. \tag{9.14}$$

Set VI:

$$c = -bk^3p - \frac{6M_1}{abkq(\gamma+2\varepsilon)}, \ k = k, \ A_0 = 0, \ A_1 = \frac{2\sqrt{-3ab^2k^4pq(\gamma+2\varepsilon)+6M_1}}{a(\gamma+2\varepsilon)},$$

$$B_1 = -\frac{2ab^2k^4qr(\gamma+2\varepsilon)\sqrt{-3ab^2k^4pq(\gamma+2\varepsilon)-18M_1}}{a^2b^2k^4q(\gamma+2\varepsilon)^2\sqrt{qr}},$$

where

$$M_1 = \sqrt{a^2b^4k^8q^3r(\gamma+2\varepsilon)^2}. \tag{9.15}$$

Set VII:

$$c = -bk^3p + \frac{6M_1}{abkq(\lambda+2\varepsilon)}, \ k = k, \ A_0 = 0, \ A_1 = -\frac{2\sqrt{-3ab^2k^4pq(\gamma+2\varepsilon)+6M_1}}{a(\gamma+2\varepsilon)},$$

$$B_1 = -\frac{2ab^2k^4qr(\gamma+2\varepsilon)\sqrt{-3ab^2k^4pq(\gamma+2\varepsilon)-18M_1}}{a^2b^2k^4q(\gamma+2\varepsilon)^2\sqrt{qr}},$$

where

$$M_1 = \sqrt{a^2b^4k^8q^3r(\gamma+2\varepsilon)^2}. \tag{9.16}$$

Set VIII:

$$c = -bk^3p + \frac{6M_1}{abkq(\lambda+2\varepsilon)}, \ k = k, \ A_0 = 0, \ A_1 = \frac{2\sqrt{-3ab^2k^4pq(\gamma+2\varepsilon)+6M_1}}{a(\gamma+2\varepsilon)},$$

$$B_1 = -\frac{2ab^2k^4qr(\gamma+2\varepsilon)\sqrt{-3ab^2k^4pq(\gamma+2\varepsilon)-18M_1}}{a^2b^2k^4q(\gamma+2\varepsilon)^2\sqrt{qr}}, \tag{9.17}$$

where

$$M_1 = \sqrt{a^2b^4k^8q^3r(\gamma+2\varepsilon)^2}.$$

A. For **Set I**, we have the following solutions:

$$\Psi_1(\zeta) = -\frac{2\sqrt{3pq}bk^2}{\sqrt{-a(\gamma+2\varepsilon)}}\phi(\zeta),$$

$$\Phi_1(\zeta) = -\frac{ak\Psi_1^2(\zeta)}{2c} = -\frac{6bk^2q}{(\gamma+2\varepsilon)}\phi^2(\zeta). \tag{9.18}$$

For different values of p, q, r, we have the following cases:

Case I:

If $\begin{cases} r=1 \\ p=-(1+m^2), \\ q=m^2 \end{cases}$ then eq. (2.86) of Chapter 2 has solution $\phi(\zeta)=sn\zeta$, then we have

the following Jacobi elliptic function solutions:

$$\Psi_{11}(\zeta) = -\frac{2mbk^2\sqrt{-3-3m^2}}{\sqrt{-a(\gamma+2\varepsilon)}}sn\zeta,$$

$$\Phi_{11}(\zeta) = -\frac{6bk^2m^2}{\gamma+2\varepsilon}(sn\zeta)^2. \tag{9.19}$$

Case II:

If $\begin{cases} r=1-m^2 \\ p=2m^2-1, \\ q=-m^2 \end{cases}$ then eq. (2.86) of Chapter 2 has solution $\phi(\zeta)=cn\zeta$, then we have the

following Jacobi elliptic function solutions:

$$\Psi_{12}(\zeta) = -\frac{2bk^2\sqrt{m^2(3-6m^2)}}{\sqrt{-a(\gamma+2\varepsilon)}}cn\zeta,$$

$$\Phi_{12}(\zeta) = \frac{6bk^2m^2}{\gamma+2\varepsilon}(cn\zeta)^2. \tag{9.20}$$

Case III:

If $\begin{cases} r=m^2 \\ p=-(1+m^2), \\ q=1 \end{cases}$ then eq. (2.86) of Chapter 2 has solution $\phi(\zeta)=ns\zeta=(sn\zeta)^{-1}$, then

we have the following Jacobi elliptic function solutions:

$$\Psi_{13}(\zeta) = -\frac{2bk^2\sqrt{-3-3m^2}}{\sqrt{-a(\gamma+2\varepsilon)}}ns\zeta,$$

$$\Phi_{13}(\zeta) = -\frac{6bk^2}{\gamma+2\varepsilon}(ns\zeta)^2. \tag{9.21}$$

Case IV:

If $\begin{cases} r = -m^2 \\ p = 2m^2 - 1, \\ q = 1 - m^2 \end{cases}$ then eq. (2.86) of Chapter 2 has solution $\phi(\zeta) = nc\zeta = (cn\zeta)^{-1}$, then we

have the following Jacobi elliptic function solutions:

$$\Psi_{14}(\zeta) = -\frac{2bk^2\sqrt{(3-3m^2)(-1+2m^2)}}{\sqrt{-a(\gamma+2\varepsilon)}} nc\zeta.$$

$$\Phi_{14}(\zeta) = \frac{6bk^2(-1+m^2)}{\gamma+2\varepsilon}(nc\zeta)^2. \tag{9.22}$$

Case V:

If $\begin{cases} r = q = \dfrac{1}{4} \\ p = \dfrac{1-2m^2}{2} \end{cases}$, then eq. (2.86) of Chapter 2 has solution $\phi(\zeta) = ns\zeta \pm cs\zeta$, then we

have the following Jacobi elliptic function solutions:

$$\Psi_{15}(\zeta) = -\frac{\sqrt{3-6m^2}\,bk^2}{\sqrt{-2a(\gamma+2\varepsilon)}} ns\zeta \pm cs\zeta,$$

$$\Phi_{15}(\zeta) = -\frac{3bk^2}{2\gamma+4\varepsilon}\left(ns\zeta \pm cs\zeta\right)^2. \tag{9.23}$$

Case VI:

If $\begin{cases} r = q = \dfrac{1-m^2}{4} \\ p = \dfrac{1+m^2}{2} \end{cases}$, then eq. (2.86) of Chapter 2 has solution $\phi(\zeta) = nc\zeta \pm sc\zeta, \dfrac{cn\zeta}{1 \pm sn\zeta}$,

then we have the following Jacobi elliptic function solutions:

$$\Psi_{16}(\zeta) = -\frac{\sqrt{3-3m^4}\,bk^2}{\sqrt{-2a(\gamma+2\varepsilon)}} nc\zeta \pm sc\zeta,$$

$$\Phi_{16}(\zeta) = -\frac{3bk^2(1-m^2)}{2(\gamma+2\varepsilon)}\left(nc\zeta \pm sc\zeta\right)^2. \tag{9.24}$$

New Exact Solutions of FCDEs by Jacobi Elliptic Function Method 237

B. For **Set V**, we have the following solutions:

$$\Psi_5(\zeta) = \frac{2\sqrt{-18M_1 - 3ab^2k^4pq(\gamma + 2\varepsilon)}(ab^2k^4qr(\gamma + 2\varepsilon) - M_1\phi^2(\zeta))}{a(\gamma + 2\varepsilon)M_1\phi(\zeta)}.$$

$$\Phi_5(\zeta) = -\frac{ak\Psi_1^2(\zeta)}{2c} = -\frac{6(ab^2k^4qr(\gamma + 2\varepsilon) - M_1\phi^2(\zeta))^2}{a^2b^3k^6q^2r(\gamma + 2\varepsilon)^3\phi(\zeta)}. \tag{9.25}$$

For different values of p, q, r, we have the following cases:

Case I:

If $\begin{cases} r = 1 \\ p = -(1 + m^2), \\ q = m^2 \end{cases}$ then eq. (2.86) of Chapter 2 has solution $\phi(\zeta) = sn\zeta$, then we have

the following Jacobi elliptic function solutions:

$$\Psi_{51}(\zeta) = \frac{2\sqrt{-18M_2 - 3ab^2k^4m^2(-1 - m^2)(\gamma + 2\varepsilon)}(ab^2k^4m^2(\gamma + 2\varepsilon) - M_2sn^2\zeta)}{a(\gamma + 2\varepsilon)M_2sn\zeta},$$

$$\Phi_{51}(\zeta) = -\frac{6(ab^2k^4m^2(\gamma + 2\varepsilon) - M_2sn^2\zeta)^2}{a^2b^3k^6m^4(\gamma + 2\varepsilon)^3sn^2\zeta}, \tag{9.26}$$

where $0 < m < 1$ and $M_2 = \sqrt{a^2b^4k^8m^6(\gamma + 2\varepsilon)^2}$.

Case II:

If $\begin{cases} r = 1 - m^2 \\ p = 2m^2 - 1, \\ q = -m^2 \end{cases}$ then eq. (2.86) of Chapter 2 has solution $\phi(\zeta) = cn\zeta$, then we have the

following Jacobi elliptic function solutions:

$$\Psi_{52}(\zeta) = \frac{2\sqrt{-18M_3 + 3ab^2k^4m^2(-1 + 2m^2)(\gamma + 2\varepsilon)}(ab^2k^4m^2(-1 + m^2)(\gamma + 2\varepsilon) - M_3cn^2\zeta)}{a(\gamma + 2\varepsilon)M_3cn\zeta},$$

$$\Phi_{52}(\zeta) = -\frac{6(ab^2k^4m^2(-1 + m^2)(\gamma + 2\varepsilon) - M_3cn^2\zeta)^2}{a^2b^3k^6m^4(-1 + m^2)(\gamma + 2\varepsilon)^3cn^2\zeta}, \tag{9.27}$$

where $0 < m < 1$ and $M_3 = \sqrt{a^2b^4k^8m^6(-1 + m^2)(\gamma + 2\varepsilon)^2}$.

Case III:

If $\begin{cases} r = m^2 \\ p = -(1+m^2), \\ q = 1 \end{cases}$ then eq. (2.86) of Chapter 2 has solution $\phi(\zeta) = ns\zeta = (sn\zeta)^{-1}$, then

we have the following Jacobi elliptic function solutions:

$$\Psi_{53}(\zeta) = \frac{2\sqrt{-18M_4 + 3ab^2k^4(-1-m^2)(\gamma+2\varepsilon)}(M_4^4 m - M_4 ns^2\zeta)}{a(\gamma+2\varepsilon)M_4 ns\zeta},$$

$$\Phi_{53}(\zeta) = -\frac{6(M_4^4 m - M_4 ns^2\zeta)^2}{abk^2 m(\gamma+2\varepsilon)M_4^4 ns^2\zeta} \tag{9.28}$$

where $0 < m < 1$ and $M_4 = \sqrt{a^2 b^4 k^8 m^2 (\gamma+2\varepsilon)^2}$.

Case IV:

If $\begin{cases} r = -m^2 \\ p = 2m^2 - 1, \\ q = 1 - m^2 \end{cases}$ then eq. (2.86) of Chapter 2 has solution $\phi(\zeta) = nc\zeta = (cn\zeta)^{-1}$, then we

have the following Jacobi elliptic function solutions:

$$\Psi_{54}(\zeta) = \frac{2\sqrt{-18M_5 + 3ab^2k^4(1-3m^2+2m^4)(\gamma+2\varepsilon)}(ab^2k^4m^2(-1+m^2)(\gamma+2\varepsilon) - M_5 nc^2\zeta)}{a(\gamma+2\varepsilon)M_5 nc\zeta},$$

$$\Phi_{54}(\zeta) = -\frac{6(ab^2k^4m^2(-1+m^2)(\gamma+2\varepsilon) - M_5 nc^2\zeta)^2}{a^2 b^3 k^6 m^4(-1+m^2)^2(\gamma+2\varepsilon)^3 nc^2\zeta}, \tag{9.29}$$

where $0 < m < 1$ and $M_5 = \sqrt{a^2 b^4 k^8 m^2 (-1+m^2)^3 (\gamma+2\varepsilon)^2}$.

Case V:

If $\begin{cases} r = q = \dfrac{1}{4} \\ p = \dfrac{1-2m^2}{2} \end{cases}$, then eq. (2.86) of Chapter 2 has solution $\phi(\zeta) = ns\zeta \pm cs\zeta$, then we

have the following Jacobi elliptic function solutions:

$$\Psi_{55}(\zeta) = \frac{\sqrt{-9M_6 + 3M_6^4(-1+2m^2)}(M_6^4 - M_6(ns\zeta \pm cs\zeta)^2)}{\sqrt{2}a(\gamma+2\varepsilon)M_6(ns\zeta \pm cs\zeta)},$$

$$\Phi_{55}(\zeta) = -\frac{3(M_6^4 - M_6 nc^2\zeta)^2}{2M_6^4 abk^2(\gamma+2\varepsilon)^2 nc^2\zeta}, \tag{9.30}$$

where $M_6 = \sqrt{a^2 b^4 k^8 (\gamma+2\varepsilon)^2}$.

New Exact Solutions of FCDEs by Jacobi Elliptic Function Method

Case VI:

If $\begin{cases} r = q = \dfrac{1-m^2}{4} \\ p = \dfrac{1+m^2}{2} \end{cases}$, then eq. (2.86) of Chapter 2 has solution $\phi(\zeta) = nc\zeta \pm sc\zeta$ or $\dfrac{cn\zeta}{1 \pm sn\zeta}$,

then we have the following Jacobi elliptic function solutions:

$$\Psi_{56}(\zeta) = \frac{\sqrt{-9M_7 + 3ab^2k^4(-1+m^4)(\gamma+2\varepsilon)}(M_7^4 - M_7(nc\zeta \pm sc\zeta)^2)}{\sqrt{2}a(\gamma+2\varepsilon)M_7(nc\zeta \pm sc\zeta)},$$

$$\Phi_{56}(\zeta) = \frac{3(M_7^4 - M_7(nc\zeta \pm sc\zeta)^2)^2}{2a^2b^3k^6m^4(-1+m^2)^3(\gamma+2\varepsilon)^3(nc\zeta \pm sc\zeta)^2} - ab^2k^2m^4(-1+m^2)(\gamma+2\varepsilon)^2, \quad (9.31)$$

where $0 < m < 1$ and $M_7 = \sqrt{a^2b^4k^8(-1+m^2)^4(\gamma+2\varepsilon)^2}$.

NOTE: Similarly, the exact solutions as presented for **Set I** and **Set V**, can be obtained for **Set II**, **Set III**, **Set IV**, **Set VI**, **Set VII**, and **Set VIII**, which are omitted here.

9.4 Numerical Simulations for Time-Fractional Coupled Drinfeld–Sokolov–Wilson Equations

In this present section, we have presented the numerical simulations of time-fractional coupled Drinfeld–Sokolov–Wilson equations by the proposed analytical method. Here, the solutions presented in eqs. (9.19), (9.20), and (9.23) have been used to draw the 3-D and the corresponding 2-D solution graphs for time-fractional coupled Drinfeld–Sokolov–Wilson equations.

9.4.1 Numerical Simulation for Time-Fractional Coupled DSW Equations Based on the Solutions Obtained by Case I of Set I (Figures 9.1–9.6)

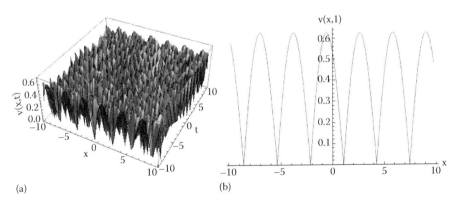

FIGURE 9.1

Case I: For $\alpha = 1$ (classical order). (a) The 3-D double periodic solution graph for $v(x,t)$ appears in eq. (9.19) as Ψ_{11} in **Case I** of **Set I**, when $a = 3$, $b = 2$, $k = 1$, $\gamma = 2$, $\varepsilon = 1$, $m = 0.3$, and $\alpha = 1$, (b) the corresponding 2-D graph for $v(x,t)$, when $t = 1$.

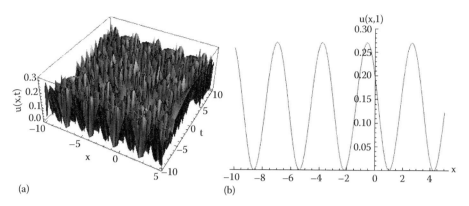

FIGURE 9.2
Case I: For $\alpha = 1$ (classical order). (a) The 3-D double periodic solution graph for $u(x,t)$ appears in eq. (9.19) as Φ_{11} in **Case I** of **Set I**, when $a = 3$, $b = 2$, $k = 1$, $\gamma = 2$, $\varepsilon = 1$, $m = 0.3$, and $\alpha = 1$, (b) the corresponding 2-D graph for $u(x,t)$, when $t = 1$.

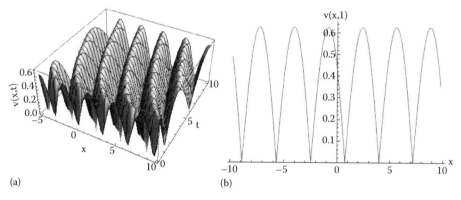

FIGURE 9.3
Case II: For $\alpha = 0.5$ (fractional order). (a) The 3-D double periodic solution graph for $v(x,t)$ appears in eq. (9.19) as Ψ_{11} in **Case I** of **Set I**, when $a = 3$, $b = 2$, $k = 1$, $\gamma = 2$, $\varepsilon = 1$, $m = 0.3$, and $\alpha = 0.5$, (b) the corresponding 2-D graph for $v(x,t)$, when $t = 1$.

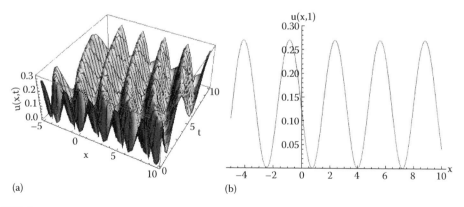

FIGURE 9.4
Case II: For $\alpha = 0.5$ (fractional order). (a) The 3-D double periodic solution graph for $u(x,t)$ appears in eq. (9.19) as Φ_{11} in **Case I** of **Set I**, when $a = 3$, $b = 2$, $k = 1$, $\gamma = 2$, $\varepsilon = 1$, $m = 0.3$, and $\alpha = 0.5$, (b) the corresponding 2-D graph for $u(x,t)$, when $t = 1$.

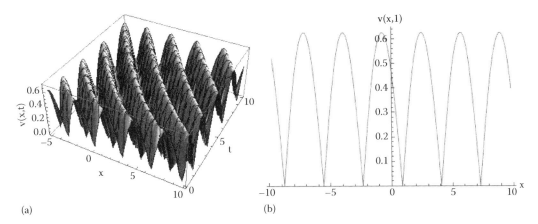

FIGURE 9.5
Case III: For $\alpha = 0.75$ (fractional order). (a) The 3-D double periodic solution graph for $v(x,t)$ appears in eq. (9.19) as Ψ_{11} in **Case I** of **Set I**, when $a = 3$, $b = 2$, $k = 1$, $\gamma = 2$, $\varepsilon = 1$, $m = 0.3$, and $\alpha = 0.75$, (b) the corresponding 2-D graph for $v(x,t)$, when $t = 1$.

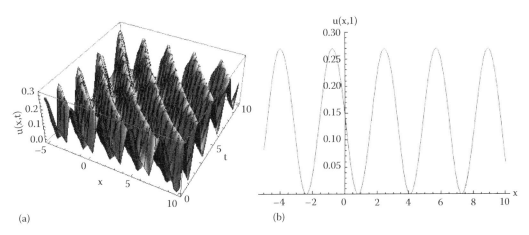

FIGURE 9.6
Case I: For $\alpha = 1$ (classical order). (a) The 3-D double periodic solution graph for $u(x,t)$ appears in eq. (9.19) as Φ_{11} in **Case I** of **Set I**, when $a = 3$, $b = 2$, $k = 1$, $\gamma = 2$, $\varepsilon = 1$, $m = 0.3$, and $\alpha = 0.75$, (b) the corresponding 2-D graph for $u(x,t)$, when $t = 1$.

9.4.2 Numerical Simulation for Time-Fractional Coupled DSW Equations Based on the Solutions Obtained by Case II of Set I (Figures 9.7–9.10)

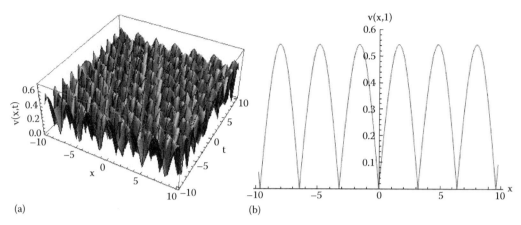

FIGURE 9.7
Case I: For $\alpha = 1$ (classical order). (a) The 3-D double periodic solution graph for $v(x,t)$ appears in eq. (9.20) as Ψ_{12} in **Case II** of **Set I**, when $a = 3, b = 2, k = 1, \gamma = 2, \varepsilon = 1, m = 0.3$, and $\alpha = 1$, (b) the corresponding 2-D graph for $v(x,t)$, when $t = 1$.

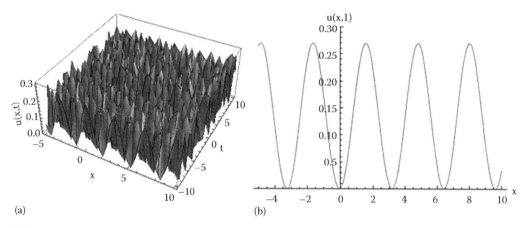

FIGURE 9.8
Case I: For $\alpha = 1$ (classical order). (a) The 3-D double periodic solution graph for $u(x,t)$ appears in eq. (9.20) as Φ_{12} in **Case II** of **Set I**, when $a = 3, b = 2, k = 1, \gamma = 2, \varepsilon = 1, m = 0.3$, and $\alpha = 1$, (b) the corresponding 2-D graph for $u(x,t)$, when $t = 1$.

New Exact Solutions of FCDEs by Jacobi Elliptic Function Method 243

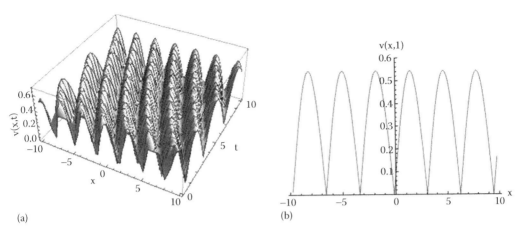

FIGURE 9.9
Case II: For $\alpha = 0.75$ (fractional order). (a) The 3-D double periodic solution graph for $v(x,t)$ appears in eq. (9.20) as Ψ_{12} in **Case II** of **Set I**, when $a = 3, b = 2, k = 1, \gamma = 2, \varepsilon = 1, m = 0.3$, and $\alpha = 0.75$, (b) the corresponding 2-D graph for $v(x,t)$, when $t = 1$.

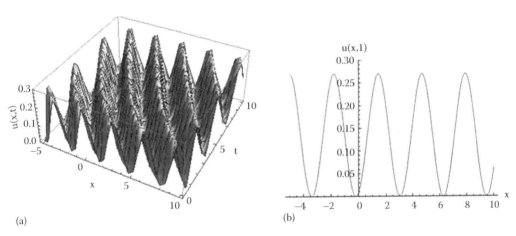

FIGURE 9.10
Case II: For $\alpha = 0.75$ (fractional order). (a) The 3-D double periodic solution graph for $u(x,t)$ appears in eq. (9.20) as Φ_{12} in **Case II** of **Set I**, when $a = 3, b = 2, k = 1, \gamma = 2, \varepsilon = 1, m = 0.3$, and $\alpha = 0.75$, (b) the corresponding 2-D graph for $u(x,t)$, when $t = 1$.

9.4.3 Numerical Simulation for Time-Fractional Coupled DSW Equations Based on the Solutions Obtained by Case V of Set I (Figures 9.11–9.14)

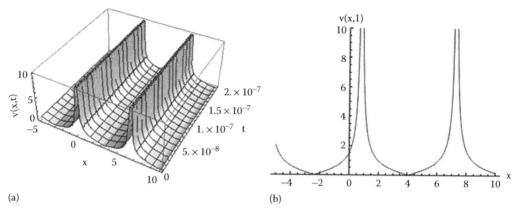

FIGURE 9.11
Case I: For $\alpha = 1$ (classical order). (a) The 3-D double periodic solution graph for $v(x,t)$ appears in eq. (9.23) as Ψ_{15} in **Case V** of **Set I**, when $a = 3, b = 2, k = 1, \gamma = 2, \varepsilon = 1, m = 0.3$, and $\alpha = 1$, (b) the corresponding 2-D graph for $v(x,t)$, when $t = 1$.

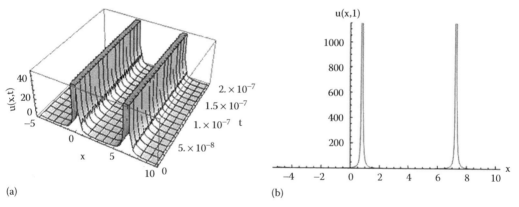

FIGURE 9.12
Case I: For $\alpha = 1$ (classical order). (a) The 3-D double periodic solution graph for $u(x,t)$ appears in eq. (9.23) as Φ_{15} in **Case V** of **Set I**, when $a = 3, b = 2, k = 1, \gamma = 2, \varepsilon = 1, m = 0.3$, and $\alpha = 1$, (b) the corresponding 2-D graph for $u(x,t)$, when $t = 1$.

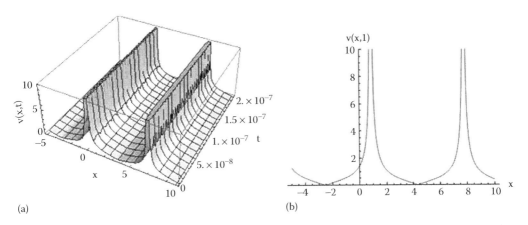

FIGURE 9.13
Case II: For $\alpha = 0.75$ (fractional order). (a) The 3-D double periodic solution graph for $v(x,t)$ appears in eq. (9.23) as Ψ_{15} in **Case V** of **Set I**, when $a = 3, b = 2, k = 1, \gamma = 2, \varepsilon = 1, m = 0.3$, and $\alpha = 0.75$, (b) the corresponding 2-D graph for $v(x,t)$, when $t = 1$.

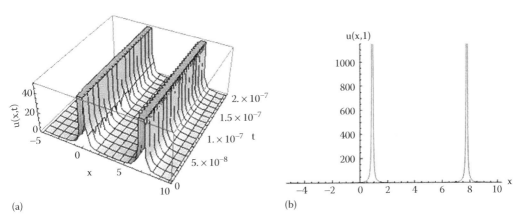

FIGURE 9.14
Case II: For $\alpha = 0.75$ (fractional order). (a) The 3-D double periodic solution graph for $u(x,t)$ appears in eq. (9.23) as Φ_{15} in **Case V** of **Set I**, when $a = 3, b = 2, k = 1, \gamma = 2, \varepsilon = 1, m = 0.3$, and $\alpha = 0.75$, (b) the corresponding 2-D graph for $u(x,t)$, when $t = 1$.

The present section contains the numerical simulations for time-fractional coupled Drinfeld–Sokolov–Wilson equations. We have presented the solution graphs for time-fractional coupled Drinfeld–Sokolov–Wilson equations for both classical and fractional order. Figures 9.1 to 9.14 show the evolution of the double periodic wave solutions for eqs. (9.1) and (9.2).

9.5 Conclusion

In this chapter, we have implemented a new analytical method for getting exact solutions of time-fractional coupled Drinfeld–Sokolov–Wilson equations. We have used here fractional complex transform for transformation of the non-linear fractional differential equations to non-linear ordinary differential equations. The most essential interest of the newly proposed method is that it takes less computation for obtaining the exact solutions. The exact solutions obtained from the proposed method have also been used here for presenting the numerical simulations. From the numerical simulations, we can analyze the nature of solution in physical form as double periodic waves. The proposed method is an effective and powerful technique for handling the fractional non-linear differential equation to obtain the explicit analytical solutions.

10

Formulation and Solutions of Fractional Continuously Variable-Order Mass-Spring Damper Systems

10.1 Introduction

The classical order differential operators play an important role in describing the physical phenomenon of dynamic systems. The differential operator is mainly used for formulating the physical system and analyzing its dynamic behavior. In view of describing the complex behavior of the system, the classical differential operator fails to describe the viscoelastic behavior which lays between the normal integer orders descriptions. The convenient way to analyze the intermediate behavior of the system, the differential operator of variable order is introduced.

By varying the order of the derivative we get the generalized functional order concept, which can be used to analyze the most complex systems in a better way. This concept of functional order is quite new and known as the variable order. Several definitions of variable order have been discussed in [64,313–319]. The application of variable-order differential operator for deformation of viscoelastic materials has been discussed in [320–322]. Due to variable-order operator, having a variable exponent in its kernel, the analytical solutions of variable-order differential equations (VODEs) are difficult to obtain. Although some analytical methods are well suited for constant order fractional differential equations [323,324]. But the numerical solutions associated with variable-order differential equations have been obtained in recent past [325–328].

The fractional calculus [1–3] has been in mathematical science from the end of seventeenth century. However, its interest is only in the past few decades or so, that it has drawn the attention of mainstream science as a way to describe the dynamics of complex phenomena with long term memory, spatial heterogeneity, along with non-stationary statistics described by Das [324], and West [329]. The fractional differential equations examples include in describing the dynamics of visco-elastic materials, turbulence, phase transition, complex networks, di-electric relaxations, control systems, and several other physical phenomena. The fractional generalization of the stress relaxation equation is phenomenological rather than fundamental. The empirical law that stress is proportional to strain for solids was provided by Hooke. For fluids, Newton proposed that stress is proportional to the first (integer order) derivative of strain. Blair et al. [330] suggested that a material with properties intermediate to that of a solid and a fluid, for example, a polymer should be modeled by fractional derivative of strain. Glöckle and Nonnenmacher [331–333] also compared the theoretical results with experimental data

247

sets obtained by stress-strain experiments carried out on poly-isobutylene and natural rubber and found agreement over more than 10 orders of magnitude. Thus, in between behavior between pure Hookian solid and pure Newtonian fluid is termed as viscoelastic material.

Here, in this chapter, the concept of dynamic variable-order fractional differential equations has been presented. The objectives of this chapter are first the mathematical formulation of fractional continuously variable-order spring-mass damping systems, and then analyzing approximate analytical solution of fractional continuous-variable-order models, in which damping is controlled by viscoelastic, viscous-viscoelastic, and viscoelastic-viscous dampers. Due to the dynamic varying nature of fractional order derivative of damper material, it is very difficult to obtain the analytic solutions of the system. The solutions for fractional continuously variable dynamic models have been newly studied in this chapter. The linear damping natures of the systems have been taken here for modeling the problems. The changing property of the guide, on which the motion takes place, results in oscillation of the systems, which are modeled here by fractional continuously variable-order q. The obtained results have been plotted for showing the nature of oscillation, with continuously variable damping order.

10.2 Outline of Present Study

In the past few years, the fractional order physical models [1,2,334,335] have seen much attention by researchers due to dynamic behavior and the viscoelastic behavior of material [336,337]. Thus, the fractional order model is remarkably used for describing the frequency distribution of the structural damping systems [324,338,339]. The fractional differential equations have been often used to model the behavior of dynamic systems. The motion of an N-degree-of-freedom system in viscoelastic material with respect to fractional damping has been studied by Ingman and Suzdalnitsky [340]. In this study, Ingman and Suzdalnitsky have transferred the system into the Volterra integral equation and then applied the iteration method for getting the numerical solution. Several authors have modeled the dynamic system based on fractional calculus. Rossikhin and Shitikova [341] have done analysis on the viscoelastic single-mass system by considering the damped vibration. Enelund and Josefson [342] used the finite element method for analysis of fractionally damped viscoelastic material. The exact solution of fractional order of 1/2 was obtained by Elshehawey et al. [343]. The Green function approach for finding the solution of the dynamic system was studied by Agrawal [344] using the Mittag–Leffler function proposed by Miller [345]. By using the fractional Green function and Laplace transform, Hong et al. [346] have obtained the solution of the single-degree freedom mass-spring system of order $0 < \alpha < 1$. The analytical solution of the fractional systems mass-spring and spring-damper system formed by using the Mittag–Leffler function was analyzed by Gomez-Aguilar et al. [347]. The fractional Maxwell model for viscous-damper model and its analytical solution was proposed by Makris and Constantinous [348] and Choudhury et al. [349]. Recently, the variable-order formation of the dynamic variable-order fractional differential equation has been described by Sahoo et al. [350].

Several analytical and numerical approaches have been used for analyzing the solutions of fractional damper systems. Such methods like Fourier transform [351–353] and

Fractional Continuously Variable-Order Mass-Spring Damper Systems 249

Laplace transform [353–356] have been proposed by researchers to find the solution of fractional damper systems. Recently Saha Ray and Bera [357] used the Adomian decomposition method to determine the analytical solution of dynamic system of order one-half and proclaim that the acquired solutions coincided with the solutions obtained through the eigenvector expansion method given by Suarez and Shokooh [358]. In this study, the Adomian decomposition method has been used by Saha Ray et al. for obtaining a solution for dynamic analysis of a single degree-of-freedom spring-mass-damper system whose damping is described by a fractional derivative of order 1/2. Naber [359] used the Caputo approach to study linear damping system. The generalization of linear oscillator to form the fractional oscillator has been studied by Stanislavsky [360]. The variable-order structure is described in Laplace domain by Das [324]. In the recent past, Saha Ray et al. [361] have presented the analytical solutions for continuous order mass-spring model by using the new recursive approach.

In this chapter, the formation of the variable-order model is established for continuous order fractional model. Also, the definitions and properties of variable-order operators given by many researchers have been reviewed. Then, the variable-order operator has been used to define the new transfer function and analyze the model of a dynamic viscoelastic oscillator. Also, the formulation and a new approach to find analytic solutions for fractional continuously variable-order dynamic models, viz., fractional continuously variable-order mass-spring damper systems have been presented. Here, the viscoelastic, viscous-viscoelastic, and viscoelastic-viscous dampers have been used for describing the damping nature of the oscillating systems, where the order of fractional derivative varies continuously.

Here, the continuous changing nature of fractional order derivative for dynamic systems has been studied for the first time. The solutions of the fractional continuously variable-order mass-spring damper systems have been presented here by using the successive recursive method and also the closed form of the solutions have been obtained. By using graphical plots, the natures of the solutions have been discussed for the different cases of continuously variable fractional order of damping force for oscillator.

10.3 Theory of Variable-Order Fractional Calculus

10.3.1 Variable-Order Differential Operator

In this section, some definitions of the variable-order operators have been presented. The variable-order differential operators are the differential operators whose order of differentiation is a function of the independent variable.

Definition 10.3.1 The generalized linear variable-order Riemann–Liouville integration operator is given as [64,317–319]:

$$
{}^{RL}_0 D_t^{-q(t)} f(t) = \int_0^t \frac{(t-\tau)^{q_e(t,\tau)-1}}{\Gamma(q_g(t,\tau))} f(\tau) d\tau, \, q(t) > 0, \tag{10.1}
$$

where $q_e(t,\tau)$ and $q_g(t,\tau)$ denote the arguments of the q's of the exponent and the gamma function, respectively, which may be different. Some examples of arguments are $q(t,\tau) \to q(t)$, $q(t,\tau) \to q(\tau)$, and $q(t,\tau) \to q(t-\tau)$, yielding nine permutations of the operators, given by:

$$
{}_0^1 D_t^{-q(t)} f(t) = \int_0^t \frac{(t-\tau)^{q(t)-1}}{\Gamma(q(t))} f(\tau) d\tau, \tag{10.2}
$$

$$
{}_0^2 D_t^{-q(t)} f(t) = \int_0^t \frac{(t-\tau)^{q(t)-1}}{\Gamma(q(\tau))} f(\tau) d\tau, \tag{10.3}
$$

$$
{}_0^3 D_t^{-q(t)} f(t) = \int_0^t \frac{(t-\tau)^{q(t)-1}}{\Gamma(q(t-\tau))} f(\tau) d\tau, \tag{10.4}
$$

$$
{}_0^4 D_t^{-q(t)} f(t) = \int_0^t \frac{(t-\tau)^{q(\tau)-1}}{\Gamma(q(t))} f(\tau) d\tau, \tag{10.5}
$$

$$
{}_0^5 D_t^{-q(t)} f(t) = \int_0^t \frac{(t-\tau)^{q(\tau)-1}}{\Gamma(q(\tau))} f(\tau) d\tau, \tag{10.6}
$$

$$
{}_0^6 D_t^{-q(t)} f(t) = \int_0^t \frac{(t-\tau)^{q(\tau)-1}}{\Gamma(q(t-\tau))} f(\tau) d\tau, \tag{10.7}
$$

$$
{}_0^7 D_t^{-q(t)} f(t) = \int_0^t \frac{(t-\tau)^{q(t-\tau)-1}}{\Gamma(q(t))} f(\tau) d\tau, \tag{10.8}
$$

$$
{}_0^8 D_t^{-q(t)} f(t) = \int_0^t \frac{(t-\tau)^{q(t-\tau)-1}}{\Gamma(q(\tau))} f(\tau) d\tau, \tag{10.9}
$$

$$
{}_0^9 D_t^{-q(t)} f(t) = \int_0^t \frac{(t-\tau)^{q(t-\tau)-1}}{\Gamma(q(t-\tau))} f(\tau) d\tau, \tag{10.10}
$$

where the superscript before the derivative symbol indicates the serial number of the operators in the permutations. Based on the behavior of the operator for different function $f(t)$, the q- arguments $q_e(t,\tau)$ and $q_g(t,\tau)$ were set equal, reducing the set of nine permutation to three, we have:

Case I: For $q(t,\tau) \to q(t)$:

$$
{}_0 D_t^{-q(t)} f(t) = \int_0^t \frac{(t-\tau)^{q(t)-1}}{\Gamma(q(t))} f(\tau) d\tau.
$$

Fractional Continuously Variable-Order Mass-Spring Damper Systems 251

Case II: For $q(t,\tau) \to q(\tau)$:

$$_0D_t^{-q(t)}f(t) = \int_0^t \frac{(t-\tau)^{q(\tau)-1}}{\Gamma(q(\tau))}\,f(\tau)\,d\tau.$$

Case III: For $q(t,\tau) \to q(t-\tau)$:

$$_0D_t^{-q(t)}f(t) = \int_0^t \frac{(t-\tau)^{q(t-\tau)-1}}{\Gamma(q(t-\tau))}\,f(\tau)\,d\tau.$$

Among these three cases, the $q(t,\tau) \to q(t-\tau)$ (Case III) is preferred for its adherence to the index rule, that is:

$$_0D_t^{-q(t)}\,_0D_t^{-\alpha(t)}f(t) = \,_0D_t^{-\alpha(t)}\,_0D_t^{-q(t)}f(t), \tag{10.11}$$

with assuming $q(t) > 0$ and $\alpha(t) \geq 0$ are real positive functions, and its full convolution form is:

$$_0D_t^{-q(t)}f(t) = \frac{t^{q(t)-1}}{\Gamma(q(t))} * f(t) = \int_0^t \frac{(t-\tau)^{q(t-\tau)-1}}{\Gamma(q(t-\tau))}\,f(\tau)\,d\tau,\ q(t) > 0. \tag{10.12}$$

This full convolution variable-order (VO) integral definition is preferred by Larenzo and Hartley, as it satisfies the index rule for certain functions and is time invariant [318,319]. The associated Riemann–Liouvillie based on the variable-order differential operator may be arrived at by sequential application of derivatives of order m ($m \in \mathbb{Z}^+$), as follows:

$$D_t^{q(t)}f(t) = D^m J_t^{m-q(t)}f(t) = \frac{d^m}{dt^m}\int_0^t \frac{(t-\tau)^{(m-q(t-\tau)-1)}}{\Gamma(m-q(t-\tau))}\,f(\tau)\,d\tau,\ m-1 < q(t) < m. \tag{10.13}$$

From eq. (10.13), the Riemann–Liouvillie derivative [321,362] can be written as:

$$D_t^{\hat{q}(t)}f(t) = \frac{d}{dt}\left(\int_0^t \frac{(t-\tau)^{\hat{q}(t,\tau)-1}}{\Gamma(\hat{q}(t,\tau))}\,f(\tau)\,d\tau\right), \tag{10.14}$$

where $\hat{q}(t) = 1 - q(t)$ and $0 < \hat{q}(t) < 1$. The above definition eq. (10.14) considered by taking the Riemann–Liouville basis. By considering the different arguments of $\hat{q}(t,\tau)$, the eq. (10.14) has the following three forms:

1. For $\hat{q}(t,\tau) \to \hat{q}(t)$, we have:

$$D_t^{\hat{q}(t)}f(t) = \frac{d}{dt}\left(\int_0^t \frac{(t-\tau)^{\hat{q}(t-1)}}{\Gamma(\hat{q}(t))}\,f(\tau)\,d\tau\right). \tag{10.15}$$

2. For $\hat{q}(t,\tau) \to \hat{q}(\tau)$, we have:

$$D_t^{\hat{q}(t)} f(t) = \frac{d}{dt} \left(\int_0^t \frac{(t-\tau)^{\hat{q}(\tau)-1}}{\Gamma(\hat{q}(\tau))} f(\tau) d\tau \right). \tag{10.16}$$

3. For $\hat{q}(t,\tau) \to \hat{q}(t-\tau)$, we have:

$$D_t^{\hat{q}(t)} f(t) = \frac{d}{dt} \left(\int_0^t \frac{(t-\tau)^{\hat{q}(t-\tau)-1}}{\Gamma(\hat{q}(t-\tau))} f(\tau) d\tau \right). \tag{10.17}$$

The Caputo type variable-order fractional derivative can be written as [327]:

$$D_t^{q(t)} f(t) = \frac{1}{\Gamma(1-q(t))} \int_0^t (t-\tau)^{-q(t)} f'(\tau) d\tau, \, 0 < q(t) < 1, \tag{10.18}$$

and by taking Caputo basis [321,362], we have the following definition:

$$D_t^{\hat{q}(t)} f(t) = \int_0^t \frac{(t-\tau)^{\hat{q}(t,\tau)-1}}{\Gamma(\hat{q}(t,\tau))} f^{(1)}(\tau) d\tau, \tag{10.19}$$

where $\hat{q}(t) = 1 - q(t)$ and $0 < \hat{q}(t) < 1$. By considering the different arguments of $\hat{q}(t,\tau)$, the eq. (10.19) has following three forms:

1. For $\hat{q}(t,\tau) \to \hat{q}(t)$, we have:

$$D_t^{\hat{q}(t)} f(t) = \int_0^t \frac{(t-\tau)^{\hat{q}(t)-1}}{\Gamma(\hat{q}(t))} f^{(1)}(\tau) d\tau. \tag{10.20}$$

2. For $\hat{q}(t,\tau) \to \hat{q}(\tau)$, we have:

$$D_t^{\hat{q}(t)} f(t) = \int_0^t \frac{(t-\tau)^{\hat{q}(\tau)-1}}{\Gamma(\hat{q}(\tau))} f^{(1)}(\tau) d\tau. \tag{10.21}$$

3. For $\hat{q}(t,\tau) \to \hat{q}(t-\tau)$, we have:

$$D_t^{\hat{q}(t)} f(t) = \int_0^t \frac{(t-\tau)^{\hat{q}(t-\tau)-1}}{\Gamma(\hat{q}(t-\tau))} f^{(1)}(\tau) d\tau. \tag{10.22}$$

Here, we discuss some definitions based on variable-order fractional operators. The variable-order Riemann–Liouville differential operator [64,317–319] is written as:

$$D_t^{q(t)} f(t) = \frac{1}{\Gamma(1-q(t))} \frac{d}{dt} \int_0^t \frac{f(\tau) d\tau}{(t-\tau)^{q(t)}}, \, 0 < q(t) < 1. \tag{10.23}$$

Fractional Continuously Variable-Order Mass-Spring Damper Systems

Variable-order operators based on other fractional derivative definition forms have also been proposed by many notable researchers. Based on the definition of the Marchaud fractional derivative, Samko and Ross proposed a variable-order fractional operator, which is discussed in [317] and given as:

$$D_t^{q(t)} f(t) = \frac{f(t)}{\Gamma(1-q(t))(t)^{q(t)}} + \frac{1}{\Gamma(1-q(t))} \int_0^t \frac{f(t)-f(\tau)}{(t-\tau)^{1+q(t)}} d\tau. \tag{10.24}$$

Using Laplace transform, Coimbra [313] derived the variable-order fractional differential operator, which is given as:

$$D_t^{q(t)} f(t) = \frac{1}{\Gamma(1-q(t))} \int_0^t (t-\tau)^{-q(t)} f^{(1)}(\tau) d\tau + \frac{(f(0_+)-f(0_-))t^{-q(t)}}{\Gamma(1-q(t))}, 0 < q(t) < 1. \tag{10.25}$$

The above definitions are defined for real derivative order between 0 and 1. Because all the physical processes are not likely to be in steady state, we can make the value of lower terminals is equal to zero. As the case of VO, no single definition is widely considered as the correct definition. Samko and Ross [317] used eq. (10.24) definition for describing the VO fractional derivative, as the operator retains the symmetry on power functions alike in the case of constant order, viz.:

$$D_{c+}^{q(t)}(t-c)^\alpha = \frac{\Gamma(\alpha+1)}{\Gamma(\alpha+1-q(t))}(t-c)^{\alpha-q(t)}, \tag{10.26}$$

for $0 < \mathrm{Re}\ q(t) < 1$ and $\alpha > -1$.

10.3.2 Analysis of Variable-Order Differential Operator

In the recent past, the researcher formulated variable-order differential operators (VODOs) using a Riemann–Liouville definition. Coimbra proposed a different approach and independently derived VODO from the Laplace transform of the Caputo fractional derivative:

$$D_t^{q(t)} f(t) = \frac{1}{\Gamma(1-q(t))} \int_0^t (t-\tau)^{-q(t)} f^{(1)}(\tau) d\tau + \frac{(f(0_+)-f(0_-))t^{-q(t)}}{\Gamma(1-q(t))}, 0 < q(t) < 1. \tag{10.27}$$

The additional term on definition (10.27) takes into account the behavior of a physical system when it departs from dynamic equilibrium giving a discontinuity at the initial point. The second term in (10.27) gives the correction for that initial point discontinuity or sudden change; which decays to zero slowly while taking derivative. A jump in the value of $f(t)$ at $t=0$ implies a Heaviside distribution, and that gets corrected as power law decay due to second term of (10.27). For a constant function, the definition returns to zero for all $q(t) > 0$. Without the extra term, the definition would return to zero for the derivative of a Heaviside distribution, just as the Caputo fractional derivative. For a constant order function, definition (10.27) reduces to the Caputo fractional derivative only if $f(0^-) = f(0^+)$. The most important characteristic of definition (10.27) is that for any time t, the operator returns the $q(t)$th-order derivative of the function $f(t)$.

In the recent past, Ingman and Suzdalnitsky [314] introduce another VO differintegral operator:

$$D_t^{q(t)} f(t) = \frac{d^m}{dt^m} \int_a^t \frac{(\tau - a)^{m-q(\tau)-1}}{\Gamma(m-q(\tau))} \left[f(t-\tau+a) - \sum_{j=0}^{m-1} \frac{f^{(j)}(a)}{j!})(t-\tau)^j \right] d\tau, \qquad (10.28)$$

where $q(t) > 0$ and $m-1 < q(t) < m$, the operator performs differentiation. If $q(t) < 0$, $m = 0$, then summation is omitted and it performs integration. During integration, the operator is equivalent to Riemann–Liouville full convolution definition proposed by Lorenzo and Hartley [318] and suggested by Ingman et al. [320]. When the operator performs integration, the presence of summation implies that the operator is not of the true Riemann–Liouville form. In fact, for $q(t) \geq 0$, the definition (10.28) is equivalent to:

$$_a D_t^{q(t)} f(t) = \int_a^t \frac{(\tau - a)^{m-q(\tau)-1}}{\Gamma(m-q(\tau))} f^{(m)}(t-\tau+a) d\tau. \qquad (10.29)$$

10.3.3 Laplace Transform of Variable-Order Operator

The derivation of Laplace transform of the variable-order integral can be done using the application of convolution theorem.

$$L\left\{ _0 D_t^{-q(t)} f(t) \right\} = \int_0^\infty e^{-st} \left(\int_a^t \frac{(t-\tau)^{q(t-\tau)-1}}{\Gamma(q(t-\tau))} f(\tau) d\tau \right) dt, \; q(t) > 0, \, t > 0. \qquad (10.30)$$

We know, from convolution theorem, that is:

$$L\left\{ h(t) * g(t) \right\} = H(s)G(s) = L\left(\int_0^t h(\tau)g(t-\tau)d\tau \right),$$

where, Laplace transform of $h(t)$ is $L\{h(t)\} = H(s)$, and for $g(t)$ is $L\{g(t)\} = G(s)$.

Now taking $h(t) = f(t)$ and $g(t) = \frac{t^{q(t)-1}}{\Gamma(q(t))}$, the convolution theorem yields:

$$L\left\{ _0 D_t^{-q(t)} f(t) \right\} = F(s)G(s) = L\{f(t)\} L\left\{ \frac{t^{q(t)-1}}{\Gamma(q(t))} \right\}. \qquad (10.31)$$

When the variable-order integral operator is considered as a dynamic system, block diagram for variable-order operator allows the introduction of new transfer function concept. The conventional transfer function as observed from Figure 10.1, relates to the Laplace transform of the output $y(t)$ to the Laplace transform of input $f(t)$, given by:

$$TF_1 = \frac{L\{y(t)\}}{L\{f(t)\}} = \frac{L\left\{ _0 D_t^{-q(t)} f(t) \right\}}{L\{f(t)\}} = L\left\{ \frac{t^{q(t)-1}}{\Gamma(q(t))} \right\}, \qquad (10.32)$$

according to eq. (10.31).

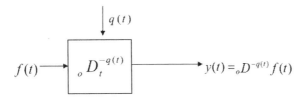

FIGURE 10.1
Block diagram for variable-order integral operator.

Now, if we consider $q(t)$ as input function, then new transfer function may be defined as:

$$TF_2 = \frac{L\{y(t)\}}{L\{q(t)\}}. \tag{10.33}$$

The relation between two transfer function (10.32) and (10.33) is given as follows. Consider $f(t)$ and $q(t)$ to be related by $g(t)$, where:

$$q(t) = \int_0^t f(\tau)g(t-\tau)d\tau.$$

So $L\{q(t)\} = L\{f(t)\}L\{g(t)\}$. (10.34)

The above assumption (10.34) is of very significant nature, meaning that evolution of the variable order is related to input via a memory kernel. This is physically impossible as on today to ascertain.

According to eq. (10.34):

$$TF_2 = \frac{L\{y(t)\}}{L\{q(t)\}} = \frac{L\{f(t)\}TF_1}{L\{f(t)\}L\{g(t)\}}.$$

So
$$\frac{TF_1}{TF_2} = L\{g(t)\}. \tag{10.35}$$

The above formulations need special analytical treatment to get Laplace transforms and its inverse, for power functions of variable exponent with time.

10.4 Analysis of Viscoelastic Oscillator of a Spring Damper System

Consider a force system composed of a mass, spring, and variable viscoelasticity friction element oscillating one-dimensionally on a guide, named as the variable viscoelasticity oscillator. The variable viscoelastic element generates a dependent force that acts on the mass as it traverses the guide. At the end of the guide, the force is purely viscous, for example, order 1 [that is $f(t) \propto D_t^1 x(t)$, meaning force proportional to derivative of displacement] and at the equilibrium position, the force is purely elastic, for example, order 0 [that is force is proportional to displacement $f(t) \propto x(t)$ or $f(t) \propto D^0 x(t)$]. In between the equilibrium positions and the end of the guide, a continuous transition exists between the viscous and elastic regimes. Therefore, in between the ends the visco-elastic force will take a form as $f(t) \propto D_t^{q(x(t))} x(t)$; where $q(x(t))$ is position dependent function varying from zero at one end to one at other end of total travel. In this situation, the nature of the viscoelasticity depends on the position $x(t)$ of the mass.

A mass-spring damper system has a dynamic equation of integer order as:

$$mD^2 x(t) + cD^1 x(t) + kx(t) = F(t), \tag{10.36}$$

where $x(t)$ is the position of mass m, $F(t)$ is the forcing function (external force) on the mass, and c, k are the friction or damping constant and spring constants, respectively. The eq. (10.36) is a simple damped harmonic oscillator; which is balancing (or reacting to) the inertial force $f_1(t) = mD^2 x(t)$ plus the damping (frictional force) $f_2(t) = cD^1 x(t)$ plus the spring force $f_3(t) = kx(t)$ to the acting external force $F(t)$.

By considering the semi differential operator, a simple viscoelastic oscillator can be written as [362,363]:

$$mD^2 x(t) + cD^{\frac{1}{2}} x(t) + kD^0 x(t) = F(t), \tag{10.37}$$

where $D^{\frac{1}{2}} x(t)$ represents the half-derivative of the displacement $x(t)$, with respect to time. A mass-spring damper system has a dynamic equation of variable order as:

$$mD^2 x(t) + cD^{q(x(t))} x(t) + kD^0 x(t) = F_0(t). \tag{10.38}$$

Consider a mass oscillating smoothly and repeatedly on a guide that is covered with a non-uniform viscoelastic film such that there is a continuous variation order of the frictional force mass translates from the viscous to the purely half-order viscoelastic part, for example, from integer order one to the half order of the guide (Figure 10.2). A smooth frictional transition between viscous to viscoelastic portion can be modeled by using the fractional behavior of the guide. At the point when the system experiences pure viscoelastic friction at high speeds and a viscous friction at low-speeds, it is said to be "viscoelastic-viscous", and when the system undergoes pure viscous friction at high speeds and a viscoelastic friction at low speeds, it is said to be "viscous- viscoelastic" [313,363].

It is assumed a smooth transition between viscoelastic and viscous regimes. Under this condition, we can model the fractional behavior as varying continuously from $c_1 D^1 x(t)$ at the viscous portion to $c_{1/2} D^{1/2} x(t)$ at the viscoelastic portion (Figure 10.2). The distance parameter x is defined for $|x| \leq L$.

We make the system dimensionless by changing the dependent variable x so that $\xi = \frac{x}{L}$ and also by taking dimensionless time τ as the independent variable. Additionally, the mass m and constants are taken to be 1. Further, it may be remembered in general $c = c(x(t), t)$.

We can model the above described problem as a constant order (CO) non-linear fractional differential equation (FDE) as described below:

$$D^2 \xi(\tau) + c_0 f(\xi) D^1 \xi(\tau) + c_0 (1 - f(\xi)) D^{\frac{1}{2}} \xi(\tau) + k_0 D^0 \xi(\tau) = F_0^*(\tau), \tag{10.39}$$

Fractional Continuously Variable-Order Mass-Spring Damper Systems

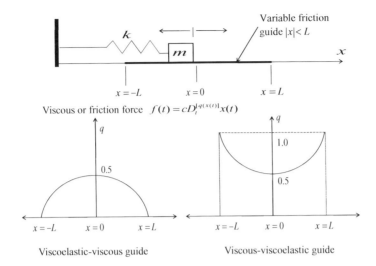

FIGURE 10.2
A mass-spring oscillator sliding on a variable-order guide.

where $f(\xi) = \xi^2$ for the viscous-viscoelastic oscillator and $f(\xi) = 1-\xi^2$ for the viscoelastic-viscous oscillator. c_0 and k_0 are dimensionless constants, also $F_0^*(t) \equiv F_0(t)/m$. Therefore, the (10.39) takes the form as:

$$D^2\xi(\tau) + c_0\xi(\tau)^2 D^1\xi(\tau) + c_0(1-\xi(\tau)^2)D^{\frac{1}{2}}\xi(\tau) + k_0 D^0\xi(\tau) = F_0^*(\tau),$$

$$D^2\xi(\tau) + c_0(1-\xi(\tau)^2)D^1\xi(\tau) + c_0\xi(\tau)^2 D^{\frac{1}{2}}\xi(\tau) + k_0 D^0\xi(\tau) = F_0^*(\tau).$$

However, the equation of motion that describes this mechanism exactly modeled by using the following variable-order differential equation:

$$D^2\xi(\tau) + c_0 D^{q(\xi(\tau))}\xi(\tau) + k_0 D^0\xi(\tau) = F_0^*(\tau), \tag{10.40}$$

where $q(\xi(\tau))$ is a viscoelastic-viscous or viscous-viscoelastic type of operator and defined as:

$$q(\xi(\tau)) = \begin{cases} \dfrac{1+\xi(\tau)^2}{2}, & \text{if viscous-viscoelastic oscillator,} \\ \dfrac{1-\xi(\tau)^2}{2}, & \text{if viscoelastic-viscous oscillator.} \end{cases}$$

It is noted that if the forcing function $F_0^*(t)$ exceeds, some definite amount of the oscillator leaves the domain of the problem for which the solution is not valid. The problem is well defined for $|\xi| < 1$, as no information about the variable viscoelastic force is known beyond the limit of the guide. Eqs. (10.39) and (10.40) can be made variable-order initial value problems by assigning suitable initial conditions. The variable-order initial value problem is well posed for the initial conditions $\xi(0) = 0$ and $D^1\xi(0) = 0$ yielding $D^2\xi(0) = \frac{f_0(0)}{m} = F_0^*(0) = A\delta(\tau)$. These are rest conditions at the start with an impulse force of strength A to excite the system and start the process of oscillation.

The analytical technique to solve the (10.40) is to be developed. However, one can resort to numerical scheme by approximating the variable-order differential structure via the Grünwald–Letnikov (GL) method [323,324]. The basic definition of GL approximation for q a real number is:

$$_aD_t^q f(t) = \lim_{N\to\infty} \frac{\left(\frac{t-a}{N}\right)^{-q}}{\Gamma(-q)} \sum_{j=0}^{N-1} \frac{\Gamma(j-q)}{\Gamma(j+1)} f\left(t-j\left[\frac{t-a}{N}\right]\right).$$

We can use this formulation for structuring $D^{q(\xi(\tau))}\xi(\tau)$ in (10.40), by following [323,324]:

$$_0D_t^{q(\xi(\tau))}\xi(\tau) = \lim_{\substack{N\to\infty\\ \Delta T\to 0}} \sum_{j=0}^{N-1} \Delta T^{-q(\tau-j\Delta T)} \frac{\Gamma(j-q(\tau-j\Delta T))}{\Gamma(-q(\tau-j\Delta T))\Gamma(j+1)} \xi(\tau - j\Delta T),$$

where $\Delta T = (\tau - 0)/N$ is the time step. The operator $D^2\xi(\tau)$ is approximated as:

$$D^2\xi(\tau) = \frac{\xi(\tau) - 2\xi(\tau - \Delta T) + \xi(\tau - 2\Delta T)}{\Delta T^2}.$$

Thus, the approximated (10.40) is:

$$\frac{\xi(\tau) - 2\xi(\tau - \Delta T) + \xi(\tau - 2\Delta T)}{\Delta T^2}$$

$$+ c_0 \left[\lim_{\substack{N\to\infty\\ \Delta T\to 0}} \sum_{j=0}^{N-1} \Delta T^{-q(\tau-j\Delta T)} \frac{\Gamma(j-q(\tau-j\Delta T))}{\Gamma(-q(\tau-j\Delta T))\Gamma(j+1)} \xi(\tau - j\Delta T) \right] + k_0\xi(\tau) = F_0^*(\tau),$$

$F_0^*(\tau) = A$ and zero for all other $\tau > 0$.

At initial point in time, we start with at $\xi = 0$, with value of q as 0.5 (Figure 10.2) with $F_0^*(\tau) = A$ $F_0^*(0) = A$ (an impulse force). We chose a suitable time step ΔT, and find the value of ξ at time $\tau = \Delta T$. As for the case of consideration (viscoelastic-viscous or viscous-viscoelastic), we chose the new value of q from q versus ξ relation, and then repeat the process to find the new value of ξ, (now with $F_0^* = 0$).

10.5 Definition of Continuous Variable Fractional Order Frictional Damping Force

Consider a case of a road section where at the start point, we have a severe oil spill making that region almost frictionless and as the road proceeds it gradually dries up. Thus, we observe in such cases we have a variable frictional system that opposes the motion differently at different locations. The Newtonian Friction force (or pure viscous force), which resists the motion is given as integer-one-order derivative with respect to time of displacement variable. For example, $F_d = -cD_t^1 x(t)$; the negative sign indicating that it is opposing the motion. The viscoelastic elements are those elements which impart

Fractional Continuously Variable-Order Mass-Spring Damper Systems 259

force of opposition that is in between pure spring force $F_s = -cx(t)$ or $F_s = -cD_t^0 x(t)$ and $F_d = -cD_t^1 x(t)$; are given as fractional order derivative $q \in [0,1]$ as $F_d = -cD_t^q x(t)$. We may have a system with viscoelastic fractional order friction where q is fixed say $q = \frac{1}{2}$ or anything between zero and one for $q = 0.5 + 0.5 \tanh(|v|)$. Similarly, we may have a condition where we have a guide of length $2L$; the end point $x = \pm L$ of which are friction-less with $q = 0$, and center of the guide has pure Newtonian viscosity, for example, at $x = 0$ we have $F_d = -cD_t^1 x(t)$; and the fractional order q varies continuously along the guide length as $q = \cos\left(\frac{\pi x(t)}{2L}\right)$. This case we call as "viscoelastic-viscous" damping. We can have "viscoelastic-viscous" damping with end points of travel guide are friction-less, and the center point of the guide having fractional order derivative as half-given as $q = \frac{1}{2} - \frac{1}{2}\left(\frac{x(t)}{L}\right)^2$. We may have a case of variable fractional friction where the end points of the guide present a pure Newtonian viscous friction $F_d = -cD_t^1 x(t)$; and the center point presents a half-order visco-elastic friction, for example, $F_d = -cD_t^{1/2} x(t)$. Here, we express in one way the function $q = \frac{1}{2} + \frac{1}{2}\left(\frac{x(t)}{L}\right)^2$ representing continuously variable frictional force as "viscous-viscoelastic" damping. There may be several types of manifestations in representing the continuous variable order; we here will develop a method to get a solution of these variable-order fractional differential equations, especially for a mass-spring-(variable order) damper system.

10.6 Formulation for Mass-Spring Damper System

The damping on a linear spring-mass system usually occurred due to the dashpot. A dashpot is considered as a mechanical device, which resists motion via viscous friction. As a result, the obtained force, which is also known as resulting force is directly proportional to the velocity and acts to the opposite direction of motion. Due to the force acting in the opposite direction, it absorbs energy and impedes the motion of the system. When the damping force is considered as viscoelastic, it follows the viscous and elastic characteristic to prevent or damp the oscillation of the system. When the system achieves a pure viscous friction at high speed and viscoelastic friction at low speed the damping force is called "viscous-viscoelastic". Similarly, when the system achieves a pure viscoelastic friction at high speed and viscous friction at low speed the damping force is called "viscoelastic-viscous". The damping force is expressed in the form of fractional derivative of position [340,364–368], with damping constant c. In this chapter, the order of fractional derivative is taken as q, which varies continuously with position on the guide where the body moves. This forms a continuously variable-order viscosity (or friction) mass-spring damper system. The damping force (or frictional force) is given by $F_d = -cD_t^q x(t)$, where the order of fractional derivative is varying continuously.

In the current section, we analyze four suitable cases for linear damping of fractional continuously variable-order mass-spring damper system with single-degree freedom, viz.:

Case I: Free oscillation with viscoelastic damping with continuously variable fractional order damping given as $q = 0.5 + 0.5 \tanh(|v|)$.

Case II: Free oscillation with viscous-viscoelastic or viscoelastic-viscous damping with continuously variable fractional order damping given as $q = \frac{1}{2} + \frac{1}{2}\left(\frac{x(t)}{L}\right)^2$.

Case III: Forced oscillation with viscous-viscoelastic with continuously variable fractional order damping given as $q = \frac{1}{2} + \frac{1}{2}\left(\frac{x(t)}{L}\right)^2$.

Case IV: Forced oscillation with viscoelastic-viscous damping with continuously variable fractional order damping given as $q = \cos\left(\frac{\pi x(t)}{2L}\right)$.

10.6.1 Free Oscillation with Viscoelastic Damping in Case I

First, we consider the free oscillation of the fractional continuously variable-order mass-spring-damper system with single-degree freedom. Here, the mass m is displaced from its equilibrium position, and then it vibrates freely without any external force F_0 (Figure 10.3).

For the displaced mass from equilibrium, the system experiences restoring force F_s due to spring constant k, opposing its displacement which is given as:

$$F_s = -kx(t), \qquad (10.41)$$

and a damping force F_d, which is viscoelastic that is described by a fractional continuously variable-order derivative due to viscoelastic of damping coefficient c and given as:

$$F_d = -cD_t^q x(t), \qquad (10.42)$$

where q is a continuously variable fractional order viscoelastic oscillator. By the Newton's second law, due to oscillation, the free body with mass m experiences a total force F_{Net} which is given as:

$$F_{Net} = ma, \qquad (10.43)$$

where a denotes the acceleration while the mass oscillates:

$$F_{Net} = mD_t^2 x(t), \qquad (10.44)$$

where $a = D_t^2 x(t)$.

The total force on the body is given as:

$$F_{Net} = F_s + F_d + F_0, \qquad (10.45)$$

which is equal to:

$$mD_t^2 x(t) = -kx(t) - cD_t^q x(t) + F_0. \qquad (10.46)$$

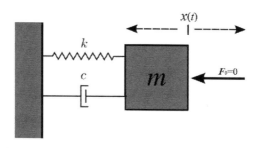

FIGURE 10.3
A mass-spring oscillator under viscoelastic damping when no external force is applied (**Case I**).

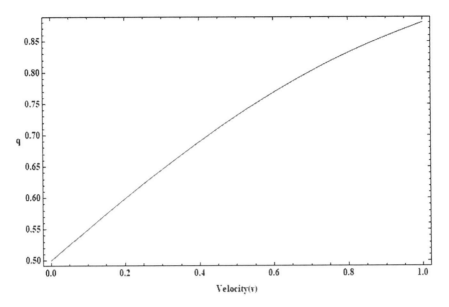

FIGURE 10.4
Figure representing continuously variable viscoelastic oscillator (**Case I**).

We can model the above described problem as a continuous-variable-order linear FDE with viscoelastic oscillator, which is described as:

$$mD_t^2 x(t) + cD_t^q x(t) + kx(t) = F_0, \quad (10.47)$$

where q is a continuously variable fractional order viscoelastic damper, let us assume that the fractional order q is continuously variable order, defined as:

$$q = 0.5 + 0.5\tanh(|v|), \quad (10.48)$$

where v is velocity possessed by the system, for example, $v(t) = \left(\frac{dx(t)}{dt}\right) = x'(t)$ and $0 < v \leq 1$. This continuously variable order is shown in Figure 10.4. In this case, the continuous-variable damper order, for example, q is maintained to be a function of velocity. For demonstration of a continuously variable fractional order damping, we have assumed the function of q varying with velocity in the form (10.48). Therefore, with this definition of q in eq. (10.48) at very low speeds, the order q tends to 0.5 and the equation of oscillator is with half-order damping, that is:

$$mD_t^2 x(t) + cD_t^{0.5} x(t) + kx(t) = F_0.$$

At high speeds $v = 1$, the fractional order is $q = 8807$, and the equation of oscillator is tending toward classical integer order damped oscillator that is described as follows:

$$mD_t^2 x(t) + cD_t^{0.8807} x(t) + kx(t) = F_0, \quad mD_t^2 x(t) + cD_t^1 x(t) + kx(t) = F_0.$$

Therefore, with the oscillation process, the fractional order of viscoelastic damping changes continuously with position, time from value half to almost unity, and that also changes the damping order of the fractional differential equation (10.46).

For free oscillation case, there is no external force, we take $F_0 = 0$. So the governing eq. (10.46) changes to:

$$mD_t^2 x(t) + cD_t^q x(t) + kx(t) = 0. \tag{10.49}$$

The eq. (10.49) can be made a continuously variable-order initial value problem by assigning suitable initial conditions. In this case, the continuous-variable-order initial value problem is well posed for the initial conditions $x(0) = 0$ and $D_t^1 x(t)\big|_{t=0} \equiv D_t^1 x(0) = x'(0) = 1$.

10.6.2 Free Oscillation with Viscous-Viscoelastic or Viscoelastic-Viscous Damping System in Case II

Let us consider a fractional continuously variable-order mass-spring-damper system with a mass m oscillating smoothly and repeatedly about its equilibrium position and vibrating freely without any external force F_0 on a variable "viscous-viscoelastic" or "viscoelastic-viscous" path of length L. Therefore, the generalized damping force that is $F_d = -cD_t^q x(t)$ will be having a continuously variable order q which depends on position $x(t)$ that also depends on where at the present instant within travel length L the system is positioned. Say we formulate the "viscous-viscoelastic" damping by $q = \frac{1}{2} + \frac{1}{2}\left(\frac{x(t)}{L}\right)^2$ and "viscoelastic-viscous" damping by $q = \frac{1}{2} - \frac{1}{2}\left(\frac{x(t)}{L}\right)^2$, given in eq. (10.50).

Thus, we have:

$$q = \begin{cases} \dfrac{1}{2} + \dfrac{1}{2}\left(\dfrac{x(t)}{L}\right)^2, & \text{viscous-viscoelastic damping,} \\[3mm] \dfrac{1}{2} - \dfrac{1}{2}\left(\dfrac{x(t)}{L}\right)^2, & \text{viscoelastic-viscous damping.} \end{cases} \tag{10.50}$$

Here, $x(t)$ denotes the displacement that is defined for a guide as $|x(t)| \le L$. The guide is of the length $2L$, where the continuous variable damping force exists and the motion of the mass is within the guide's limit. The guide length is thus $2L$, where $x(t)$ varies from $x(t) = -L$ to $x(t) = +L$. With q is a "viscoelastic-viscous" or "viscous-viscoelastic" type of friction as defined in expression (10.50); depicted in Figure 10.5.

For the case of "viscous-viscoelastic" damping, at the beginning position, that is $x(t) = \pm L$, the system starts with a pure "viscous" friction (Newtonian type) with order of derivative as $q = 1$; the damping force is Newtonian in nature $F_d = -cx'(t)$. Here, at $x(t) = \pm L$ velocity, for example, $v(\pm L) = x'(\pm L) = 0$, the velocity is zero; whereas at $x(t) = 0$, the velocity is maximum; and damping is "viscoelastic" friction with order of its derivative $q = \frac{1}{2}$ and damping force is $F_d = -cD_t^{1/2} x(t)$, at $x(t) = 0$. For the case of "viscoelastic-viscous" damping at zero speeds, for example, at position $x(t) = \pm L$, the order of derivative is zero, that is there is no friction, damping force is just $F_d = -c(x(t))$, whereas at the maximum speed, for example, at $x(t) = 0$, the order of the damping is half; and damping force is $F_d = -cD_t^{1/2} x(t)$. The Figure 10.5 shows the variation of fractional order for (10.50). In this chapter, we will solve the case with "viscous-viscoelastic" damping with $q = \frac{1}{2} + \frac{1}{2}\left(\frac{x(t)}{L}\right)^2$ and have shown in Figure 10.6.

Instead of solving for the case with $q = \frac{1}{2} - \frac{1}{2}\left(\frac{x(t)}{L}\right)^2$; we modify the order of fractional derivative given as in eq. (10.50) for "visco-elastic viscous" guide and write as $q = \cos\left(\frac{\pi x(t)}{2L}\right)$. This implies that at $x(t) = \pm L$, the guide behaves as a friction-less system, and at $x(t) = 0$ the guide behaves as integer order ($q = 1$), for example, a Newtonian system with classical

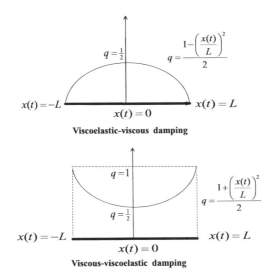

FIGURE 10.5
Viscous-viscoelastic and viscoelastic-viscous oscillators.

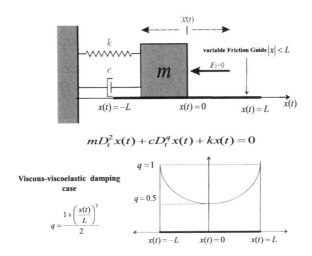

FIGURE 10.6
A mass-spring oscillator under "viscous-viscoelastic" guide when no external force is applied (**Case II**).

frictional damping. In between the travel length, the guide behaves as a fractional order frictional system. In this case, thus, we have a guide where the end points are friction-less and the viscoelasticity (or friction) goes to classical integer order (Newtonian) at the central point of the guide, with continuously variable fractional order $0 \leq q \leq 1$ when $-L \leq x(t) \leq 0$ or $0 \leq x(t) \leq L$. This case we solve and call as Case IV; and have shown in Figure 10.9.

For the displaced mass from equilibrium, the system experiences restoring force F_s due to spring constant k, opposing its displacement which is given as:

$$F_s = -kx(t), \qquad (10.51)$$

and a damping force F_d, which is viscoelastic that is described by a fractional continuously variable-order derivative due to viscoelastic of damping coefficient c and given as:

$$F_d = -cD_t^q x(t), \qquad (10.52)$$

where q is a continuously variable fractional order viscoelastic oscillator. By the Newton's second law, due to oscillation, the free body with mass m experiences a total force F_{Net} which is given as:

$$F_{Net} = ma, \qquad (10.53)$$

where $a = x''(t) = D_t^2 x(t)$ denotes the acceleration while the mass oscillates.

$$F_{Net} = mD_t^2 x(t). \qquad (10.54)$$

The total force on the body is given as:

$$F_{Net} = F_s + F_d + F_0, \qquad (10.55)$$

which is equal to:

$$mD_t^2 x(t) = -kx(t) - cD_t^q x(t) + F_0. \qquad (10.56)$$

We can model the above described problem as a continuous-variable-order linear FDE with viscoelastic oscillator, which is described as:

$$mD_t^2 x(t) + cD_t^q x(t) + kx(t) = F_0, \qquad (10.57)$$

where q is a continuously variable fractional order "viscoelastic" damper, let us assume that the fractional order q is continuously variable order be defined as:

$$q = \frac{1}{2}\left(1 + \left(\frac{x(t)}{L}\right)^2\right), \qquad (10.58)$$

This continuously variable order is shown in Figure 10.6. In this case, the continuous-variable damper order that is q is maintained to be a function of velocity. For demonstration of a continuously variable fractional order damping, we have assumed the function of q varying with velocity in the form (10.58). Therefore, with this definition of q in eq.(10.58) at zero speed, the order q tends to 1 (at $x(t) = \pm L$) and the equation of oscillator is with one (integer)-order Newtonian damping. At maximum speed, for example, at $x(t) = 0$, the fractional order is $q = 0.5$, and the equation of oscillator is tending toward half order damped oscillator that is described as follows:

$$mD_t^2 x(t) + cD_t^{\frac{1}{2}} x(t) + kx(t) = F_0, \, x(t) = 0,$$

$$mD_t^2 x(t) + cD_t^1 x(t) + kx(t) = F_0, \, x(t) = \pm L,$$

$$mD_t^2 x(t) + cD_t^q x(t) + kx(t) = F_0, \, q \in \left(\frac{1}{2}, 1\right), 0 \le |x(t)| \le L. \qquad (10.59)$$

Fractional Continuously Variable-Order Mass-Spring Damper Systems 265

Therefore, with the oscillation process, the fractional order of "viscoelastic" damping changes continuously with position (time) from value half to unity, and that also changes the damping order of the fractional differential equation (10.56). This example we will solve subsequently.

For free oscillation case, there is no external force, we take $F_0 = 0$. So the governing eq. (10.56) changes to:

$$mD_t^2 x(t) + cD_t^q x(t) + kx(t) = 0. \tag{10.60}$$

The problem is well defined for $\left|\frac{x(t)}{L}\right| < 1$ and no information about the viscoelastic force is known beyond the limit of the guide. The eq. (10.60) can be made a continuously variable-order initial value problem by assigning suitable initial conditions. In this case, the continuous-variable-order initial value problem is well posed for the initial conditions $x(0) = 0$ and $D_t^1 x(t)\big|_{t=0} \equiv D_t^1 x(0) = x'(0) = 1$.

10.6.3 Forced Oscillation with Viscous-Viscoelastic or Viscoelastic-Viscous Damping in Case III

Consider a fractional continuously variable-order mass-spring damper system with a mass m oscillating smoothly and repeatedly about its equilibrium position and vibrating freely with external force F_0 on a variable viscous-viscoelastic or viscoelastic-viscous path of length L. Therefore, the generalized damping force that is $F_d = -cD_t^q x(t)$, will be having a continuously variable order q which is depending on position $x(t)$ that is also depending on where at the present instant within travel length L the system is positioned. Say we formulate the viscous-viscoelastic damping by $q = \frac{1}{2}\left[1 + \left(\frac{x(t)}{L}\right)^2\right]$, and viscoelastic-viscous damping by $q = \frac{1}{2}\left[1 - \left(\frac{x(t)}{L}\right)^2\right]$, given in (10.50). Here, $x(t)$ denotes the displacement that is defined for a guide $|x| \leq L$. The guide is the length where the continuous variable damping force exists, and the motion of the mass is within the guide's limit. The guide length is thus $2L$, from $x = -L$ to $x = +L$. Here, q is a viscoelastic-viscous or viscous-viscoelastic type of function and defined in (10.50).

For the case of viscous-viscoelastic damping, at the beginning position, that is $x(t) = \pm L$, the system starts with a pure viscous friction with order of derivative as $q = 1$ and here $v(\pm L) = x'(\pm L) = 0$ the velocity is low; whereas at high speed that is at $x(t) = 0$, the damping is viscoelastic friction with order of its derivative $q = \frac{1}{2}$. For the case of viscoelastic-viscous damping at very low speeds, for example, at position $x(t) = \pm L$, the order of derivative is zero, that is there is no friction, whereas at the high speed at $x(t) = 0$, the order of the damping is half. In this chapter, we will solve the case with viscous-viscoelastic damping (Figure 10.7).

The Figure 10.8 gives the difference in the two cases.

Suppose the body has been acted upon by small impulsive force $F_0(t)$, which is applied externally and is given by:

$$F_0(t) = \delta(t), \tag{10.61}$$

where $\delta(t)$ is defined as $\delta(t) = \lim_{\varepsilon \to 0} \frac{1}{2\varepsilon}$ for $-\varepsilon \leq t \leq \varepsilon$, else $\delta(t) = 0$, which is known as Dirac delta function.

So the equation of motion is:

$$mD_t^2 x(t) + cD_t^q x(t) + kx(t) = \delta(t). \tag{10.62}$$

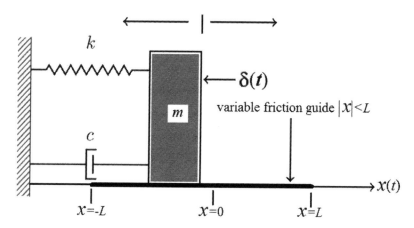

FIGURE 10.7
A mass-spring oscillator sliding on a continuous order guide when external force is applied (**Case III**).

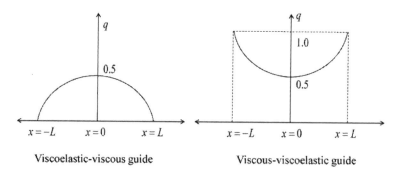

FIGURE 10.8
Viscous-viscoelastic and viscoelastic-viscous oscillators.

The problem is well defined for $\left|\frac{x(t)}{L}\right| < 1$ and no information about the viscoelastic force is known beyond the limit of the guide. The eq. (10.62) can be made a continuous-variable-order initial value problem by assigning suitable initial conditions. The continuous-order initial value problem is well posed for the initial conditions $x(0) = 0$ and $D_t^1 x(t)\big|_{t=0} = 0$. These are equilibrium state at the start with an impulse force $\delta(t)$.

10.6.4 Forced Oscillation with Viscoelastic-Viscous Damping in Case IV

Here, for the "viscoelastic-viscous" guide, we have taken $q = \cos\left(\frac{\pi x(t)}{2L}\right)$. This implies that at $x(t) = \pm L$, the guide behaves as a friction-less system, and at $x(t) = 0$, the guide behaves as an integer order ($q = 1$) system with classical frictional damping. In this case, thus, we have a guide where the end points are friction-less and the visco-elasticity (or friction) goes to classical integer order at the central point of the guide, with continuously variable fractional order $0 \leq q \leq 1$ when $-L \leq x(t) \leq 0$ or $0 \leq x(t) \leq L$. This problem we have solved for Case IV (Figure 10.9).

Fractional Continuously Variable-Order Mass-Spring Damper Systems

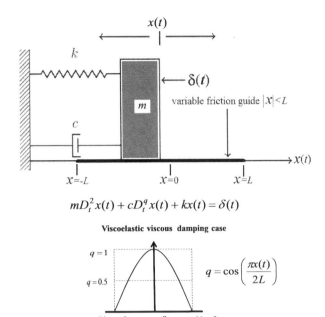

FIGURE 10.9
A mass-spring oscillator sliding on a continuous order guide with viscoelastic damping when external force is applied (**Case IV**).

Let us suppose the body has been acted upon by small impulsive force $F_0(t)$, which is applied externally and is given by:

$$F_0(t) = \delta(t), \tag{10.63}$$

where $\delta(t)$ is defined as $\delta(t) = \lim_{\varepsilon \to 0} \frac{1}{2\varepsilon}$ for $-\varepsilon \leq t \leq \varepsilon$, else $\delta(t) = 0$, which is known as Dirac delta function. So the equation of motion is:

$$mD_t^2 x(t) + cD_t^q x(t) + kx(t) = \delta(t). \tag{10.64}$$

The eq. (10.64) can be made a continuous-variable-order initial value problem by assigning suitable initial conditions. The continuous-order initial value problem is well posed for the initial conditions $x(0) = 0$ and $D_t^1 x(t)\big|_{t=0} = 0$. These are equilibrium state at the start with an impulse force $\delta(t)$.

10.7 Application of Proposed Successive Recursion Method for Solution of Fractional Continuously Variable-Order Mass-Spring Damper System

The present section includes the analytic solutions (obtained via successive recursion method proposed in Section 2.14 of Chapter 2) for fractional continuously variable-order spring-mass damper systems for free oscillation with viscoelastic, viscous-viscoelastic

268   *Generalized Fractional Order Differential Equations Arising in Physical Models*

damping, and force oscillation with viscous-viscoelastic, viscoelastic-viscous damping. The successive recursion method has been implemented here for finding the analytical solutions for proposed systems.

10.7.1 Implementation of Successive Recursion Method for Free Oscillation of Mass-Spring Viscoelastic Damping System in Case I

Consider eq. (10.49) with initial conditions, given as:

$$x(0) = 0 \text{ and } D_t^1 x(t)\big|_{t=0} = x'(0) = 1, \tag{10.65}$$

which tend to the equilibrium states of the proposed dynamic system at the beginning process.

The eq. (10.49) can be written in the following form:

$$D_t^2 x(t) + \frac{c}{m} D_t^q x(t) + \frac{k}{m} x(t) = 0, \tag{10.66}$$

where q is the continuously variable order of a viscous-viscoelastic oscillator, which is defined as in (10.48) that we re-write again as following with discretized $t = n\Delta t$:

$$q = 0.5 + 0.5 \tanh\left(|v|\right)\big|_{t=n\Delta t}. \tag{10.67}$$

Here, the q changes with the small change in time say Δt and velocity, where $n \in \mathbb{N}$ in (10.67). This has been depicted in Figure 10.4. So for each q, we have the new solution, which continuously changes throughout the oscillation period. In the successive recursion method that we will use subsequently, we use $D_t x(0)$, which implies, initial velocity that is:

$$D_t x(t)\big|_{t=0} \equiv D_t^1 x(t)\big|_{t=0} \equiv \tfrac{dx}{dt}\big|_{t=0} \equiv x'(0).$$

By using the successive recursion method eq. (10.66) can be written as:

$$x(t) = x(0) + t D_t x(0) - \frac{k}{m} L^{-1} x(t) - \frac{c}{m} L^{-1} D_t^q x(t). \tag{10.68}$$

Here, the inverse linear operator is taken as $L^{-1} = D_t^{-2}$, that is a two-fold definite integration operation from 0 to t. The eq. (10.68) can be rewritten as:

$$x(t) = x(0) + t D_t x(0) - \frac{k}{m} D_t^{-2} x(t) - \frac{c}{m} D_t^{-2}\left(D_t^q x(t)\right)$$

$$= x(0) + t D_t x(0) - \frac{k}{m} D_t^{-2} x(t) - \frac{c}{m} D_t^{-2+q} x(t). \tag{10.69}$$

By using initial conditions (10.65), we can calculate initially for first recursion q as:

$$q = 0.5 + 0.5 \tanh\left(|v|\right)\big|_{t=0}$$

$$= 0.5 + 0.5 \tanh(1) = 0.8807. \tag{10.70}$$

Fractional Continuously Variable-Order Mass-Spring Damper Systems

10.7.1.1 First Recursion

Therefore eq. (10.69) can be written as:

$$x(t) = x(0) + tD_t x(0) - \frac{k}{m} D_t^{-2} x(t) - \frac{c}{m} D_t^{-1.1192} x(t). \tag{10.71}$$

So by using the successive recursion method, we have the following successive terms as follows:

$$x_0(t) = x(0) + tD_t x(0) = t,$$

$$x_1(t) = -\frac{k}{m} D_t^{-2} x_0(t) - \frac{c}{m} D_t^{-1.1192} x_0(t)$$

$$= -\frac{k}{m} \frac{t^3}{\Gamma(4)} - \frac{c}{m} \frac{t^{2.1192}}{\Gamma(3.1192)},$$

$$x_2(t) = -\frac{k}{m} D_t^{-2} x_1(t) - \frac{c}{m} D_t^{-1.1192} x_1(t)$$

$$= \frac{k^2}{m^2} \frac{t^5}{\Gamma(6)} + \frac{2kc}{m^2} \frac{t^{4.1192}}{\Gamma(5.1192)} + \frac{c^2}{m^2} \frac{t^{3.2384}}{\Gamma(4.2384)},$$

$$x_3(t) = -\frac{k}{m} D_t^{-2} x_2(t) - \frac{c}{m} D_t^{-1.1192} x_2(t)$$

$$= -\frac{k^3}{m^3} \frac{t^7}{\Gamma(8)} - \frac{3k^2 c}{m^3} \frac{t^{6.1192}}{\Gamma(7.1192)} - \frac{3kc^2}{m^3} \frac{t^{5.3284}}{\Gamma(6.2384)} - \frac{c^3}{m^3} \frac{t^{4.3576}}{\Gamma(5.3576)},$$

and so on. So the solution of eq. (10.49) for $q = 0.880797$ is given as:

$$x(t) = x_0(t) + x_1(t) + x_2(t) + x_3(t)$$

$$= t - \frac{k}{m} \frac{t^3}{\Gamma(4)} - \frac{c}{m} \frac{t^{2.1192}}{\Gamma(3.1192)} + \frac{k^2}{m^2} \frac{t^5}{\Gamma(6)} + \frac{2kc}{m^2} \frac{t^{4.1192}}{\Gamma(5.1192)} + \frac{c^2}{m^2} \frac{t^{3.2384}}{\Gamma(4.2384)}. \tag{10.72}$$

$$- \frac{k^3}{m^3} \frac{t^7}{\Gamma(8)} - \frac{3k^2 c}{m^3} \frac{t^{6.1192}}{\Gamma(7.1192)} - \frac{3kc^2}{m^3} \frac{t^{5.2384}}{\Gamma(6.2384)} - \frac{c^3}{m^3} \frac{t^{4.3576}}{\Gamma(5.3576)}$$

10.7.1.2 Second Recursion

By differentiation, we obtain from (10.72) $v(t) = x'(t)$ that is velocity. Taking the values $\frac{k}{m} = \omega_n^2$, $\frac{c}{m} = 2\eta\omega_n^{3/2}$, $\omega_n = 2$, $\eta = 0.5$ in eq. (10.72); this substitution is done to make the equation of motion in similar lines that of a classical damped oscillator equation (Elshehaway et al. [343], Gómez-Aguilar et al. [347], Bagley et al. [354], Saha Ray et al. [357], and Torvik et al. [368]). We have also solved for other values of η as we report in graphical plots in a subsequent section. With this, we have by putting $\Delta t = 0.01$, that is for the next time step $n = 1$, $t = 0.001$, we obtain $x(0.01) = 0.0099267$ and $x'(0.01) = v(0.01) = 0.984444$.

270   *Generalized Fractional Order Differential Equations Arising in Physical Models*

From the expression (10.48) of variable order of viscoelastic element, we obtain the next value of the variable order that is:

$$q = 0.5 + 0.5 \tanh\left(\left|v\right|\right)\Big|_{t=0.01}$$

$$= 0.5 + 0.5 \tanh\left(0.984444\right) = 0.877492.$$

Here, we note that initially at $t = 0$, we started with order of viscoelastic element with fractional order $q = 0.8807$ and in next recursion, for $t = 0.01$ with change in time, we have the new value of fractional order of viscoelastic damper that is $q = 0.877492$. So eq. (10.69) becomes with changed fractional order of damping as described below:

$$x(t) = x(0) + tD_t x(0) - \frac{k}{m} D_t^{-2} x(t) - \frac{c}{m} D_t^{-1.122508} x(t). \tag{10.73}$$

So by using the successive recursion method, we have:

$$x_0(t) = x(0) + tD_t x(0) = t,$$

$$x_1(t) = -\frac{k}{m} D_t^{-2} x_0(t) - \frac{c}{m} D_t^{-1.122508} x_0(t)$$

$$= -\frac{k}{m} \frac{t^3}{\Gamma(4)} - \frac{c}{m} \frac{t^{2.122508}}{\Gamma(3.122508)},$$

$$x_2(t) = -\frac{k}{m} D_t^{-2} x_1(t) - \frac{c}{m} D_t^{-1.122508} x_1(t)$$

$$= \frac{k^2}{m^2} \frac{t^5}{\Gamma(6)} + \frac{2kc}{m^2} \frac{t^{4.122508}}{\Gamma(5.122508)} + \frac{c^2}{m^2} \frac{t^{3.45016}}{\Gamma(4.45016)},$$

$$x_3(t) = -\frac{k}{m} D_t^{-2} x_2(t) - \frac{c}{m} D_t^{-1.122508} x_2(t)$$

$$= -\frac{k^3}{m^3} \frac{t^7}{\Gamma(8)} - \frac{3k^2 c}{m^3} \frac{t^{6.122508}}{\Gamma(7.122508)} - \frac{3kc^2}{m^3} \frac{t^{5.45016}}{\Gamma(6.45016)} - \frac{c^3}{m^3} \frac{t^{4.67524}}{\Gamma(5.67524)},$$

and so on.

So the solution of eq. (10.49) for $q = 0.877492$ is given as:

$$x(t) = x_0(t) + x_1(t) + x_2(t) + x_3(t)$$

$$= t - \frac{k}{m} \frac{t^3}{\Gamma(4)} - \frac{c}{m} \frac{t^{2.122508}}{\Gamma(3.122508)} + \frac{k^2}{m^2} \frac{t^5}{\Gamma(6)} + \frac{2kc}{m^2} \frac{t^{4.122508}}{\Gamma(5.122508)} + \frac{c^2}{m^2} \frac{t^{3.45016}}{\Gamma(4.45016)}. \tag{10.74}$$

$$- \frac{k^3}{m^3} \frac{t^7}{\Gamma(8)} - \frac{3k^2 c}{m^3} \frac{t^{6.122508}}{\Gamma(7.122508)} - \frac{3kc^2}{m^3} \frac{t^{5.45016}}{\Gamma(6.45016)} - \frac{c^3}{m^3} \frac{t^{4.67524}}{\Gamma(5.67524)}$$

Fractional Continuously Variable-Order Mass-Spring Damper Systems

Similarly, by taking ω_n and using eq. (10.67), the fractional order q for next all recursion can be calculated. Here, $\Delta t = 0.01$, for example, the time step is 0.01 second and $n = 2, 3, 4, \ldots$.

By generalizing the solution by the successive recursion method, we have the following, where q is time dependent and can be taken as $q_{n\Delta t}$.

$$x_0(t) = x(0) + t D_t x(0) = t,$$

$$x_1(t) = -\frac{k}{m} D_t^{-2} x_0(t) - \frac{c}{m} D_t^{-2+q_{n\Delta t}} x_0(t)$$

$$= -\frac{k}{m}\frac{t^3}{\Gamma(4)} - \frac{c}{m}\frac{t^{3-q_{n\Delta t}}}{\Gamma(4 - q_{n\Delta t})},$$

$$x_2(t) = -\frac{k}{m} D_t^{-2} x_1(t) - \frac{c}{m} D_t^{-2+q_{n\Delta t}} x_1(t)$$

$$= \frac{k^2}{m^2}\frac{t^5}{\Gamma(6)} + \frac{2kc}{m^2}\frac{t^{5-q_{n\Delta t}}}{\Gamma(6 - q_{n\Delta t})} + \frac{c^2}{m^2}\frac{t^{5-2q_{n\Delta t}}}{\Gamma(6 - 2q_{n\Delta t})},$$

$$x_3(t) = -\frac{k}{m} D_t^{-2} x_2(t) - \frac{c}{m} D_t^{-2+q_{n\Delta t}} x_2(t)$$

$$= -\frac{k^3}{m^3}\frac{t^7}{\Gamma(8)} - \frac{3k^2 c}{m^3}\frac{t^{7-q_{n\Delta t}}}{\Gamma(8 - q_{n\Delta t})} - \frac{3kc^2}{m^3}\frac{t^{7-2q_{n\Delta t}}}{\Gamma(8 - 2q_{n\Delta t})} - \frac{c^3}{m^3}\frac{t^{7-3q_{n\Delta t}}}{\Gamma(8 - 3q_{n\Delta t})},$$

and so on.

Thus, the cumulative solution is therefore:

$$x(t) = x_0(t) + x_1(t) + x_2(t) + x_3(t) + \ldots$$

$$= t - \frac{k}{m}\frac{t^3}{\Gamma(4)} - \frac{c}{m}\frac{t^{3-q_{n\Delta t}}}{\Gamma(4 - q_{n\Delta t})} + \frac{k^2}{m^2}\frac{t^5}{\Gamma(6)} + \frac{2kc}{m^2}\frac{t^{5-q_{n\Delta t}}}{\Gamma(6 - q_{n\Delta t})} + \frac{c^2}{m^2}\frac{t^{5-2q_{n\Delta t}}}{\Gamma(6 - 2q_{n\Delta t})}.$$

$$- \frac{k^3}{m^3}\frac{t^7}{\Gamma(8)} - \frac{3k^2 c}{m^3}\frac{t^{7-q_{n\Delta t}}}{\Gamma(8 - q_{n\Delta t})} - \frac{3kc^2}{m^3}\frac{t^{7-2q_{n\Delta t}}}{\Gamma(8 - 2q_{n\Delta t})} - \frac{c^3}{m^3}\frac{t^{7-3q_{n\Delta t}}}{\Gamma(8 - 3q_{n\Delta t})} + \ldots$$

The value of q is variable and that in this case depends on velocity or $x'(t)$ and also with t given by expression (10.48). With rearrangement in above series solution, we can write the above series in following compact form as following:

$$x(t) = \sum_{r=0}^{\infty}\frac{(-k/m)^r}{r!} t^{2r+1} \sum_{j=0}^{\infty}\frac{(-c/m)^j (j+r)! t^{(2-q_{n\Delta t})j}}{j!\,\Gamma((2 - q_{n\Delta t})j + 2r + 2)}$$

$$= \sum_{r=0}^{\infty}\frac{(-k/m)^r}{r!} t^{2r+1} E_{2-q_{n\Delta t},\,q_{n\Delta t}r+2}^{(r)}\left((-c/m)t^{(2-q_{n\Delta t})}\right), \tag{10.75}$$

$$= \sum_{r=0}^{\infty}\frac{(-\omega_n^2)^r}{r!} t^{2r+1} E_{2-q_{n\Delta t},\,q_{n\Delta t}r+2}^{(r)}\left((-2\eta\omega_n^{3/2})t^{(2-q_{n\Delta t})}\right)$$

where $E_{\lambda,\mu}(.)$ defines the Mittag–Leffler function in two parameter given by:

$$E_{\lambda,\mu}(y) = \sum_{j=0}^{\infty} \frac{y^j}{\Gamma(\lambda j + \mu)} \text{ for } \lambda > 0 \text{ and } \mu > 0,$$

and the r th-derivative of the two parameter Mittag–Leffler function is defined as:

$$E_{\lambda,\mu}^{(r)}(y) = \frac{d^r}{dy^r} E_{\lambda,\mu}(y) = \sum_{j=0}^{\infty} \frac{(j+r)! y^j}{j! \Gamma(\lambda j + \lambda r + \mu)}, (r = 0,1,2,...).$$

The choice of time step of 0.01 is for convenience. Ideally, it should be as small as possible. A smaller value of time step that is 0.01 gives a very large time to obtain the solution in computer and a larger value of time step gives inaccurate results. The idea is to simulate the results for a continuously variable order; and we found the 0.01 time step to be convenient for our 600 steps recursion, which are plotted in the graphs.

10.7.2 Implementation of Successive Recursion Method for Free Oscillation of Mass-Spring Viscous-Viscoelastic Damping System in Case II

Let us consider eq. (10.60) with the initial conditions, presented as:

$$x(0) = 0 \text{ and } D_t^1 x(t)\big|_{t=0} = x'(0) = 1, \tag{10.76}$$

which correspond to the equilibrium states at the beginning of the proposed dynamic process.

Now, the eq. (10.60) can be written as:

$$D_t^2 x(t) + \frac{c}{m} D_t^q x(t) + \frac{k}{m} x(t) = 0, \tag{10.77}$$

where q is the continuously variable order of a viscous-viscoelastic oscillator, defined as in eq. (10.58). The eq. (10.58) can be written as following with discretized $t = n\Delta t$.

$$q = \frac{1 + \left(\dfrac{x(t)}{L}\right)^2 \bigg|_{t=n\Delta t}}{2}. \tag{10.78}$$

Here, the q changes with the small change in time say Δt and displacement, where $n \in N$ in (10.78). This has been depicted in Figure 10.6. So, by eq. (10.78), we will get a new solution for each q, which continuously changes throughout the oscillation period.

By using the successive recursion method in eq. (10.77), we have:

$$x(t) = x(0) + t D_t x(0) - \frac{k}{m} L^{-1} x(t) - \frac{c}{m} L^{-1} D_t^q x(t), \tag{10.79}$$

where L^{-1} denotes the inverse linear operator. For this case, it is taken as $L^{-1} = D_t^{-2}$, that is a two-fold definite integration operation from 0 to t.

Fractional Continuously Variable-Order Mass-Spring Damper Systems 273

So, by replacing $L^{-1} = D_t^{-2}$, the eq. (10.79) can be written as:

$$x(t) = x(0) + tD_t x(0) - \frac{k}{m} D_t^{-2} x(t) - \frac{c}{m} D_t^{-2} \left(D_t^q x(t) \right).$$

$$= x(0) + tD_t x(0) - \frac{k}{m} D_t^{-2} x(t) - \frac{c}{m} D_t^{-2+q} x(t)$$

(10.80)

Then, using initial conditions (10.76) and by taking unit length of regime, for example, $L = 1$, the q can be calculated for initial step, which is given as:

$$q = \frac{1 + \left(\dfrac{x(0)}{L} \right)^2}{2} = 0.5.$$

(10.81)

10.7.2.1 First Recursion

By using the fractional order $q = 0.5$, the eq. (10.80) can be written as:

$$x(t) = x(0) + tD_t x(0) - \frac{k}{m} D_t^{-2} x(t) - \frac{c}{m} D_t^{-1.5} x(t).$$

(10.82)

So by using the successive recursion method, we can get the following successive terms for first recursion, those are given as:

$$x_0(t) = x(0) + tD_t x(0) = t,$$

$$x_1(t) = -\frac{k}{m} D_t^{-2} x_0(t) - \frac{c}{m} D_t^{-1.5} x_0(t)$$

$$= -\frac{k}{m} \frac{t^3}{\Gamma(4)} - \frac{c}{m} \frac{t^{2.5}}{\Gamma(3.5)},$$

$$x_2(t) = -\frac{k}{m} D_t^{-2} x_1(t) - \frac{c}{m} D_t^{-1.5} x_1(t)$$

$$= \frac{k^2}{m^2} \frac{t^5}{\Gamma(6)} + \frac{2kc}{m^2} \frac{t^{4.5}}{\Gamma(5.5)} + \frac{c^2}{m^2} \frac{t^4}{\Gamma(5)},$$

$$x_3(t) = -\frac{k}{m} D_t^{-2} x_2(t) - \frac{c}{m} D_t^{-1.5} x_2(t)$$

$$= -\frac{k^3}{m^3} \frac{t^7}{\Gamma(8)} - \frac{3k^2 c}{m^3} \frac{t^{6.5}}{\Gamma(7.5)} - \frac{3kc^2}{m^3} \frac{t^6}{\Gamma(7)} - \frac{c^3}{m^3} \frac{t^{5.5}}{\Gamma(6.5)},$$

and so on. So the solution of eq. (10.58) for $q(x(t))\big|_{t=0} = 0.5$ is given as:

$$x(t) = x_0(t) + x_1(t) + x_2(t) + x_3(t)$$

$$= t - \frac{k}{m} \frac{t^3}{\Gamma(4)} - \frac{c}{m} \frac{t^{2.5}}{\Gamma(3.5)} + \frac{k^2}{m^2} \frac{t^5}{\Gamma(6)} + \frac{2kc}{m^2} \frac{t^{4.5}}{\Gamma(5.5)} + \frac{c^2}{m^2} \frac{t^4}{\Gamma(5)}.$$

$$- \frac{k^3}{m^3} \frac{t^7}{\Gamma(8)} - \frac{3k^2c}{m^3} \frac{t^{6.5}}{\Gamma(7.5)} - \frac{3kc^2}{m^3} \frac{t^6}{\Gamma(7)} - \frac{c^3}{m^3} \frac{t^{5.5}}{\Gamma(6.5)}$$

(10.83)

10.7.2.2 Second Recursion

By putting $\Delta t = 0.01$, that is for $n = 1$, and taking $m = 1$, $\frac{k}{m} = \omega_n^2$, $\frac{c}{m} = 2\eta\omega_n^{3/2}$, $\omega_n = 2$, and $\eta = 0.5$ in eq. (10.83), we have $x(0.01) = 0.00999$. For other values of η, we have also presented graphical plots in a subsequent section. By taking unit length, for example, $L = 1$ and $x(0.01) = 0.00999$, the q for second recursion can be calculated, which is given as:

$$q = \frac{1 + \left(\dfrac{x(0.01)}{L}\right)^2}{2} = 0.50005.$$

The eq. (10.80) can be written as:

$$x(t) = x(0) + tD_t x(0) + D_t^{-2} \frac{\delta(t)}{m} - \frac{k}{m} D_t^{-2} x(t) - \frac{c}{m} D_t^{-1.49995} x(t).$$

(10.84)

So by using the successive recursion method, we have:

$$x_0(t) = x(0) + tD_t x(0) = t,$$

$$x_1(t) = -\frac{k}{m} D_t^{-2} x_0(t) - \frac{c}{m} D_t^{-1.49995} x_0(t)$$

$$= -\frac{k}{m} \frac{t^3}{\Gamma(4)} - \frac{c}{m} \frac{t^{2.49995}}{\Gamma(3.49995)},$$

$$x_2(t) = -\frac{k}{m} D_t^{-2} x_1(t) - \frac{c}{m} D_t^{-1.49995} x_1(t)$$

$$= \frac{k^2}{m^2} \frac{t^5}{\Gamma(6)} + \frac{2kc}{m^2} \frac{t^{4.49995}}{\Gamma(5.49995)} + \frac{c^2}{m^2} \frac{t^{3.9999}}{\Gamma(4.9999)},$$

$$x_3(t) = -\frac{k}{m} D_t^{-2} x_2(t) - \frac{c}{m} D_t^{-1.49995} x_2(t)$$

$$= -\frac{k^3}{m^3} \frac{t^7}{\Gamma(8)} - \frac{3k^2c}{m^3} \frac{t^{6.49995}}{\Gamma(7.49995)} - \frac{3kc^2}{m^3} \frac{t^{5.9999}}{\Gamma(6.9999)} - \frac{c^3}{m^3} \frac{t^{5.49985}}{\Gamma(6.49985)},$$

Fractional Continuously Variable-Order Mass-Spring Damper Systems 275

and so on. So the solution of eq. (10.60) for $q = 0.50005$ is given as:

$$x(t) = x_0(t) + x_1(t) + x_2(t) + x_3(t)$$

$$= t - \frac{k}{m} \frac{t^3}{\Gamma(4)} - \frac{c}{m} \frac{t^{2.49995}}{\Gamma(3.49995)} + \frac{k^2}{m^2} \frac{t^5}{\Gamma(6)} + \frac{2kc}{m^2} \frac{t^{4.49995}}{\Gamma(5.49995)} + \frac{c^2}{m^2} \frac{t^{3.9999}}{\Gamma(4.9999)} \quad (10.85)$$

$$- \frac{k^3}{m^3} \frac{t^7}{\Gamma(8)} - \frac{3k^2c}{m^3} \frac{t^{6.49995}}{\Gamma(7.49995)} - \frac{3kc^2}{m^3} \frac{t^{5.9999}}{\Gamma(6.9999)} - \frac{c^3}{m^3} \frac{t^{5.49985}}{\Gamma(6.49985)}$$

Similarly, by taking ω_n and using eq. (10.78), the fractional order q for next all recursion can be calculated. Here, $\Delta t = 0.01$, for example, the time step is 0.01 seconds and $n = 2, 3, 4, \dots$.

By generalizing the solution by the successive recursion method, we have the following, where q is time dependent too, calling that $q_{n\Delta t}$.

$$x_0(t) = x(0) + tD_t x(0) = t,$$

$$x_1(t) = -\frac{k}{m} D_t^{-2} x_0(t) - \frac{c}{m} D_t^{-2 + q_{n\Delta t}} x_0(t)$$

$$= -\frac{k}{m} \frac{t^3}{\Gamma(4)} - \frac{c}{m} \frac{t^{3 - q_{n\Delta t}}}{\Gamma(4 - q_{n\Delta t})},$$

$$x_2(t) = -\frac{k}{m} D_t^{-2} x_1(t) - \frac{c}{m} D_t^{-2 + q_{n\Delta t}} x_1(t)$$

$$= \frac{k^2}{m^2} \frac{t^5}{\Gamma(6)} + \frac{2kc}{m^2} \frac{t^{5 - q_{n\Delta t}}}{\Gamma(6 - q_{n\Delta t})} + \frac{c^2}{m^2} \frac{t^{5 - 2q_{n\Delta t}}}{\Gamma(6 - 2q_{n\Delta t})},$$

$$x_3(t) = -\frac{k}{m} D_t^{-2} x_2(t) - \frac{c}{m} D_t^{-2 + q_{n\Delta t}} x_2(t)$$

$$= -\frac{k^3}{m^3} \frac{t^7}{\Gamma(8)} - \frac{3k^2c}{m^3} \frac{t^{7 - q_{n\Delta t}}}{\Gamma(8 - q_{n\Delta t})} - \frac{3kc^2}{m^3} \frac{t^{7 - 2q_{n\Delta t}}}{\Gamma(8 - 2q_{n\Delta t})} - \frac{c^3}{m^3} \frac{t^{7 - 3q_{n\Delta t}}}{\Gamma(8 - 3q_{n\Delta t})},$$

and so on.

Thus, the cumulative solution is therefore:

$$x(t) = x_0(t) + x_1(t) + x_2(t) + x_3(t) + \dots$$

$$= t - \frac{k}{m} \frac{t^3}{\Gamma(4)} - \frac{c}{m} \frac{t^{3 - q_{n\Delta t}}}{\Gamma(4 - q_{n\Delta t})} + \frac{k^2}{m^2} \frac{t^5}{\Gamma(6)} + \frac{2kc}{m^2} \frac{t^{5 - q_{n\Delta t}}}{\Gamma(6 - q_{n\Delta t})} + \frac{c^2}{m^2} \frac{t^{5 - 2q_{n\Delta t}}}{\Gamma(6 - 2q_{n\Delta t})}.$$

$$- \frac{k^3}{m^3} \frac{t^7}{\Gamma(8)} - \frac{3k^2c}{m^3} \frac{t^{7 - q_{n\Delta t}}}{\Gamma(8 - q_{n\Delta t})} - \frac{3kc^2}{m^3} \frac{t^{7 - 2q_{n\Delta t}}}{\Gamma(8 - 2q_{n\Delta t})} - \frac{c^3}{m^3} \frac{t^{7 - 3q_{n\Delta t}}}{\Gamma(8 - 3q_{n\Delta t})} + \dots$$

The value of q is variable and that in this case depends on velocity or $x'(t)$ and also with t given by expression (10.57). With rearrangement in above series solution, we can write the above series in the following compact form as following:

$$x(t) = \sum_{r=0}^{\infty} \frac{(-k/m)^r}{r!} t^{2r+1} \sum_{j=0}^{\infty} \frac{(-c/m)^j (j+r)! t^{(2-q_{n\Delta t})j}}{j! \Gamma((2-q_{n\Delta t})j + 2r + 2)}$$

$$= \sum_{r=0}^{\infty} \frac{(-k/m)^r}{r!} t^{2r+1} E^{(r)}_{2-q_{n\Delta t}, q_{n\Delta t}r+2} \left((-c/m) t^{(2-q_{n\Delta t})} \right), \qquad (10.86)$$

$$= \sum_{r=0}^{\infty} \frac{(-\omega_n^2)^r}{r!} t^{2r+1} E^{(r)}_{2-q_{n\Delta t}, q_{n\Delta t}r+2} \left((-2\eta\omega_n^{3/2}) t^{(2-q_{n\Delta t})} \right)$$

where $E_{\lambda,\mu}(.)$ defines the Mittag–Leffler function in two parameter given by:

$$E_{\lambda,\mu}(y) = \sum_{j=0}^{\infty} \frac{y^j}{\Gamma(\lambda j + \mu)} \text{ for } \lambda > 0 \text{ and } \mu > 0,$$

and the r th-derivative of the two parameter Mittag–Leffler function is defined as:

$$E^{(r)}_{\lambda,\mu}(y) = \frac{d^r}{dy^r} E_{\lambda,\mu}(y) = \sum_{j=0}^{\infty} \frac{(j+r)! y^j}{j! \Gamma(\lambda j + \lambda r + \mu)}, (r = 0, 1, 2, \ldots).$$

The choice of time step of 0.01 is for convenience. Ideally, it should be as small as possible. A smaller value of time step that is 0.01, gives a very large time to obtain the solution in computer; and a larger value of time step gives inaccurate results. The idea is to simulate the results for a continuously variable order; and we found the 0.01 time step to be convenient for our 300 steps recursion, which are plotted in the graphs.

10.7.3 Application of Successive Recursion Method for Forced Oscillation of Spring-Mass Viscous-Viscoelastic Damping System in Case III

Let us consider eq. (10.62) with the homogenous initial conditions:

$$x(0) = 0 \text{ and } D_t^1 x(t)\big|_{t=0} = 0, \qquad (10.87)$$

which is at the equilibrium states of the dynamic system at the beginning process. The eq. (10.62) can be written in the following form:

$$D_t^2 x(t) + \frac{c}{m} D_t^q x(t) + \frac{k}{m} x(t) = \frac{\delta(t)}{m}. \qquad (10.88)$$

Fractional Continuously Variable-Order Mass-Spring Damper Systems 277

Here, q is the fractional continuously variable-order viscous-viscoelastic oscillator, which is defined as:

$$q = \frac{1 + \left(\dfrac{x(t)}{L}\right)^2\Big|_{t=n\Delta t}}{2}. \tag{10.89}$$

Here, the q changes with the small change in time say Δt and $n \in N$ in (10.89). So for each $t = n\Delta t$, we have the new fractional differential equation and its solution, which continuously changes throughout the period of oscillation. By using the successive recursion method, eq. (10.88) can be written as:

$$x(t) = x(0) + tD_t x(0) + L^{-1}\frac{\delta(t)}{m} - \frac{k}{m}L^{-1}x(t) - \frac{c}{m}L^{-1}D_t^q x(t). \tag{10.90}$$

Here, the inverse linear operator is taken as $L^{-1} = D_t^{-2}$. The eq. (10.90) can be rewritten as:

$$x(t) = x(0) + tD_t x(0) + D_t^{-2}\frac{\delta(t)}{m} - \frac{k}{m}D_t^{-2}x(t) - \frac{c}{m}D_t^{-2}\left(D_t^q x(t)\right)$$

$$= x(0) + tD_t x(0) + D_t^{-2}\frac{\delta(t)}{m} - \frac{k}{m}D_t^{-2}x(t) - \frac{c}{m}D_t^{-2+q}x(t). \tag{10.91}$$

By using initial conditions (10.87) and by taking unit length of regime, for example, $L = 1$, we can calculate initially for first recursion q as:

$$q = \frac{1 + \left(\dfrac{x(0)}{L}\right)^2}{2} = \frac{1}{2}. \tag{10.92}$$

10.7.3.1 First Recursion

Therefore, eq. (10.91) can be written as:

$$x(t) = x(0) + tD_t x(0) + D_t^{-2}\frac{\delta(t)}{m} - \frac{k}{m}D_t^{-2}x(t) - \frac{c}{m}D_t^{-1.5}x(t). \tag{10.93}$$

So by using the successive recursion method, we have:

$$x_0(t) = x(0) + tD_t x(0) + D_t^{-2}\frac{\delta(t)}{m} = \frac{t}{m},$$

$$x_1(t) = -\frac{k}{m}D_t^{-2}x_0(t) - \frac{c}{m}D_t^{-1.5}x_0(t)$$

$$= -\frac{k}{m^2}\frac{t^3}{\Gamma(4)} - \frac{c}{m^2}\frac{t^{2.5}}{\Gamma(3.5)},$$

$$x_2(t) = -\frac{k}{m}D_t^{-2}x_1(t) - \frac{c}{m}D_t^{-1.5}x_1(t)$$

$$= \frac{k^2}{m^3}\frac{t^5}{\Gamma(6)} + \frac{2kc}{m^3}\frac{t^{4.5}}{\Gamma(5.5)} + \frac{c^2}{m^3}\frac{t^4}{\Gamma(5)},$$

$$x_3(t) = -\frac{k}{m}D_t^{-2}x_2(t) - \frac{c}{m}D_t^{-1.5}x_2(t)$$

$$= -\frac{k^3}{m^4}\frac{t^7}{\Gamma(8)} - \frac{3k^2c}{m^4}\frac{t^{6.5}}{\Gamma(7.5)} - \frac{3kc^2}{m^4}\frac{t^6}{\Gamma(7)} - \frac{c^3}{m^4}\frac{t^{5.5}}{\Gamma(6.5)}.$$

and so on. So the solution of eq. (10.62) for $q(x(t))\big|_{t=0} = \frac{1}{2}$ is given as:

$$x(t) = x_0(t) + x_1(t) + x_2(t) + x_3(t)$$

$$= \frac{t}{m} - \frac{k}{m^2}\frac{t^3}{\Gamma(4)} - \frac{c}{m^2}\frac{t^{2.5}}{\Gamma(3.5)} + \frac{k^2}{m^3}\frac{t^5}{\Gamma(6)} + \frac{2kc}{m^3}\frac{t^{4.5}}{\Gamma(5.5)} + \frac{c^2}{m^3}\frac{t^4}{\Gamma(5)}. \tag{10.94}$$

$$- \frac{k^3}{m^4}\frac{t^7}{\Gamma(8)} - \frac{3k^2c}{m^4}\frac{t^{6.5}}{\Gamma(7.5)} - \frac{3kc^2}{m^4}\frac{t^6}{\Gamma(7)} - \frac{c^3}{m^4}\frac{t^{5.5}}{\Gamma(6.5)}$$

10.7.3.2 Second Recursion

By putting $\Delta t = 0.01$, that is for $n = 1$, and taking $m = 1$, $\frac{k}{m} = \omega_n^2$, $\frac{c}{m} = 2\eta\omega_n^{3/2}$, $\omega_n = 2$, and $\eta = 0.5$ in eq. (10.94), we have $x(0.01) = 0.0100078$. We have also solved for other values of η, as we report in graphical plots in a subsequent section. By taking unit length, for example, $L = 1$ and $x(0.01) = 0.0100078$, we find q for second recursion and calculated as follows:

$$q = \frac{1 + \left(\dfrac{x(0.01)}{L}\right)^2}{2} = 0.50005.$$

The eq. (10.91) can be written as:

$$x(t) = x(0) + tD_tx(0) + D_t^{-2}\frac{\delta(t)}{m} - \frac{k}{m}D_t^{-2}x(t) - \frac{c}{m}D_t^{-1.49995}x(t). \tag{10.95}$$

So by using the successive recursion method, we have:

$$x_0(t) = x(0) + tD_tx(0) + D_t^{-2}\frac{\delta(t)}{m} = \frac{t}{m},$$

$$x_1(t) = -\frac{k}{m}D_t^{-2}x_0(t) - \frac{c}{m}D_t^{-1.49995}x_0(t)$$

$$= -\frac{k}{m^2}\frac{t^3}{\Gamma(4)} - \frac{c}{m^2}\frac{t^{2.49995}}{\Gamma(3.49995)},$$

Fractional Continuously Variable-Order Mass-Spring Damper Systems

$$x_2(t) = -\frac{k}{m} D_t^{-2} x_1(t) - \frac{c}{m} D_t^{-1.49995} x_1(t)$$

$$= \frac{k^2}{m^3} \frac{t^5}{\Gamma(6)} + \frac{2kc}{m^3} \frac{t^{4.49995}}{\Gamma(5.49995)} + \frac{c^2}{m^3} \frac{t^{3.9999}}{\Gamma(4.9999)},$$

$$x_3(t) = -\frac{k}{m} D_t^{-2} x_2(t) - \frac{c}{m} D_t^{-1.49995} x_2(t)$$

$$= -\frac{k^3}{m^4} \frac{t^7}{\Gamma(8)} - \frac{3k^2 c}{m^4} \frac{t^{6.49995}}{\Gamma(7.49995)} - \frac{3kc^2}{m^4} \frac{t^{5.9999}}{\Gamma(6.9999)} - \frac{c^3}{m^4} \frac{t^{5.49985}}{\Gamma(6.49985)},$$

and so on. So the solution of eq. (10.62) for $q = 0.50005$ is given as:

$$x(t) = x_0(t) + x_1(t) + x_2(t) + x_3(t)$$

$$= \frac{t}{m} - \frac{k}{m^2} \frac{t^3}{\Gamma(4)} - \frac{c}{m^2} \frac{t^{2.49995}}{\Gamma(3.49995)} + \frac{k^2}{m^3} \frac{t^5}{\Gamma(6)} + \frac{2kc}{m^3} \frac{t^{4.49995}}{\Gamma(5.49995)} + \frac{c^2}{m^3} \frac{t^{3.9999}}{\Gamma(4.9999)}. \qquad (10.96)$$

$$- \frac{k^3}{m^4} \frac{t^7}{\Gamma(8)} - \frac{3k^2 c}{m^4} \frac{t^{6.49995}}{\Gamma(7.49995)} - \frac{3kc^2}{m^4} \frac{t^{5.9999}}{\Gamma(6.9999)} - \frac{c^3}{m^4} \frac{t^{5.49985}}{\Gamma(6.49985)}$$

Similarly, by taking $t = n\Delta t$ and using eq. (10.89), the fractional order q for next all recursions can be calculated. Here, $\Delta t = 0.01$, for example, the time step is 0.01 seconds and $n = 2, 3, 4, \dots$.

By generalizing the solution by the successive recursion method, we have the following:

$$x_0(t) = x(0) + t D_t x(0) + D_t^{-2} \frac{\delta(t)}{m} = \frac{t}{m},$$

$$x_1(t) = -\frac{k}{m} D_t^{-2} x_0(t) - \frac{c}{m} D_t^{-2 + q_{n\Delta t}} x_0(t)$$

$$= -\frac{k}{m^2} \frac{t^3}{\Gamma(4)} - \frac{c}{m^2} \frac{t^{3 - q_{n\Delta t}}}{\Gamma(4 - q_{n\Delta t})},$$

$$x_2(t) = -\frac{k}{m} D_t^{-2} x_1(t) - \frac{c}{m} D_t^{-2 + q_{n\Delta t}} x_1(t)$$

$$= \frac{k^2}{m^3} \frac{t^5}{\Gamma(6)} + \frac{2kc}{m^3} \frac{t^{5 - q_{n\Delta t}}}{\Gamma(6 - q_{n\Delta t})} + \frac{c^2}{m^3} \frac{t^{5 - 2q_{n\Delta t}}}{\Gamma(6 - 2q_{n\Delta t})},$$

$$x_3(t) = -\frac{k}{m} D_t^{-2} x_2(t) - \frac{c}{m} D_t^{-2 + q_{n\Delta t}} x_2(t)$$

$$= -\frac{k^3}{m^4} \frac{t^7}{\Gamma(8)} - \frac{3k^2 c}{m^4} \frac{t^{7 - q_{n\Delta t}}}{\Gamma(8 - q_{n\Delta t})} - \frac{3kc^2}{m^4} \frac{t^{7 - 2q_{n\Delta t}}}{\Gamma(8 - 2q_{n\Delta t})} - \frac{c^3}{m^4} \frac{t^{7 - 3q_{n\Delta t}}}{\Gamma(8 - 3q_{n\Delta t})},$$

and so on.

Thus the cumulative solution is:

$$x(t) = x_0(t) + x_1(t) + x_2(t) + x_3(t) + \dots$$

$$= \frac{t}{m} - \frac{k}{m^2}\frac{t^3}{\Gamma(4)} - \frac{c}{m^2}\frac{t^{3-q_{n\Delta t}}}{\Gamma(4-q_{n\Delta t})} + \frac{k^2}{m^3}\frac{t^5}{\Gamma(6)} + \frac{2kc}{m^3}\frac{t^{5-q_{n\Delta t}}}{\Gamma(6-q_{n\Delta t})} + \frac{c^2}{m^3}\frac{t^{5-2q_{n\Delta t}}}{\Gamma(6-2q_{n\Delta t})}.$$

$$- \frac{k^3}{m^4}\frac{t^7}{\Gamma(8)} - \frac{3k^2c}{m^4}\frac{t^{7-q_{n\Delta t}}}{\Gamma(8-q_{n\Delta t})} - \frac{3kc^2}{m^4}\frac{t^{7-2q_{n\Delta t}}}{\Gamma(8-2q_{n\Delta t})} - \frac{c^3}{m^4}\frac{t^{7-3q_{n\Delta t}}}{\Gamma(8-3q_{n\Delta t})} + \dots$$

The value of q is variable and that in this case depends on position $x(t)$ given by expression (10.89). With rearrangement in the above series solution, we can write the above series in following compact form as following:

$$x(t) = \frac{1}{m}\sum_{r=0}^{\infty}\frac{(-k/m)^r}{r!}t^{2r+1}\sum_{j=0}^{\infty}\frac{(-c/m)^j(j+r)!t^{(2-q_{n\Delta t})j}}{j!\Gamma((2-q_{n\Delta t})j+2r+2)}$$

$$= \frac{1}{m}\sum_{r=0}^{\infty}\frac{(-k/m)^r}{r!}t^{2r+1}E_{2-q_{n\Delta t},q_{n\Delta t}r+2}^{(r)}\left((-c/m)t^{(2-q_{n\Delta t})}\right), \qquad (10.97)$$

$$= \frac{1}{m}\sum_{r=0}^{\infty}\frac{(-\omega_n^2)^r}{r!}t^{2r+1}E_{2-q_{n\Delta t},q_{n\Delta t}r+2}^{(r)}\left((-2\eta\omega_n^{3/2})t^{(2-q_{n\Delta t})}\right)$$

where $E_{\lambda,\mu}(.)$ defines the Mittag–Leffler function in two parameter and the r th-derivative of two-parameter Mittag–Leffler function is defined as:

$$E_{\lambda,\mu}^{(r)}(y) = \frac{d^r}{dy^r}E_{\lambda,\mu}(y) = \sum_{j=0}^{\infty}\frac{(j+r)!y^j}{j!\Gamma(\lambda j+\lambda r+\mu)}, \quad (r=0,1,2,\dots).$$

10.7.4 Application of Successive Recursion Method for Forced Oscillation of Spring-Mass Viscoelastic-Viscous Damping System in Case IV

Let us consider eq. (10.64) with the homogenous initial conditions:

$$x(0) = 0 \text{ and } D_t^1 x(t)\big|_{t=0} = 0, \qquad (10.98)$$

which is at the equilibrium states of the dynamic system at the beginning process. The eq. (10.64) can be written in the following form:

$$D_t^2 x(t) + \frac{c}{m}D_t^q x(t) + \frac{k}{m}x(t) = \frac{\delta(t)}{m}, \qquad (10.99)$$

where q is the fractional continuously variable-order viscous-viscoelastic oscillator, which is defined as:

$$q = \cos\left(\frac{\pi x(t)}{2L}\right)\bigg|_{t=n\Delta t}. \qquad (10.100)$$

Fractional Continuously Variable-Order Mass-Spring Damper Systems 281

Here, the q changes with the small change in time say Δt and $n \in \mathbb{N}$ in (10.100). So for each $t = n\Delta t$, we have the new fractional differential equation and its solution, which continuously changes throughout the period of oscillation. By using the successive recursion method, eq. (10.99) can be written as:

$$x(t) = x(0) + tD_t x(0) + L^{-1}\frac{\delta(t)}{m} - \frac{k}{m}L^{-1}x(t) - \frac{c}{m}L^{-1}D_t^q x(t).$$ (10.101)

Here, the inverse linear operator is taken as $L^{-1} = D_t^{-2}$. The eq. (10.101) can be rewritten as:

$$x(t) = x(0) + tD_t x(0) + D_t^{-2}\frac{\delta(t)}{m} - \frac{k}{m}D_t^{-2}x(t) - \frac{c}{m}D_t^{-2}\left(D_t^q x(t)\right)$$

$$= x(0) + tD_t x(0) + D_t^{-2}\frac{\delta(t)}{m} - \frac{k}{m}D_t^{-2}x(t) - \frac{c}{m}D_t^{-2+q}x(t).$$ (10.102)

By using initial conditions (10.98) and by taking unit length of regime, for example, $L = 1$, we can calculate initially for first recursion q as:

$$q = \cos\left(\frac{\pi x(t)}{2L}\right)\Bigg|_{t=0} = \cos\left(\frac{\pi x(0)}{2L}\right) = 1.$$ (10.103)

10.7.4.1 First Recursion

Therefore, the eq. (10.102) can be written as:

$$x(t) = x(0) + tD_t x(0) + D_t^{-2}\frac{\delta(t)}{m} - \frac{k}{m}D_t^{-2}x(t) - \frac{c}{m}D_t^{-1}x(t).$$ (10.104)

So by using the successive recursion method, we have:

$$x_0(t) = x(0) + tD_t x(0) + D_t^{-2}\frac{\delta(t)}{m} = \frac{t}{m},$$

$$x_1(t) = -\frac{k}{m}D_t^{-2}x_0(t) - \frac{c}{m}D_t^{-1}x_0(t)$$

$$= -\frac{k}{m^2}\frac{t^3}{\Gamma(4)} - \frac{c}{m^2}\frac{t^2}{\Gamma(3)},$$

$$x_2(t) = -\frac{k}{m}D_t^{-2}x_1(t) - \frac{c}{m}D_t^{-1}x_1(t)$$

$$= \frac{k^2}{m^3}\frac{t^5}{\Gamma(6)} + \frac{2kc}{m^3}\frac{t^4}{\Gamma(5)} + \frac{c^2}{m^3}\frac{t^3}{\Gamma(4)},$$

$$x_3(t) = -\frac{k}{m}D_t^{-2}x_2(t) - \frac{c}{m}D_t^{-1}x_2(t)$$

$$= -\frac{k^3}{m^4}\frac{t^7}{\Gamma(8)} - \frac{3k^2c}{m^4}\frac{t^6}{\Gamma(7)} - \frac{3kc^2}{m^4}\frac{t^5}{\Gamma(6)} - \frac{c^3}{m^4}\frac{t^4}{\Gamma(5)},$$

and so on. So the solution of eq. (10.64) for $q(x(t))\big|_{t=0} = 1$ is given as:

$$x(t) = x_0(t) + x_1(t) + x_2(t) + x_3(t)$$

$$= \frac{t}{m} - \frac{k}{m^2}\frac{t^3}{\Gamma(4)} - \frac{c}{m^2}\frac{t^2}{\Gamma(3)} + \frac{k^2}{m^3}\frac{t^5}{\Gamma(6)} + \frac{2kc}{m^3}\frac{t^4}{\Gamma(5)} + \frac{c^2}{m^3}\frac{t^3}{\Gamma(4)}.$$

$$- \frac{k^3}{m^4}\frac{t^7}{\Gamma(8)} - \frac{3k^2c}{m^4}\frac{t^6}{\Gamma(7)} - \frac{3kc^2}{m^4}\frac{t^5}{\Gamma(6)} - \frac{c^3}{m^4}\frac{t^4}{\Gamma(5)}$$

(10.105)

10.7.4.2 Second Recursion

By putting $\Delta t = 0.01$, that is for $n = 1$, and taking $m = 1$, $\frac{k}{m} = \omega_n^2$, $\frac{c}{m} = 2\eta\omega_n^{3/2}$, $\omega_n = 2$, and $\eta = 0.5$ in eq. (10.105), we have $x(0.01) = 0.00985925$. We have also presented in graphical plots for other values of η, in a subsequent section. By taking unit length, for example, $L = 1$ and $x(0.01) = 0.00985925$, the q for second recursion can be calculated, which is given as:

$$q = \cos\left(\frac{\pi\, x(0.01)}{2L}\right) = 0.99988.$$

(10.106)

The eq. (10.102) can be written as:

$$x(t) = x(0) + tD_t x(0) + D_t^{-2}\frac{\delta(t)}{m} - \frac{k}{m}D_t^{-2}x(t) - \frac{c}{m}D_t^{-1.00012}x(t).$$

(10.107)

So by using the successive recursion method, we have:

$$x_0(t) = x(0) + tD_t x(0) + D_t^{-2}\frac{\delta(t)}{m} = \frac{t}{m},$$

$$x_1(t) = -\frac{k}{m}D_t^{-2}x_0(t) - \frac{c}{m}D_t^{-1.00012}x_0(t)$$

$$= -\frac{k}{m^2}\frac{t^3}{\Gamma(4)} - \frac{c}{m^2}\frac{t^{2.00012}}{\Gamma(3.00012)},$$

$$x_2(t) = -\frac{k}{m}D_t^{-2}x_1(t) - \frac{c}{m}D_t^{-1.00012}x_1(t)$$

$$= \frac{k^2}{m^3}\frac{t^5}{\Gamma(6)} + \frac{2kc}{m^3}\frac{t^{4.00012}}{\Gamma(5.00012)} + \frac{c^2}{m^3}\frac{t^{3.00024}}{\Gamma(4.00024)},$$

$$x_3(t) = -\frac{k}{m}D_t^{-2}x_2(t) - \frac{c}{m}D_t^{-1.00012}x_2(t)$$

$$= -\frac{k^3}{m^4}\frac{t^7}{\Gamma(8)} - \frac{3k^2c}{m^4}\frac{t^{6.00012}}{\Gamma(7.00012)} - \frac{3kc^2}{m^4}\frac{t^{5.00024}}{\Gamma(6.00024)} - \frac{c^3}{m^4}\frac{t^{4.00036}}{\Gamma(5.00036)},$$

Fractional Continuously Variable-Order Mass-Spring Damper Systems 283

and so on. So the solution of eq. (10.64) for $q = 0.99988$ is given as:

$$x(t) = x_0(t) + x_1(t) + x_2(t) + x_3(t)$$

$$= \frac{t}{m} - \frac{k}{m^2} \frac{t^3}{\Gamma(4)} - \frac{c}{m^2} \frac{t^{2.00012}}{\Gamma(3.00012)} + \frac{k^2}{m^3} \frac{t^5}{\Gamma(6)} + \frac{2kc}{m^3} \frac{t^{4.00012}}{\Gamma(5.00012)} + \frac{c^2}{m^3} \frac{t^{3.00024}}{\Gamma(4.00024)} .$$

$$- \frac{k^3}{m^4} \frac{t^7}{\Gamma(8)} - \frac{3k^2c}{m^4} \frac{t^{6.00012}}{\Gamma(7.00012)} - \frac{3kc^2}{m^4} \frac{t^{5.00024}}{\Gamma(6.00024)} - \frac{c^3}{m^4} \frac{t^{4.00036}}{\Gamma(5.00036)}$$

(10.108)

Similarly, by taking $t = n\Delta t$ and using eq. (10.100), the fractional order q for next all recursions can be calculated. Here, $\Delta t = 0.01$, for example, the time step is 0.01 second and $n = 2, 3, 4, \ldots$.

By generalizing the solution by the successive recursion method, we have the following:

$$x_0(t) = x(0) + t D_t x(0) + D_t^{-2} \frac{\delta(t)}{m} = \frac{t}{m},$$

$$x_1(t) = -\frac{k}{m} D_t^{-2} x_0(t) - \frac{c}{m} D_t^{-2 + q_{n\Delta t}} x_0(t)$$

$$= -\frac{k}{m^2} \frac{t^3}{\Gamma(4)} - \frac{c}{m^2} \frac{t^{3 - q_{n\Delta t}}}{\Gamma(4 - q_{n\Delta t})},$$

$$x_2(t) = -\frac{k}{m} D_t^{-2} x_1(t) - \frac{c}{m} D_t^{-2 + q_{n\Delta t}} x_1(t)$$

$$= \frac{k^2}{m^3} \frac{t^5}{\Gamma(6)} + \frac{2kc}{m^3} \frac{t^{5 - q_{n\Delta t}}}{\Gamma(6 - q_{n\Delta t})} + \frac{c^2}{m^3} \frac{t^{5 - 2q_{n\Delta t}}}{\Gamma(6 - 2q_{n\Delta t})},$$

$$x_3(t) = -\frac{k}{m} D_t^{-2} x_2(t) - \frac{c}{m} D_t^{-2 + q_{n\Delta t}} x_2(t)$$

$$= -\frac{k^3}{m^4} \frac{t^7}{\Gamma(8)} - \frac{3k^2c}{m^4} \frac{t^{7 - q_{n\Delta t}}}{\Gamma(8 - q_{n\Delta t})} - \frac{3kc^2}{m^4} \frac{t^{7 - 2q_{n\Delta t}}}{\Gamma(8 - 2q_{n\Delta t})} - \frac{c^3}{m^4} \frac{t^{7 - 3q_{n\Delta t}}}{\Gamma(8 - 3q_{n\Delta t})}$$

and so on.

Thus the cumulative solution is:

$$x(t) = x_0(t) + x_1(t) + x_2(t) + x_3(t) + \ldots$$

$$= \frac{t}{m} - \frac{k}{m^2} \frac{t^3}{\Gamma(4)} - \frac{c}{m^2} \frac{t^{3 - q_{n\Delta t}}}{\Gamma(4 - q_{n\Delta t})} + \frac{k^2}{m^3} \frac{t^5}{\Gamma(6)} + \frac{2kc}{m^3} \frac{t^{5 - q_{n\Delta t}}}{\Gamma(6 - q_{n\Delta t})} + \frac{c^2}{m^3} \frac{t^{5 - 2q_{n\Delta t}}}{\Gamma(6 - 2q_{n\Delta t})} .$$

$$- \frac{k^3}{m^4} \frac{t^7}{\Gamma(8)} - \frac{3k^2c}{m^4} \frac{t^{7 - q_{n\Delta t}}}{\Gamma(8 - q_{n\Delta t})} - \frac{3kc^2}{m^4} \frac{t^{7 - 2q_{n\Delta t}}}{\Gamma(8 - 2q_{n\Delta t})} - \frac{c^3}{m^4} \frac{t^{7 - 3q_{n\Delta t}}}{\Gamma(8 - 3q_{n\Delta t})} + \ldots$$

The value of q is variable and that in this case depends on position $x(t)$ given by expression (10.100). With rearrangement in the above series solution, we can write the above series in following compact form as following:

$$x(t) = \frac{1}{m}\sum_{r=0}^{\infty}\frac{(-k/m)^r}{r!}t^{2r+1}\sum_{j=0}^{\infty}\frac{(-c/m)^j(j+r)!t^{(2-q_{n\Delta t})j}}{j!\Gamma((2-q_{n\Delta t})j+2r+2)}$$

$$= \frac{1}{m}\sum_{r=0}^{\infty}\frac{(-k/m)^r}{r!}t^{2r+1}E_{2-q_{n\Delta t},q_{n\Delta t}r+2}^{(r)}\left((-c/m)t^{(2-q_{n\Delta t})}\right), \qquad (10.109)$$

$$= \frac{1}{m}\sum_{r=0}^{\infty}\frac{(-\omega_n^2)^r}{r!}t^{2r+1}E_{2-q_{n\Delta t},q_{n\Delta t}r+2}^{(r)}\left((-2\eta\omega_n^{3/2})t^{(2-q_{n\Delta t})}\right),$$

where $E_{\lambda,\mu}(.)$ defines the Mittag–Leffler function in two parameter, and the rth-derivative of two-parameter Mittag–Leffler function is defined as:

$$E_{\lambda,\mu}^{(r)}(y) = \frac{d^r}{dy^r}E_{\lambda,\mu}(y) = \sum_{j=0}^{\infty}\frac{(j+r)!y^j}{j!\Gamma(\lambda j+\lambda r+\mu)}, \quad (r=0,1,2,\ldots).$$

10.8 Numerical Simulations and Discussion

The solution to the oscillator problem with continuously variable damping order q, defined in the first case as viscoelastic damping via expression (10.67), in the second case as viscous-viscoelastic damping via expression (10.78) governed by fractional differential equation $D_t^2 x(t) + \frac{c}{m}D_t^q x(t) + \frac{k}{m}x(t) = 0$, the third case as viscous-viscoelastic damping via expression (10.89), and the fourth case as viscoelastic-viscous damping via expression (10.100) governed by fractional differential equation $D_t^2 x(t) + \frac{c}{m}D_t^q x(t) + \frac{k}{m}x(t) = \delta(t)$, respectively, gives new type of analytical solutions in compact form expressed in eqs. (10.75), (10.86), (10.97), and (10.109), respectively. In the first and second cases, though, we are not having external forcing function, but the initial condition at $t = 0$ is $x(0) = 0$ and $D_t^1 x(t)\big|_{t=0} = x'(0) = 1$, that is the system has initial velocity of unity magnitude. In the third and fourth cases, we have Dirac delta function as forcing function that is applied at $t = 0$ with system initially at rest, for example, at $t = 0$, we have $x(0) = 0$ and $D_t x(t)\big|_{t=0} = x'(0) = 0$. In the third and fourth cases, the forcing function starts the oscillations. All discussed are cases of free running damped oscillator, with continuously varying order of damping, that is q throughout the oscillation period.

In the present section, the displacement-time graphs have been presented for fractional continuously variable-order mass-spring damper systems for free oscillation with viscoelastic damping, viscous-viscoelastic damping and forced oscillation with viscous-viscoelastic damping, and viscoelastic-viscous damping. Thus, the two fractional differential equations with continuous variable order q are following; with the initial conditions stated as in the above sections.

$$D_t^2 x(t) + 2\eta\omega_n^{3/2}D_t^q x(t) + \omega_n^2 x(t) = 0, \ D_t^2 x(t) + 2\eta\omega_n^{3/2}D_t^q x(t) + \omega_n^2 x(t) = \delta(t).$$

Here, ω_n is called the natural or angular frequency and η is called the damping ratio of the system.

10.8.1 Numerical Simulation of Fractional Continuously Variable-Order Mass-Spring Damper System for Free Oscillation with Viscoelastic Damping-Case I

For Case I, the body oscillates without implementation of any external forces with the continuous change of the fractional continuously variable-order viscoelastic damper q, with $q(x'(t)) = 0.5 + 0.5\tanh(|x'(t)|)$. Here, we have taken $0 \le t \le 6$ and 600 recursions for plotting the time response of the fractional continuous order damping system with displacement $x(t)$ and velocity $v(t)$ (Figures 10.10 and 10.11). Also in the solution to eq. (10.75), we have taken mass $m = 1$ unit, natural or angular frequency $\omega_n = 2\,\text{rad/sec}$, and the damping ratio $\eta = 0.05, 0.5, 1, \sqrt{\pi}$.

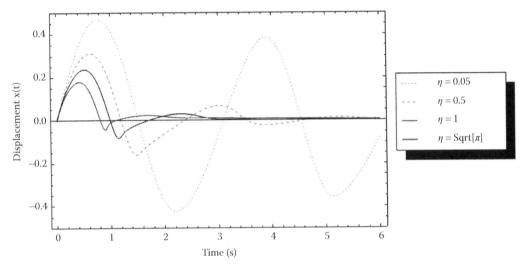

FIGURE 10.10
The displacement-time graph for fractional continuous order spring-mass damper model for free oscillation with viscoelastic damping plots for **Case I**.

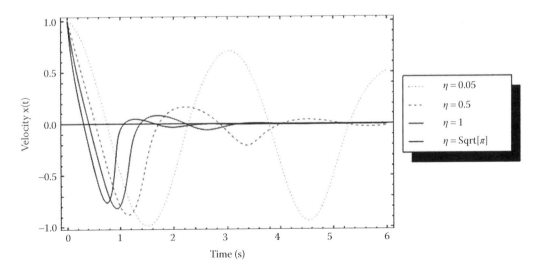

FIGURE 10.11
The velocity-time graph for fractional continuous order spring-mass damper model for free oscillation with viscoelastic damping plots for **Case I**.

10.8.2 Numerical Simulation of Fractional Continuously Variable-Order Mass-Spring Damper System for Free Oscillation with Viscous-Viscoelastic Damping-Case II

For Case II, the body oscillates without implementation of any external forces with the continuous change of the fractional continuously variable fractional order of "viscous-viscoelastic" damper, for example, q, with $q(x(t)) = \frac{1}{2} + \frac{1}{2}\left(\frac{x(t)}{L}\right)^2$. Here, we have taken $0 \leq t \leq 3$ and 300 recursions for plotting the time response of the fractional continuous order damping system with displacement $x(t)$ and velocity $v(t)$. Also in the solution to eq. (10.86), we have taken mass $m = 1$ unit, natural or angular frequency $\omega_n = 2\,\text{rad/sec}$, and the damping ratio $\eta = 0.05, 0.5, 1, \sqrt{\pi}$ (Figures 10.12 and 10.13).

10.8.3 Numerical Simulation of Fractional Continuously Variable-Order Mass-Spring Damper System for Forced Oscillation with Viscous-Viscoelastic Damping-Case III

It is important to mention here that, in the system in Case III, oscillation takes place with the small external impulse force $\delta(t)$ with the continuous change of the fractional order viscous-viscoelastic oscillator q, with $q(x(t)) = \frac{1}{2}\left[1 + \left(\frac{x(t)}{L}\right)^2\right]$ which ranges between 0.5 to 1, for example, $0.5 \leq q \leq 1$. Here, we have taken $0 \leq t \leq 6$ and 600 recursions for plotting the time response of the fractional continuous order damping system with displacement $x(t)$. Also in the solution to eq.(10.97), we have taken mass $m = 1$ unit, natural or angular frequency $\omega_n = 2\,\text{rad/sec}$, and the damping ratio $\eta = 0.05, 0.5, 1, \sqrt{\pi}$ (Figure 10.14).

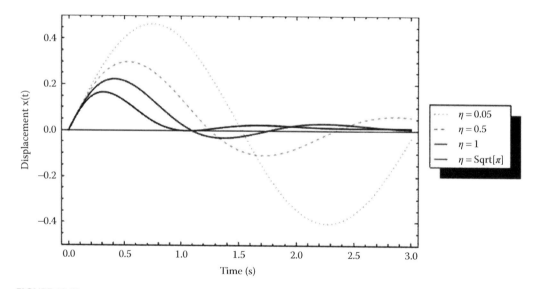

FIGURE 10.12
The displacement-time graph for fractional continuous order spring-mass damper model for free oscillation with viscous-viscoelastic damping plots for **Case II**.

Fractional Continuously Variable-Order Mass-Spring Damper Systems

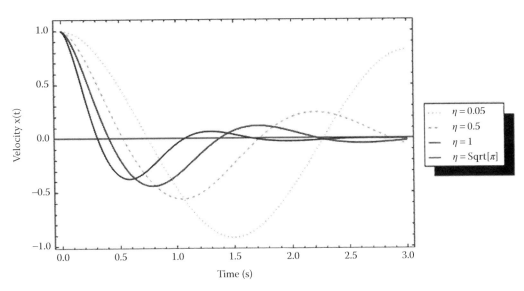

FIGURE 10.13
The velocity-time graph for fractional continuous order spring-mass damper model for free oscillation with viscous-viscoelastic damping plots for **Case II**.

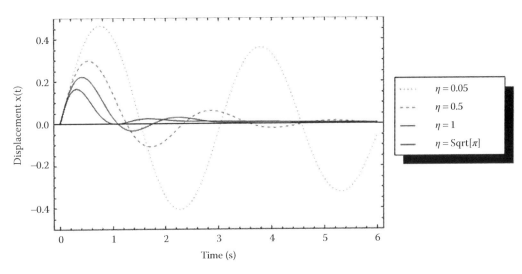

FIGURE 10.14
The displacement-time graph for fractional continuous order mass-spring damper model for forced oscillation with viscous-viscoelastic damping plots for **Case III**.

10.8.4 Numerical Simulation of Fractional Continuously Variable-Order Mass-Spring-Damper System for Forced Oscillation with Viscoelastic-Viscous Damping-Case IV

It is important to mention here that, in the system in Case IV, oscillation takes place with the small external impulse force $\delta(t)$ with the continuous change of the fractional order of "viscous-viscoelastic" damping, for example, q, with $q(x(t)) = \cos\left(\pi \frac{x(t)}{2L}\right)$. Here, we have taken $0 \le t \le 3$ and 300 recursions for plotting the time response of the fractional continuous order damping system with displacement $x(t)$. Also in the solution to eq. (10.109), we have taken mass $m = 1$ unit, natural or angular frequency $\omega_n = 2\,\text{rad/sec}$, and the damping ratio $\eta = 0.05, 0.5, 1, \sqrt{\pi}$ (Figure 10.15).

10.8.5 Physical Interpretations

Figures 10.10, 10.12, 10.14, and 10.15 represent the displacement-time graphs for Cases I to IV; and Figures 10.11 and 10.13 present the velocity-time graph (for Case I and Case II) of the fractional continuously variable-order mass-spring damper models for the different values of damping ratio η. The Figures 10.10, 10.12, 10.14, and 10.15 clearly bring out the difference in oscillatory character for the two different cases of continuously variable fractional derivative order in the system, for higher values of $\eta \gg 0$. The Case I and Case II show presence of abrupt change in rate of position, such as a kink, whereas the Case III and Case IV have rather smooth decaying oscillation. It is noted here that, the initial position state, for example, $x(t)$ with increase in time tend to asymptotic with the equilibrium state that is $x = 0$. From the above figures, we observe the following:

1. When $\eta = 0.05$ or near to or equal zero, the system oscillates at its natural frequency ω_n and the system is called "un-damped". That means the oscillation will continue almost forever, like simple harmonic motion.

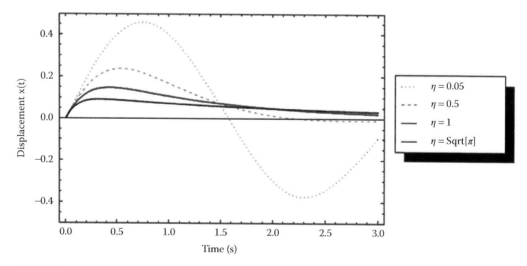

FIGURE 10.15
The displacement-time graph for fractional continuous order mass-spring damper model for forced oscillation with viscoelastic-viscous damping plots for **Case IV**.

Fractional Continuously Variable-Order Mass-Spring Damper Systems

2. When $\eta = 0.5$, the system oscillates at higher than the natural frequency and the system is called "under-damped", like the classical damped integer order oscillator. In this situation, the oscillation gradually tends to zero. However, relation of damped natural frequency say ω_d to ω_n in the case of the continuously variable-order damping case is to be developed; like we have for integer order systems, for example, given by $\omega_d = \omega_n \sqrt{1 - \eta^2}$.

3. When $\eta = 1$, the system oscillates quickly and it converges to zero as quickly as possible, and the system is similar as called "critical damped", like in integer order damped oscillators. In this situation, the oscillation returns to equilibrium in the shortest period of time.

4. When $\eta = \sqrt{\pi}$, the system oscillates a little as compared to critical damping, and it converges to zero slowly, and the system is similar as called "over damped", like in integer order damped oscillators.

We have pointed out here the definitions regarding natural frequency, damped-frequency, under-damped oscillation, critically damped oscillation, and over-damped oscillations in the continuously variable fractional order damped oscillator that we have developed. However, it needs to be re-defined with respect to the variable fractional order q of the system. But here we have drawn similarity with the classical integer order damped oscillator system.

10.9 Conclusion

In this chapter, we have analyzed several definitions of variable-order operators and concluded that the variable operator must be able to return the intermediate values between 0 and 1, which corresponds to the continuous order argument of the operator in any case. The variable-order operators defined in different forms by the independent researchers satisfy the criteria for modeling the dynamic systems. As an instance, we have also analyzed the dynamic nature of a variable-order operator in modeling the viscoelastic oscillator, viz., the mass-spring damping system.

Also, new analytical approximate solutions have been presented for fractional continuously variable-order mass-spring damper systems for free oscillation with viscous-viscoelastic damping and forced oscillation with viscoelastic-viscous damping. Also, we modeled the fractional continuously variable-order mass-spring damper systems for free oscillation with viscoelastic damping, viscous-viscoelastic damping and forced oscillation with viscous-viscoelastic damping, and viscoelastic-viscous damping. The approach is new in the sense of changing of behavior of guide continuously with the small change of time Δt with respect to both viscoelastic and viscous-viscoelastic oscillators of order q. We also find the analytical solutions of the fractional continuously variable-order mass-spring damper systems by the successive recursion method. From the above analysis, it is concluded that the proposed analytical method is exceptionally powerful and effective for solving the fractional dynamic models; with continuously varying fractional order of fractional derivative term.

The graphical plots also have been presented for different values of damping ratio. From the graph, we have given the conclusion for the nature of damping, viz., undamped, underdamped, critical damping, and over damping of the system, similar to those for the integer order classical damped oscillator system, however, though these parameters in context of continuously variable-order damping oscillator systems need to be developed. From this new method developed in this chapter, we conclude that the proposed method is highly effective for finding the solution for the fractional dynamic model, where the fractional order of damping changes continuously. This development has immense potential in study of various physical dynamic systems.

References

1. I. Podlubny, *Fractional Differential Equation*, Academic Press, New York, 1999.
2. K. S. Miller, and B. Ross, *An Introduction to the Fractional Calculus and Fractional Differential Equations*, Wiley, New York, 1993.
3. S. G. Samko, A. A. Kilbas, and O. I. Marichev, *Fractional Integrals and Derivatives: Theory and Applications*, Taylor & Francis Group, London, UK, 2002.
4. A. A. Kilbas, H. M. Srivastava, and J. J. Trujillo, *Theory and Applications of Fractional Differential Equations*, Elsevier, New York, 2006.
5. S. Saha Ray, *Fractional Calculus with Applications for Nuclear Reactor Dynamics*, CRC Press, Boca Raton, FL, 2015.
6. R. Herrmann, *Fractional Calculus: An Introduction for Physicists*, World Scientific, Singapore, 2011.
7. H. J. Haubold, A. M. Mathai, and R. K. Saxena, "Mittag-Leffler functions and their applications," *Journal of Applied Mathematics*, vol. 2011, pp. 298628-1-51, 2011.
8. M. Kurulay, and M. Bayram, "Some properties of the Mittag-Leffler functions and their relation with the Wright functions," *Advances in Difference Equations*, vol. 2012, no. 181, pp. 1–8, 2012.
9. A. M. Mathai, and H. J. Haubold, *Special Functions For Applied Scientists*, Springer-Verlag, New York, 2008.
10. E. M. Wright, "On the coefficients of power series having exponential singularities," *Journal of the London Mathematical Society*, vol. 8, no. 1, pp. 71–79, 1933.
11. C. Fox, "The G and H functions as symmetrical Fourier kernels," *Transactions of the American Mathematical Society*, vol. 98, no. 3, pp. 395–429, 1961.
12. J. H. He, "A tutorial review on fractal spacetime and fractional calculus," *International Journal of Theoretical Physics*, vol. 53, no. 11, pp. 3698–3718, 2014.
13. S. Shen, F. Liu, and V. Anh, "Fundamental solution and discrete random walk model for a time-space fractional diffusion equation of distributed order," *Journal of Applied Mathematics and Computing*, vol. 28, no. 1–2, pp. 147–164, 2008.
14. Q. Yang, "Novel analytical and numerical methods for solving fractional dynamical system," Queensland university of technology, Brisbane, Australia, 2010.
15. B. Al-Saqabi, L. Boyadjiev, and Y. Luchko, "Comments on employing the Riesz-Feller derivative in the Schrödinger equation," *The European Physical Journal Special Topics*, vol. 222, no. 8, pp. 1779–1794, 2013.
16. Y. Zhang, "Time-fractional Camassa-Holm equation: Formulation and solution using variational methods," *ASME Journal of Computational and Nonlinear Dynamics*, vol. 8, no. 4, pp. 041020-1-8, 2013.
17. G. Jumarie, "Modified Riemann-Liouville derivative and fractional Taylor series of nondifferentiable functions further results," *Computers & Mathematics with Applications*, vol. 51, no. 9–10, pp. 1367–1376, 2006.
18. G. Jumarie, "On the representation of fractional Brownian motion as an integral with respect to $(dt)^a$," *Applied Mathematics Letters*, vol. 18, pp. 739–748, 2005.
19. X. J. Yang, *Advanced Local Fractional Calculus and Its Applications*, World Science Publisher, New York, 2012.
20. X. J. Yang, "A short note on local fractional calculus of function of one variable," *Journal of Applied Library and Information Science*, vol. 1, no. 1, pp. 1–13, 2012.
21. X. J. Yang, "The zero-mass renormalization group differential equations and limit cycles in non-smooth initial value problems," *Prespacetime Journal*, vol. 3, no. 9, pp. 913–923, 2012.
22. G. Jumarie, "Table of some basic fractional calculus formulae derived from a modified Riemann–Liouville derivative for non-differentiable functions," *Applied Mathematics Letters*, vol. 22, no. 3, pp. 378–385, 2009.

23. K. M. Kolwankar, and A. D. Gangal, "Fractional differentiability of nowhere differentiable functions and dimensions," *Chaos: An Interdisciplinary Journal of Nonlinear Science*, vol. 6, no. 4, pp. 505–513, 1996.
24. M. S. Hu, D. Baleanu, and X. J. Yang, "One-Phase problems for discontinuous heat transfer in fractal media," *Mathematical Problems in Engineering*, vol. 2013, pp. 358473-1-3, 2013.
25. J. H. He, "Homotopy perturbation technique," *Computer Methods in Applied Mechanics and Engineering*, vol. 178, no. 3–4, pp. 257–262, 1999.
26. L. M. Yan, "Modified homotopy perturbation method coupled with Laplace transform for fractional heat transfer and porous media equations," *Thermal Science*, vol. 17, no. 5, pp. 1409–1414, 2013.
27. J. H. He, "A coupling method of a homotopy technique and a perturbation technique for nonlinear problems," *International Journal of Non-Linear Mechanics*, vol. 35, no. 1, pp. 37–43, 2000.
28. J. H. He, "Comparison of homotopy perturbation method and homotopy analysis method," *Applied Mathematics and Computation*, vol. 156, no. 2, pp. 527–539, 2004.
29. J. H. He, "Some asymptotic methods for strongly nonlinear equations," *International Journal of Modern Physics B*, vol. 20, no. 10, pp. 1141–1199, 2006.
30. J. H. He, "New interpretation of homotopy perturbation method," *International Journal of Modern Physics B*, vol. 20, pp. 2561–2568, 2006.
31. A. Saadatmandi, M. Dehghan, and A. Eftekhari, "Application of he's homotopy perturbation method for non-linear system of second-order boundary value problems," *Nonlinear Analysis: Real World Applications*, vol. 10, no. 3, pp. 1912–1922, 2009.
32. F. Shakeri, and M. Dehghan, "Numerical solution of a biological population model using He's variational iteration method," *Computers & Mathematics with Applications*, vol. 54, no. 7–8, pp. 1197–1209, 2007.
33. J. Biazar, and H. Ghazvini, "Convergence of the homotopy perturbation method for partial differential equations," *Nonlinear Analysis: Real World Applications*, vol. 10, pp. 2633–2640, 2009.
34. A. K. Gupta, and S. Saha Ray, "Comparison between homotopy perturbation method and optimal homotopy asymptotic method for the soliton solutions of Boussinesq–Burger equations," *Computers & Fluids*, vol. 103, pp. 34–41, 2014.
35. S. Saha Ray, "An application of the modified decomposition method for the solution of the coupled Klein–Gordon–Schrödinger equation," *Communications in Nonlinear Science and Numerical Simulation*, vol. 13, no. 7, pp. 1311–1317, 2008.
36. H. Jafari, S. Ghasempoor, and C. M. Khalique, "A comparison between Adomian's polynomials and He's polynomials for nonlinear functional equations," *Mathematical Problems in Engineering*, vol. 2013, pp. 1–4, 2013.
37. H. Jafari, S. Ghasempour, and C. M. Khalique, "Comments on "He's homotopy perturbation method for calculating Adomian polynomials"," *International Journal of Nonlinear Sciences and Numerical Simulation*, vol. 14, no. 6, pp. 339–343, 2013.
38. Y. Khan, and Q. Wu, "Homotopy perturbation transform method for nonlinear equations using He's polynomials," *Computers & Mathematics with Applications*, vol. 61, no. 8, pp. 1963–1967, 2011.
39. Y. Liu, "Approximate solutions of fractional nonlinear equations using homotopy perturbation transformation method," *Abstract and Applied Analysis*, vol. 2012, pp. 752869-1-14, 2012.
40. M. Madani, M. Fathizadeh, Y. Khan, and A. Yildirim, "On the coupling of the homotopy perturbation method and Laplace transformation," *Mathematical and Computer Modelling*, vol. 53, no. 9–10, pp. 1937–1945, 2011.
41. S. Liao, "The proposed homotopy analysis techniques for the solution of nonlinear problems," Shanghai Jiao Tong University, Shanghai, China 1992.
42. S. Liao, *Beyond Perturbation: Introduction to the Homotopy Analysis Method*, Chapman & Hall/CRC Press, Boca Raton, FL, 2003.
43. S. Liao, "On the homotopy analysis method for nonlinear problems," *Applied Mathematics and Computation*, vol. 147, no. 2, pp. 499–513, 2004.

References

44. S. Liao, *Homotopy Analysis Method in Nonlinear Differential Equations*, Springer, London, UK, 2012.
45. J. Wang, L. Biao, and Y. Wang-Chuan, "Approximate solution for the Klein–Gordon–Schrödinger equation by the homotopy analysis method," *Chinese Physics B*, vol. 19, no. 3, pp. 030401-1-7, 2010.
46. S. Saha Ray, and A. Patra, "Application of homotopy analysis method and Adomian decomposition method for the solution of neutron diffusion equation in the hemisphere and cylindrical reactors," *Journal of Nuclear Engineering & Technology*, vol. 1, no. 2–3, pp. 1–12, 2011.
47. M. G. Sakar, and F. Erdogan, "The homotopy analysis method for solving the time-fractional Fornberg–Whitham equation and comparison with Adomian's decomposition method," *Applied Mathematical Modelling*, vol. 37, no. 20–21, pp. 8876–8885, 2013.
48. M. Zurigat, S. Momani, Z. Odibat, and A. Alawneh, "The homotopy analysis method for handling systems of fractional differential equations," *Applied Mathematical Modelling*, vol. 34, no. 1, pp. 24–35, 2010.
49. G. Adomian, *Solving Frontier Problems of Physics: The Decomposition Method*, Kluwer Academic Publishers, Boston, MA, 1994.
50. S. Guo, L. Mei, Y. Li, and Y. Sun, "The improved fractional sub-equation method and its applications to the space–time fractional differential equations in fluid mechanics," *Physics Letters A*, vol. 376, no. 4, pp. 407–411, 2012.
51. S. Zhang, and H. Q. Zhang, "Fractional sub-equation method and its applications to nonlinear fractional PDEs," *Physics Letters A*, vol. 375, no. 7, pp. 1069–1073, 2011.
52. B. Zheng, and C. Wen, "Exact solutions for fractional partial differential equations by a new fractional sub-equation method," *Advances in Difference Equations*, vol. 2013, no. 199, pp. 1–12, 2013.
53. H. Jafari, H. Tajadodi, D. Baleanu, A. A. Al-Zahrani, Y. A. Alhamed, and A. H. Zahid, "Fractional sub-equation method for the fractional generalized reaction duffing model and nonlinear fractional Sharma-Tasso-Olver equation," *Central European Journal of Physics*, vol. 11, no. 10, pp. 1482–1486, 2013.
54. E. K. El-Shewy, A. A. Mahmoud, A. M. Tawfik, E. M. Abulwafa, and A. Elgarayhi, "Space time fractional KdV Burgers equation for dust acoustic shock waves in dusty plasma with non-thermal ions," *Chinese Physics B*, vol. 23, no. 7, pp. 070505-1–7, 2014.
55. S. Zhang, Q. A. Zong, D. Liu, and Q. Gao, "A generalized exp-function method for fractional Riccati differential equations," *Communications in Fractional Calculus*, vol. 1, no. 1, pp. 48–52, 2010.
56. G. W. Wang, T. Z. Xu, and T. Feng, "Lie symmetry analysis and explicit solutions of the time fractional fifth-order KdV equation," *PLoS One*, vol. 9, no. 2, pp. e88336-1-6, 2014.
57. A. Akgül, A. Kılıçman, and M. Inc, "Improved (G'/G)-expansion method for the space and time fractional foam drainage and KdV equations," *Abstract and Applied Analysis*, vol. 2013, pp. 414353-1-7, 2013.
58. G. W. Wang, X. Q. Liu, and Y. Y. Zhang, "New explicit solutions of the generalized $(2 + 1)$-dimensional Zakharov-Kuznetsov equation," *Applied Mathematics*, vol. 3, no. 6, pp. 523–527, 2012.
59. Z. Bin, "(G'/G)-expansion method for solving fractional partial differential equations in the theory of mathematical physics," *Communications in Theoretical Physics*, vol. 58, no. 5, pp. 623–630, 2012.
60. A. Bekir, Ö. Güner, and A. C. Cevikel, "Fractional complex transform and exp-function methods for fractional differential equations," *Abstract and Applied Analysis*, vol. 2013, pp. 426462-1-8, 2013.
61. O. Güner, A. Bekir, and A. C. Cevikel, "A variety of exact solutions for the time fractional Cahn-Allen equation," *The European Physical Journal Plus*, vol. 130, no. 7, pp. 146-1-13, 2015.
62. J. H. He, S. K. Elagan, and Z. B. Li, "Geometrical explanation of the fractional complex transform and derivative chain rule for fractional calculus," *Physics Letters A*, vol. 376, no. 4, pp. 257–259, 2012.

63. W. H. Su, X. J. Yang, H. Jafari, and D. Baleanu, "Fractional complex transform method for wave equations on Cantor sets within local fractional differential operator," *Advances in Difference Equations*, vol. 2013, no. 97, pp. 1–8, 2013.

64. X. J. Yang, D. Baleanu, and H. M. Srivastava, *Local Fractional Integral Transforms and their Applications*, Academic Press, New York, 2015.

65. X. Lanlan, and C. Huaitang, "New (G'/G)-expansion method and its applications to nonlinear PDE," *Physical Review Research International*, vol. 3, no. 4, pp. 407–415, 2013.

66. L. Debnath, *Nonlinear Partial Differential Equations for Scientists and Engineers*, Birkhäusher, Boston, MA 2012.

67. W. Malfliet, "Solitary wave solutions of nonlinear wave equations," *American Journal of Physics*, vol. 60, no. 7, pp. 650–654, 1992.

68. A. K. Gupta, and S. S. Ray, "The comparison of two reliable methods for accurate solution of time-fractional Kaup-Kupershmidt equation arising in capillary gravity waves," *Mathematical Methods in the Applied Sciences*, vol. 39, no. 3, pp. 583–592, 2016.

69. A. M. Wazwaz, "The tanh method: Solitons and periodic solutions for the Dodd–Bullough–Mikhailov and the Tzitzeica–Dodd–Bullough equations," *Chaos, Solitons & Fractals*, vol. 25, no. 1, pp. 55–63, 2005.

70. A. M. Wazwaz, *Partial Differential Equations and Solitary Waves Theory*, Springer, New York, 2009.

71. S. Ege, and E. Misirli, "The modified Kudryashov method for solving some fractional-order nonlinear equations," *Advances in Difference Equations*, vol. 2014, no. 135, pp. 1–13, 2014.

72. S. Saha Ray, "New analytical exact solutions of time fractional KdV–KZK equation by Kudryashov methods," *Chinese Physics B*, vol. 25, no. 4, pp. 040204-1-7, 2016.

73. Y. Pandir, "Symmetric Fibonacci function solutions of some nonlinear partial differential equations," *Applied Mathematics & Information Sciences*, vol. 8, no. 5, pp. 2237–2241, 2014.

74. A. Stakhov, and B. Rozin, "On a new class of hyperbolic functions," *Chaos, Solitons & Fractals*, vol. 23, no. 2, pp. 379–389, 2005.

75. N. A. Kudryashov, "Logistic function as solution of many nonlinear differential equations," *Applied Mathematical Modelling*, vol. 39, no. 18, pp. 5733–5742, 2015.

76. N. A. Kudryashov, "One method for finding exact solutions of nonlinear differential equations," *Communications in Nonlinear Science and Numerical Simulation*, vol. 17, no. 6, pp. 2248–2253, 2012.

77. N. A. Kudryashov, "Polynomials in logistic function and solitary waves of nonlinear differential equations," *Applied Mathematics and Computation*, vol. 219, no. 17, pp. 9245–9253, 2013.

78. N. A. Kudryashov, "Quasi-exact solutions of the dissipative Kuramoto–Sivashinsky equation," *Applied Mathematics and Computation*, vol. 219, no. 17, pp. 9213–9218, 2013.

79. M. F. El-Sabbagh, and A. T. Ali, "New generalized Jacobi elliptic function expansion method," *Communications in Nonlinear Science and Numerical Simulation*, vol. 13, no. 9, pp. 1758–1766, 2008.

80. C. Huai-Tang, and Z. Hong-Qing, "New double periodic and multiple soliton solutions of the generalized $(2 + 1)$-dimensional Boussinesq equation," *Chaos, Solitons & Fractals*, vol. 20, no. 4, pp. 765–769, 2004.

81. E. M. E. Zayed, and K. A. E. Alurrfi, "A new Jacobi elliptic function expansion method for solving a nonlinear PDE describing the nonlinear low-pass electrical lines," *Chaos, Solitons & Fractals*, vol. 78, pp. 148–155, 2015.

82. B. Zheng, "A new fractional Jacobi elliptic equation method for solving fractional partial differential equations," *Advances in Difference Equations*, vol. 2014, no. 1, pp. 228-1-11, 2014.

83. Q. Zhou, D. Yao, and F. Chen, "Analytical study of optical solitons in media with Kerr and parabolic-law nonlinearities," *Journal of Modern Optics*, vol. 60, no. 19, pp. 1652–1657, 2013.

84. Q. Zhou, D. Yao, X. Liu, F. Chen, S. Ding, Y. Zhang, and F. Chen, "Exact solitons in three-dimensional weakly nonlocal nonlinear time-modulated parabolic law media," *Optics & Laser Technology*, vol. 51, pp. 32–35, 2013.

References

85. Q. Zhou, Q. Zhu, H. Yu, Y. Liu, C. Wei, P. Yao, A. H. Bhrawy, and A. Biswas, "Bright, dark and singular optical solitons in a cascaded system," *Laser Physics*, vol. 25, no. 2, pp. 025402-1-9, 2014.

86. M. Dehghan, M. Nasri, and M. R. Razvan, "Global stability of a deterministic model for HIV infection in vivo," *Chaos, Solitons & Fractals*, vol. 34, no. 4, pp. 1225–1238, 2007.

87. F. Shakeri, and M. Dehghan, "The finite volume spectral element method to solve Turing models in the biological pattern formation," *Computers & Mathematics with Applications*, vol. 62, no. 12, pp. 4322–4336, 2011.

88. M. Dehghan, and R. Salehi, "Solution of a nonlinear time-delay model in biology via semi-analytical approaches," *Computer Physics Communications*, vol. 181, no. 7, pp. 1255–1265, 2010.

89. H. N. Hassan, and M. A. El-Tawil, "Series solution for continuous population models for single and interacting species by the homotopy analysis method," *Communications in Numerical Analysis*, vol. 2012, pp. 1–21, 2012.

90. S. Pamuk, and N. Pamuk, "He's homotopy perturbation method for continuous population models for single and interacting species," *Computers & Mathematics with Applications*, vol. 59, no. 2, pp. 612–621, 2010.

91. J. D. Murray, *Mathematical Biology I. An Introduction*, Springer, Berlin, Germany 1993.

92. M. Rafei, D. D. Ganji, and H. Daniali, "Solution of the epidemic model by homotopy perturbation method," *Applied Mathematics and Computation*, vol. 187, no. 2, pp. 1056–1062, 2007.

93. F. Shakeri, and M. Dehghan, "Solution of a model describing biological species living together using the variational iteration method," *Mathematical and Computer Modelling*, vol. 48, no. 5–6, pp. 685–699, 2008.

94. H. Liang, "Linearly implicit conservative schemes for long-term numerical simulation of Klein–Gordon–Schrödinger equations," *Applied Mathematics and Computation*, vol. 238, pp. 475–484, 2014.

95. A. Darwish, and E. G. Fan, "A series of new explicit exact solutions for the coupled Klein–Gordon–Schrödinger equations," *Chaos, Solitons & Fractals*, vol. 20, no. 3, pp. 609–617, 2004.

96. S. Liu, Z. Fu, S. Liu, and Z. Wang, "The periodic solutions for a class of coupled nonlinear Klein–Gordon equations," *Physics Letters A*, vol. 323, no. 5–6, pp. 415–420, 2004.

97. J. Xia, S. Han, and M. Wang, "The exact solitary wave solutions for the Klein-Gordon-Schrödinger equations," *Applied Mathematics and Mechanics*, vol. 23, no. 1, pp. 58–64, 2002.

98. F. T. Hioe, "Periodic solitary waves for two coupled nonlinear Klein–Gordon and Schrdinger equations," *Journal of Physics A: Mathematical and General*, vol. 36, no. 26, pp. 7307–7330, 2003.

99. W. Bao, and L. Yang, "Efficient and accurate numerical methods for the Klein–Gordon–Schrödinger equations," *Journal of Computational Physics*, vol. 225, no. 2, pp. 1863–1893, 2007.

100. M. Naber, "Time fractional Schrödinger equation," *Journal of Mathematical Physics*, vol. 45, no. 8, pp. 3339–3352, 2004.

101. S. Z. Rida, H. M. El-Sherbiny, and A. A. M. Arafa, "On the solution of the fractional nonlinear Schrödinger equation," *Physics Letters A*, vol. 372, no. 5, pp. 553–558, 2008.

102. T. Wang, J. Chen, and L. Zhang, "Conservative difference methods for the Klein–Gordon–Zakharov equations," *Journal of Computational and Applied Mathematics*, vol. 205, no. 1, pp. 430–452, 2007.

103. M. Dehghan, and A. Nikpour, "The solitary wave solution of coupled Klein–Gordon–Zakharov equations via two different numerical methods," *Computer Physics Communications*, vol. 184, no. 9, pp. 2145–2158, 2013.

104. K. R. Khusnutdinova, and D. E. Pelinovsky, "On the exchange of energy in coupled Klein–Gordon equations," *Wave Motion*, vol. 38, no. 1, pp. 1–10, 2003.

105. O. M. Braun, and Y. S. Kivshar, "Nonlinear dynamics of the Frenkel–Kontorova model," *Physics Reports*, vol. 306, no. 1–2, pp. 1–108, 1998.

106. T. Kontorova, and J. Frenkel, "On the theory of plastic deformation and twinning. II," *Zhurnal Eksperimentalnoii Teoreticheskoi Fiziki*, vol. 8, no. 89–95, pp. 1340–1368, 1938.

107. S. Yomosa, "Soliton excitations in deoxyribonucleic acid (DNA) double helices," *Physical Review A*, vol. 27, no. 4, pp. 2120–2125, 1983.

108. S. Saha Ray, "A numerical solution of the coupled Sine-Gordon equation using the modified decomposition method," *Applied Mathematics and Computation*, vol. 175, no. 2, pp. 1046–1054, 2006.
109. D. Kaya, "A numerical solution of the sine-Gordon equation using the modified decomposition method," *Applied Mathematics and Computation*, vol. 143, no. 2–3, pp. 309–317, 2003.
110. L. Cai, X. Li, M. Ghosh, and B. Guo, "Stability analysis of an HIV/AIDS epidemic model with treatment," *Journal of Computational and Applied Mathematics*, vol. 229, no. 1, pp. 313–323, 2009.
111. S. Das, and P. K. Gupta, "A mathematical model on fractional Lotka–Volterra equations," *Journal of Theoretical Biology*, vol. 277, no. 1, pp. 1–6, 2011.
112. P. V. D. Driessche, and J. Watmough, "Reproduction numbers and sub-threshold endemic equilibria for compartmental models of disease transmission," *Mathematical Biosciences*, vol. 180, no. 1–2, pp. 29–48, 2002.
113. C. Kuttler, "Mathematical models in biology II," 2009. http://www-m6.ma.tum.de/~kuttler/script_current.pdf.
114. Q. Yang, F. Liu, and I. Turner, "Numerical methods for fractional partial differential equations with Riesz space fractional derivatives," *Applied Mathematical Modelling*, vol. 34, no. 1, pp. 200–218, 2010.
115. A. I. Saichev, and G. M. Zaslavsky, "Fractional kinetic equations: Solutions and applications," *Chaos: An Interdisciplinary Journal of Nonlinear Science*, vol. 7, no. 4, pp. 753–764, 1997.
116. G. M. Zaslavsky, "Chaos, fractional kinetics, and anomalous transport," *Physics Reports*, vol. 371, no. 6, pp. 461–580, 2002.
117. M. M. Meerschaert, and C. Tadjeran, "Finite difference approximations for fractional advection–dispersion flow equations," *Journal of Computational and Applied Mathematics*, vol. 172, no. 1, pp. 65–77, 2004.
118. F. Liu, V. Anh, and I. Turner, "Numerical solution of the space fractional Fokker-Planck equation," *Journal of Computational and Applied Mathematics*, vol. 166, no. 1, pp. 209–219, 2004.
119. S. Saha Ray, "Exact solutions for time-fractional diffusion-wave equations by decomposition method," *Physica Scripta*, vol. 75, no. 1, pp. 53–61, 2007.
120. S. Shen, F. Liu, V. Anh, and I. Turner, "The fundamental solution and numerical solution of the Riesz fractional advection-dispersion equation," *IMA Journal of Applied Mathematics*, vol. 73, no. 6, pp. 850–872, 2008.
121. J. P. Boyd, "Peakons and coshoidal waves: Traveling wave solutions of the Camassa-Holm equation," *Applied Mathematics and Computation*, vol. 81, no. 2–3, pp. 173–187, 1997.
122. R. Camassa, and D. D. Holm, "An integrable shallow water equation with peaked solitons," *Physical Review Letters*, vol. 71, no. 11, pp. 1661–1664, 1993.
123. Z. Liu, and T. Qian, "Peakons and their bifurcation in a generalized Camassa-Holm equation," *International Journal of Bifurcation and Chaos*, vol. 11, no. 3, pp. 781–792, 2001.
124. Z.-R. Liu, R.-Q. Wang, and Z.-J. Jing, "Peaked wave solutions of Camassa–Holm equation," *Chaos, Solitons & Fractals*, vol. 19, no. 1, pp. 77–92, 2004.
125. T. Qian, and M. Tang, "Peakons and periodic cusp waves in a generalized Camassa-Holm equation," *Chaos, Solitons & Fractals*, vol. 12, no. 7, pp. 1347–1360, 2001.
126. L. Tian, and X. Song, "New peaked solitary wave solutions of the generalized Camassa-Holm equation," *Chaos, Solitons & Fractals*, vol. 19, no. 3, pp. 621–637, 2004.
127. H. Kalisch, "Stability of solitary waves for a nonlinearly dispersive equation," *Discrete and Continuous Dynamical Systems. Series A*, vol. 10, no. 3, pp. 709–717, 2004.
128. Z. Liu, and Z. Ouyang, "A note on solitary waves for modified forms of Camassa-Holm and Degasperis-Procesi equations," *Physics Letters A*, vol. 366, no. 4–5, pp. 377–381, 2007.
129. R. Camassa, D. D. Holm, and J. M. Hyman, "A new integrable Shallow water equation," *Advances in Applied Mechanics*, vol. 31, pp. 1–33, 1994.
130. F. Cooper, and H. Shepard, "Solitons in the Camassa-Holm shallow water equation," *Physics Letters A*, vol. 194, no. 4, pp. 246–250, 1994.
131. B. He, W. Rui, C. Chen, and S. Li, "Exact travelling wave solutions of a generalized Camassa-Holm equation using the integral bifurcation method," *Applied Mathematics and Computation*, vol. 206, no. 1, pp. 141–149, 2008.

References

132. A. M. Wazwaz, "Solitary wave solutions for modified forms of Degasperis-Procesi and Camassa-Holm equations," *Physics Letters A*, vol. 352, no. 6, pp. 500–504, 2006.

133. A. M. Wazwaz, "New solitary wave solutions to the modified forms of Degasperis-Procesi and Camassa-Holm equations," *Applied Mathematics and Computation*, vol. 186, no. 1, pp. 130–141, 2007.

134. L. Tian, and J. Yin, "New compacton solutions and solitary wave solutions of fully nonlinear generalized Camassa-Holm equations," *Chaos, Solitons & Fractals*, vol. 20, no. 2, pp. 289–299, 2004.

135. Q. Wang, and M. Tang, "New exact solutions for two nonlinear equations," *Physics Letters A*, vol. 372, no. 17, pp. 2995–3000, 2008.

136. E. Yomba, "The sub-ODE method for finding exact travelling wave solutions of generalized nonlinear Camassa-Holm, and generalized nonlinear Schrödinger equations," *Physics Letters A*, vol. 372, no. 3, pp. 215–222, 2008.

137. E. Yomba, "A generalized auxiliary equation method and its application to nonlinear Klein-Gordon and generalized nonlinear Camassa-Holm equations," *Physics Letters A*, vol. 372, no. 7, pp. 1048–1060, 2008.

138. Z. Liu, and J. Pan, "Coexistence of multifarious explicit nonlinear wave solutions for modified forms of Camassa-Holm and Degaperis-Procesi equations," *International Journal of Bifurcation and Chaos*, vol. 19, no. 7, pp. 2267–2282, 2009.

139. Z. Liu, and Y. Liang, "The explicit nonlinear wave solutions and their bifurcations of the generalized Camassa-Holm equation," *International Journal of Bifurcation and Chaos*, vol. 21, no. 11, pp. 3119–3136, 2011.

140. E. J. Parkes, and V. O. Vakhnenko, "Explicit solutions of the Camassa-Holm equation," *Chaos, Solitons & Fractals*, vol. 26, no. 5, pp. 1309–1316, 2005.

141. A. Jafarian, P. Ghaderi, A. K. Golmankhaneh, and D. Baleanu, "Analytical treatment of system of Abel integral equations by homotopy analysis method," *Romanian Reports in Physics*, vol. 66, no. 3, pp. 603–611, 2014.

142. A. Jafarian, P. Ghaderi, A. K. Golmankhaneh, and D. Baleanu, "Analytical approximate solutions of the Zakharov-Kuznetsov equations," *Romanian Reports in Physics*, vol. 66, no. 2, pp. 296–306, 2014.

143. A. R. Seadawy, "Stability analysis for Zakharov–Kuznetsov equation of weakly nonlinear ion-acoustic waves in a plasma," *Computers & Mathematics with Applications*, vol. 67, no. 1, pp. 172–180, 2014.

144. A. M. Wazwaz, "Exact solutions with solitons and periodic structures for the Zakharov–Kuznetsov (ZK) equation and its modified form," *Communications in Nonlinear Science and Numerical Simulation*, vol. 10, no. 6, pp. 597–606, 2005.

145. V. E. Zakharov, and E. A. Kuznetsov, "Three-dimensional solutions," *Soviet Physics JETP*, vol. 39, pp. 285–286, 1974.

146. B. B. Kadomtsev, and V. I. Petviashvilli, "On the stability of solitary waves in weakly dispersing media," *Soviet Physics Doklady*, vol. 15, pp. 539–541, 1970.

147. S. Guo, and L. Mei, "The fractional variational iteration method using He's polynomials," *Physics Letters A*, vol. 375, no. 3, pp. 309–313, 2011.

148. G. C. Wu, and E. W. M. Lee, "Fractional variational iteration method and its application," *Physics Letters A*, vol. 374, no. 25, pp. 2506–2509, 2010.

149. R. Y. Molliq, M. S. M. Noorani, I. Hashim, and R. R. Ahmad, "Approximate solutions of fractional Zakharov–Kuznetsov equations by VIM," *Journal of Computational and Applied Mathematics*, vol. 233, no. 2, pp. 103–108, 2009.

150. A. Yıldırım, and Y. Gülkanat, "Analytical approach to fractional Zakharov–Kuznetsov equations by He's homotopy perturbation method," *Communications in Theoretical Physics*, vol. 53, no. 6, pp. 1005–1010, 2010.

151. A. Biswas, and E. Zerrad, "Solitary wave solution of the Zakharov–Kuznetsov equation in plasmas with power law nonlinearity," *Nonlinear Analysis: Real World Applications*, vol. 11, no. 4, pp. 3272–3274, 2010.

152. A. Mushtaq, and H. A. Shah, "Nonlinear Zakharov–Kuznetsov equation for obliquely propagating two-dimensional ion-acoustic solitary waves in a relativistic, rotating magnetized electron-positron-ion plasma," *Physics of Plasmas*, vol. 12, no. 7, pp. 072306-1-8, 2005.

153. I. Kourakis, W. M. Moslem, U. M. Abdelsalam, R. Sabry, and P. K. Shukla, "Nonlinear dynamics of rotating multi-component pair plasmas and epi plasmas," *Plasma and Fusion Research*, vol. 4, pp. 018-1-11, 2009.

154. M. A. Johnson, "The transverse instability of periodic waves in Zakharov-Kuznetsov type equations," *Studies in Applied Mathematics*, vol. 124, no. 4, pp. 323–345, 2010.

155. X. Li, "The improved Riccati equation method and exact solutions to mZK equation," *International Journal of Differential Equations*, vol. 2012, pp. 1–11, 2012.

156. S. Munro, and E. J. Parkes, "The derivation of a modified Zakharov–Kuznetsov equation and the stability of its solutions," *Journal of Plasma Physics*, vol. 62, no. 3, pp. 305–317, 1999.

157. S. Munro, and E. J. Parkes, "Stability of solitary-wave solutions to a modified Zakharov-Kuznetsov equation," *Journal of Plasma Physics*, vol. 64, no. 4, pp. 411–426, 2000.

158. A. de Bouard, "Stability and instability of some nonlinear dispersive solitary waves in higher dimension," *Proceedings of the Royal Society of Edinburgh: Section A Mathematics*, vol. 126, no. 1, pp. 89–112, 1996.

159. R. L. Mace, and M. A. Hellberg, "The Korteweg–de Vries–Zakharov–Kuznetsov equation for electron-acoustic waves," *Physics of Plasmas*, vol. 8, no. 6, pp. 2649–2656, 2001.

160. S. A. El-Tantawy, and W. M. Moslem, "Nonlinear structures of the Korteweg-de Vries and modified Korteweg-de Vries equations in non-Maxwellian electron-positron-ion plasma: Solitons collision and rogue waves," *Physics of Plasmas*, vol. 21, no. 5, pp. 052112-1-10, 2014.

161. S. Guo, H. Wang, and L. Mei, "(3 + 1)-dimensional cylindrical Korteweg-de Vries equation for nonextensive dust acoustic waves: Symbolic computation and exact solutions," *Physics of Plasmas*, vol. 19, no. 6, pp. 063701-1-7, 2012.

162. M. H. Islam, K. Khan, M. A. Akbar, and M. A. Salam, "Exact traveling wave solutions of modified KdV–Zakharov–Kuznetsov equation and viscous Burgers equation," *SpringerPlus*, vol. 3, no. 105, pp. 1–9, 2014.

163. H. Naher, F. A. Abdullah, and M. A. Akbar, "New traveling wave solutions of the higher dimensional nonlinear partial differential equation by the exp-function method," *Journal of Applied Mathematics*, vol. 2012, pp. 575387-1-14, 2012.

164. H. Naher, F. A. Abdullah, and M. A. Akbar, "Generalized and improved (G'/G)-expansion method for (3 + 1)-dimensional modified KdV-Zakharov-Kuznetsev equation," *PLoS One*, vol. 8, no. 5, pp. e64618-1-7, 2013.

165. F. Demontis, "Exact solutions of the modified Korteweg-de Vries equation," *Theoretical and Mathematical Physics*, vol. 168, no. 1, pp. 886–897, 2011.

166. İ. Aslan, "Exact solutions of a fractional-type differential-difference equation related to discrete MKdV equation," *Communications in Theoretical Physics*, vol. 61, no. 5, pp. 595–599, 2014.

167. A. Biswas, "Solitary wave solution for the generalized Kawahara equation," *Applied Mathematics Letters*, vol. 22, no. 2, pp. 208–210, 2009.

168. N. Bibi, S. I. A. Tirmizi, and S. Haq, "Meshless method of lines for numerical solution of Kawahara type equations," *Applied Mathematics*, vol. 2, pp. 608–618, 2011.

169. B. S. Kashkari, "Numerical solution of Kawahara equations by using Laplace homotope perturbations method," *Applied Mathematical Sciences*, vol. 8, pp. 3243–3254, 2014.

170. H. Hasimoto, "Water waves," *Kagaku (Japanese)*, vol. 40, pp. 401–408, 1970.

171. T. Kakutani, and H. Ono, "Weak non-linear hydromagnetic waves in a cold collision-free plasma," *Journal of the Physical Society of Japan*, vol. 26, no. 5, pp. 1305–1318, 1969.

172. T. Kawahara, "Oscillatory solitary waves in dispersive media," *Journal of the Physical Society of Japan*, vol. 33, no. 1, pp. 260–264, 1972.

173. T. J. Bridges, and G. Derks, "Linear instability of solitary wave solutions of the Kawahara equation and its generalizations," *SIAM Journal on Mathematical Analysis*, vol. 33, no. 6, pp. 1356–1378, 2002.

References 299

174. J. K. Hunter, and J. Scheurle, "Existence of perturbed solitary wave solutions to a model equation for water waves," *Physica D: Nonlinear Phenomena*, vol. 32, no. 2, pp. 253–268, 1988.

175. A. Atangana, and D. Baleanu, "Nonlinear fractional Jaulent-Miodek and Whitham-Broer-Kaup equations within Sumudu transform," *Abstract and Applied Analysis*, vol. 2013, pp. 1–8, 2013.

176. E. Fan, "Uniformly constructing a series of explicit exact solutions to nonlinear equations in mathematical physics," *Chaos, Solitons & Fractals*, vol. 16, no. 5, pp. 819–839, 2003.

177. M. Jaulent, and I. Miodek, "Nonlinear evolution equations associated with 'energy–dependent Schrödinger potentials'"*Letters in Mathematical Physics*, vol. 1, no. 3, pp. 243–250, 1976.

178. Y. Matsuno, "Reduction of dispersionless coupled Korteweg–de Vries equations to the Euler–Darboux equation," *Journal of Mathematical Physics*, vol. 42, no. 4, pp. 1744–1760, 2001.

179. D. Kaya, and S. M. El-Sayed, "A numerical method for solving Jaulent–Miodek equation," *Physics Letters A*, vol. 318, no. 4–5, pp. 345–353, 2003.

180. R. Zhou, "The finite-band solution of the Jaulent–Miodek equation," *Journal of Mathematical Physics*, vol. 38, no. 5, pp. 2535–2546, 1997.

181. G. Xu, "N-fold Darboux transformation of the Jaulent-Miodek equation," *Applied Mathematics*, vol. 5, no. 17, pp. 2657–2663, 2014.

182. H. Ruan, and S. Lou, "New symmetries of the Jaulent-Miodek hierarchy," *Journal of the Physical Society of Japan*, vol. 62, no. 6, pp. 1917–1921, 1993.

183. A. K. Gupta, and S. Saha Ray, "An investigation with hermite wavelets for accurate solution of fractional Jaulent–Miodek equation associated with energy-dependent Schrödinger potential," *Applied Mathematics and Computation*, vol. 270, pp. 458–471, 2015.

184. H. Wang, and T.-C. Xia, "The fractional supertrace identity and its application to the super Jaulent–Miodek hierarchy," *Communications in Nonlinear Science and Numerical Simulation*, vol. 18, no. 10, pp. 2859–2867, 2013.

185. S. Saha Ray, and S. Sahoo, "A novel analytical method with fractional complex transform for new exact solutions of time-fractional fifth-order Sawada–Kotera equation," *Reports on Mathematical Physics*, vol. 75, no. 1, pp. 63–72, 2015.

186. S. Saha Ray, and S. Sahoo, "New exact solutions of fractional Zakharov-Kuznetsov and modified Zakharov-kuznetsov equations using fractional sub-equation method," *Communications in Theoretical Physics*, vol. 63, no. 1, pp. 25–30, 2015.

187. A. M. Wazwaz, "The tanh–coth and the sech methods for exact solutions of the Jaulent–Miodek equation," *Physics Letters A*, vol. 366, no. 1–2, pp. 85–90, 2007.

188. E. M. E. Zayed, and H. M. A. Rahman, "The extended tanh-method for finding traveling wave solutions of nonlinear evolution equations," *Applied Mathematics*, vol. 10, pp. 235–245, 2010.

189. M. M. Rashidi, G. Domairry, and S. Dinarvand, "The homotopy analysis method for explicit analytical solutions of Jaulent-Miodek equations," *Numerical Methods for Partial Differential Equations*, vol. 25, no. 2, pp. 430–439, 2009.

190. W. M. Taha, and M. S. M. Noorani, "Exact solutions of equation generated by the Jaulent-Miodek hierarchy by (G'/G)-expansion method," *Mathematical Problems in Engineering*, vol. 2013, pp. 1–7, 2013.

191. J. H. He, and M. A. Abdou, "New periodic solutions for nonlinear evolution equations using Exp-function method," *Chaos, Solitons & Fractals*, vol. 34, no. 5, pp. 1421–1429, 2007.

192. J. H. He, and L. N. Zhang, "Generalized solitary solution and compacton-like solution of the Jaulent–Miodek equations using the Exp-function method," *Physics Letters A*, vol. 372, no. 7, pp. 1044–1047, 2008.

193. A. Yildirim, and A. Kelleci, "Numerical simulation of the Jaulent-Miodek equation by He's homotopy perturbation method," *World Applied Sciences Journal*, vol. 7, pp. 84–89, 2009.

194. J. M. Dye, and A. Parker, "A bidirectional Kaup–Kupershmidt equation and directionally dependent solitons," *Journal of Mathematical Physics*, vol. 43, no. 10, pp. 4921–4949, 2002.

195. A. H. Bhrawy, A. Biswas, M. Javidi, W. X. Ma, Z. Pınar, and A. Yıldırım, "New solutions for (1 + 1)-dimensional and (2 + 1)-dimensional Kaup–Kupershmidt equations," *Results in Mathematics*, vol. 63, no. 1–2, pp. 675–686, 2013.

196. D. J. Kaup, "On the inverse scattering problem for cubic eigenvalue problems of the class $\psi_{xxx} + 6Q\psi_x + 6R_\psi = \lambda\psi$," *Studies in Applied Mathematics*, vol. 62, no. 3, pp. 189–216, 1980.
197. E. Date, M. Jimbo, M. Kashiwara, and T. Miwa, "KP hierarchies of orthogonal and symplectic type–transformation groups for soliton equations VI–," *Journal of the Physical Society of Japan*, vol. 50, no. 11, pp. 3813–3818, 1981.
198. W. Hereman, and A. Nuseir, "Symbolic methods to construct exact solutions of nonlinear partial differential equations," *Mathematics and Computers in Simulation*, vol. 43, no. 1, pp. 13–27, 1997.
199. M. Musette, and R. Conte, "Bäcklund transformation of partial differential equations from the Painlevé–Gambier classification. I. Kaup–Kupershmidt equation," *Journal of Mathematical Physics*, vol. 39, no. 10, pp. 5617–5630, 1998.
200. R. A. Zait, "Bäcklund transformations, cnoidal wave and travelling wave solutions of the SK and KK equations," *Chaos, Solitons & Fractals*, vol. 15, no. 4, pp. 673–678, 2003.
201. A. Parker, "On soliton solutions of the Kaup–Kupershmidt equation. I. Direct bilinearisation and solitary wave," *Physica D*, vol. 137, no. 1–2, pp. 25–33, 2000.
202. A. Parker, "On soliton solutions of the Kaup-Kupershmidt equation. II: Anomalous N-soliton solutions," *Physica D*, vol. 137, pp. 34–48, 2000.
203. M. Musette, and C. Verhoeven, "Nonlinear superposition formula for the Kaup–Kupershmidt partial differential equation," *Physica D: Nonlinear Phenomena*, vol. 144, no. 1–2, pp. 211–220, 2000.
204. E. G. Reyes, "Nonlocal symmetries and the Kaup-Kupershmidt equation," *Journal of Mathematical Physics*, vol. 46, no. 7, pp. 073507-1-19, 2005.
205. E. G. Reyes, and G. Sanchez, "Explicit solutions to the Kaup–Kupershmidt equation via nonlocal symmetries," *International Journal of Bifurcation and Chaos*, vol. 17, no. 8, pp. 2749–2763, 2007.
206. H. Bulut, Y. Pandir, and S. T. Demiray, "Exact solutions of time-fractional KdV equations by using generalized Kudryashov method," *International Journal of Modeling and Optimization*, vol. 4, no. 4, pp. 315–320, 2014.
207. A. Akbulut, and F. Taşcan, "Lie symmetries, symmetry reductions and conservation laws of time fractional modified Korteweg–de Vries (mkdv) equation," *Chaos, Solitons & Fractals*, vol. 100, pp. 1–6, 2017.
208. L. Debnath, and D. Bhatta, *Integral Transforms and Their Applications, Third Edition*, CRC/Chapman & Hall Press, Boca Raton, FL, 2015.
209. D. Kaya, "An application for the higher order modified KdV equation by decomposition method," *Communications in Nonlinear Science and Numerical Simulation*, vol. 10, no. 6, pp. 693–702, 2005.
210. R. M. Miura, C. S. Gardner, and M. D. Kruskal, "Korteweg-de Vries equation and generalizations. II. existence of conservation laws and constants of motion," *Journal of Mathematical Physics*, vol. 9, no. 8, pp. 1204–1209, 1968.
211. C. S. Gardner, J. M. Greene, M. D. Kruskal, and R. M. Miura, "Method for solving the Korteweg-de Vries equation," *Physical Review Letters*, vol. 19, pp. 1095–1097, 1967.
212. H. Ono, "Soliton fission in an harmonic lattices with reflectionless inhomogeneity," *Journal of the Physical Society of Japan*, vol. 61, no. 12, pp. 4336–4343, 1992.
213. K. Konno, and Y. H. Ichikawa, "A modified Korteweg de Vries equation for ion acoustic waves," *Journal of the Physical Society of Japan*, vol. 37, no. 6, pp. 1631–1636, 1974.
214. T. Nagatani, "TDGL and mKdV equations for jamming transition in the lattice models of traffic," *Physica A: Statistical Mechanics and its Applications*, vol. 264, no. 3–4, pp. 581–592, 1999.
215. V. Ziegler, J. Dinkel, C. Setzer, and K. E. Lonngren, "On the propagation of nonlinear solitary waves in a distributed Schottky barrier diode transmission line," *Chaos, Solitons & Fractals*, vol. 12, no. 9, pp. 1719–1728, 2001.
216. M. J. Ablowitz, and H. Segur, *Solitons and the Inverse Scattering Transform*, Society for Industrial & Applied Mathematics (SIAM), Philadelphia, PA, 1981.
217. P. G. Kevrekidis, A. Khare, and A. Saxena, "Breather lattice and its stabilization for the modified Korteweg–de Vries equation," *Physical Review E*, vol. 68, no. 4, pp. 047701-1-4, 2003.
218. Z. Fu, S. Liu, and S. Liu, "New solutions to mKdV equation," *Physics Letters A*, vol. 326, no. 5–6, pp. 364–374, 2004.

References

219. Z. Yan, "New jacobian elliptic function solutions to modified KdV equation: I," *Communications in Theoretical Physics*, vol. 38, no. 2, pp. 143–146, 2002.
220. M. Wadati, "The exact solution of the modified Korteweg-de Vries equation," *Journal of the Physical Society of Japan*, vol. 32, no. 6, pp. 1681–1689, 1972.
221. S. Watanabe, "Ion acoustic soliton in plasma with negative ion," *Journal of the Physical Society of Japan*, vol. 53, no. 3, pp. 950–956, 1984.
222. A. G. Johnpillai, C. M. Khalique, and A. Biswas, "Exact solutions of the mKdV equation with time-dependent coefficients," *Mathematical Communications*, vol. 16, pp. 509–518, 2011.
223. Y. X. Yu, Q. Wang, and H. Q. Zhang, "New explicit rational solitary wave solutions for discretized mKdV lattice equation," *Communications in Theoretical Physics*, vol. 44, no. 6, pp. 1011–1014, 2005.
224. S. R. Mousavian, H. Jafari, C. M. Khalique, and S. A. Karimi, "New exact-analytical solutions for the mKdV equation," *The Journal of Mathematics and Computer Science*, vol. 2, no. 3, pp. 413–416, 2011.
225. A. Biswas, "Solitary wave solution for the generalized KdV equation with time-dependent damping and dispersion," *Communications in Nonlinear Science and Numerical Simulation*, vol. 14, no. 9–10, pp. 3503–3506, 2009.
226. B. M. Vaganan, and M. S. Kumaran, "Exact linearization and invariant solutions of the generalized Burgers equation with linear damping and variable viscosity," *Studies in Applied Mathematics*, vol. 117, no. 2, pp. 95–108, 2006.
227. T. Xiao-Yan, H. Fei, and L. Sen-Yue, "Variable coefficient KdV equation and the analytical diagnoses of a dipole blocking life cycle," *Chinese Physics Letters*, vol. 23, no. 4, pp. 887–890, 2006.
228. R. Hirota, "Exact solution of the modified Korteweg-de Vries equation for multiple collisions of solitons," *Journal of the Physical Society of Japan*, vol. 33, no. 5, pp. 1456–1458, 1972.
229. M. Wadati, "The modified Korteweg-de Vries equation," *Journal of the Physical Society of Japan*, vol. 34, no. 5, pp. 1289–1296, 1973.
230. S. Tanaka, "Modified Korteweg-de Vries equation and scattering theory," *Proceedings of the Japan Academy*, vol. 48, no. 7, pp. 466–469, 1972.
231. F. Gesztesy, W. Schweiger, and B. Simon, "Commutation methods applied to the mKdV-equation," *Transactions of the American Mathematical Society*, vol. 324, no. 2, pp. 465–525, 1991.
232. J. Satsuma, "A Wronskian representation of N-Soliton solutions of nonlinear evolution equations," *Journal of the Physical Society of Japan*, vol. 46, no. 1, pp. 359–360, 1979.
233. S. Saha Ray, "Soliton solutions for time fractional coupled modified KdV equations using new coupled fractional reduced differential transform method," *Journal of Mathematical Chemistry*, vol. 51, no. 8, pp. 2214–2229, 2013.
234. K. Sawada, and T. Kotera, "A method for finding N-soliton solutions of the KdV equation and KdV-like equation," *Progress of Theoretical Physics*, vol. 51, no. 5, pp. 1355–1367, 1974.
235. Ü. Göktas, and W. Hereman, "Symbolic computation of conserved densities for systems of nonlinear evolution equations," *Journal of Symbolic Computation*, vol. 24, no. 5, pp. 591–622, 1997.
236. P. J. Caudrey, R. K. Dodd, and J. D. Gibbon, "A new hierarchy of Korteweg–de Vries equations," *Proceedings of the Royal Society A: Mathematical, Physical and Engineering Sciences*, vol. 351, pp. 407–422, 1976.
237. X. G. Geng, "Darboux transformation of the two-dimensional Sawada–Kotera equation," *Applied Mathematics-A Journal of Chinese Universities*, vol. 4, pp. 494–497, 1989.
238. J. Satsuma, and D. J. Kaup, "A Bäcklund transformation for a higher order Korteweg–de Vries equation," *Journal of the Physical Society of Japan*, vol. 43, no. 2, pp. 692–697, 1977.
239. L. Wu, G. He, and X. Geng, "Algebro-geometric solutions to the modified Sawada-Kotera hierarchy," *Journal of Mathematical Physics*, vol. 53, no. 12, pp. 123513-1-23, 2012.
240. H. G. Liang, and G. X. Guo, "An extension of the modified Sawada-Kotera equation and conservation laws," *Chinese Physics B*, vol. 21, no. 7, pp. 070205-1-6, 2012.
241. A. K. Gupta, and S. Saha Ray, "Numerical treatment for the solution of fractional fifth-order Sawada–Kotera equation using second kind Chebyshev wavelet method," *Applied Mathematical Modelling*, vol. 39, no. 17, pp. 5121–5130, 2015.

242. S. T. Mohyud-Din, F. J. Awan, J. Ahmad, and S. M. Hassan, "Differential transform method with complex transforms to some nonlinear fractional problems in mathematical physics," *Mathematical Problems in Engineering*, vol. 2015, pp. 364853-1-9, 2015.

243. N. Shang, and B. Zheng, "Exact solutions for three fractional partial differential equations by the method," *International Journal of Applied Mathematics*, vol. 43, no. 3, pp. 1–6, 2013.

244. A. H. Khater, and R. S. Temsah, "Numerical solutions of the generalized Kuramoto-Sivashinsky equation by Chebyshev spectral collocation methods," *Computers & Mathematics with Applications*, vol. 56, no. 6, pp. 1465–1472, 2008.

245. M. Kurulay, A. Secer, and M. A. Akinlar, "A new approximate analytical solution of Kuramoto–Sivashinsky equation using homotopy analysis method," *Applied Mathematics & Information Sciences*, vol. 7, no. 1, pp. 267–271, 2013.

246. A. M. Wazwaz, "New solitary wave solutions to the Kuramoto-Sivashinsky and the Kawahara equations," *Applied Mathematics and Computation*, vol. 182, no. 2, pp. 1642–1650, 2006.

247. R. Conte, Exact solutions of nonlinear partial differential equations by singularity analysis, In *Direct and Inverse Methods in Nonlinear Evolution Equations*. Springer, Berlin, Germany, pp. 1–83, 2003.

248. J. D. M. Rademacher, and R. W. Wittenberg, "Viscous shocks in the destabilized Kuramoto-Sivashinsky equation," *Journal of Computational and Nonlinear Dynamics*, vol. 1, no. 4, pp. 336–347, 2006.

249. S. Sahoo, and S. Saha Ray, "Improved fractional sub-equation method for (3 + 1)-dimensional generalized fractional KdV–Zakharov–Kuznetsov equations," *Computers & Mathematics with Applications*, vol. 70, no. 2, pp. 158–166, 2015.

250. S. Saha Ray, and A. K. Gupta, "An approach with Haar wavelet collocation method for numerical simulations of modified KdV and modified Burgers equations," *Computer Modeling in Engineering and Sciences*, vol. 103, no. 5, pp. 315–341, 2014.

251. M. Nagasawa, *Schrödinger Equations and Diffusion Theory*, Springer Birkhäuser, New York 2012.

252. P. Popelier, *Solving The Schrödinger Equation: Has Everything been Tried?* Imperial College Press, London, UK, 2011.

253. A. Hasegawa, *Plasma Instabilities and Nonlinear Effects*, Springer-Verlag Berlin, Germany, 1975.

254. J. Boussinesq, "Essai sur la theorie des eaux courantes," *Memoires Presentes Par Divers Savants a l'Academie Des Sciences de l'Institut National de France*, vol. 23, pp. 1–770, 1877.

255. D. J. Korteweg, and G. de Vries, "XLI. On the change of form of long waves advancing in a rectangular canal, and on a new type of long stationary waves," *Philosophical Magazine Series*, 5, vol. 39, no. 240, pp. 422–443, 1895.

256. J. M. Burgers, "A mathematical model illustrating the theory of turbulence," *Advances in Applied Mechanics*, pp. 171–199, 1948.

257. J. D. Cole, "On a quasi-linear parabolic equation occurring in aerodynamics," *Quarterly of Applied Mathematics*, vol. 9, no. 3, pp. 225–236, 1951.

258. Q. Wang, "Homotopy perturbation method for fractional KdV-Burgers equation," *Chaos, Solitons & Fractals*, vol. 35, no. 5, pp. 843–850, 2008.

259. S. T. Demiray, Y. Pandir, and H. Bulut, "Generalized Kudryashov method for time-fractional differential equations," *Abstract and Applied Analysis*, vol. 2014, pp. 901540-1-13, 2014.

260. A. El-Ajou, O. A. Arqub, and S. Momani, "Approximate analytical solution of the nonlinear fractional KdV–Burgers equation: A new iterative algorithm," *Journal of Computational Physics*, vol. 293, pp. 81–95, 2015.

261. S. I. Zaki, "A quintic B-spline finite elements scheme for the KdVB equation," *Computer Methods in Applied Mechanics and Engineering*, vol. 188, no. 1–3, pp. 121–134, 2000.

262. D. Kaya, "An application of the decomposition method for the KdVB equation," *Applied Mathematics and Computation*, vol. 152, no. 1, pp. 279–288, 2004.

263. X. Wang, Z. Feng, L. Debnath, and D. Y. Gao, "The Korteweg-de Vries-Burgers equation and its approximate solution," *International Journal of Computer Mathematics*, vol. 85, no. 6, pp. 853–863, 2008.

264. Z. Feng, "Exact solution in terms of elliptic functions for the Burgers-Korteweg-de Vries equation," *Wave Motion*, vol. 38, no. 2, pp. 109–115, 2003.

References

303

265. A. Jeffrey, and S. Xu, "Exact solutions to the Korteweg-de Vries-Burgers equation," *Wave Motion*, vol. 11, no. 6, pp. 559–564, 1989.

266. R. F. Bikbaev, "Shock waves in the modified Korteweg-de Vries-Burgers equation," *Journal of Nonlinear Science*, vol. 5, no. 1, pp. 1–10, 1995.

267. J. L. Bona, and M. E. Schonbek, "Travelling-wave solutions to the Korteweg-de Vries-Burgers equation," *Proceedings of the Royal Society of Edinburgh: Section A Mathematics*, vol. 101, no. 3–4, pp. 207–226, 1985.

268. Z. Feng, and R. Knobel, "Traveling waves to a Burgers-Korteweg-de Vries-type equation with higher-order nonlinearities," *Journal of Mathematical Analysis and Applications*, vol. 328, no. 2, pp. 1435–1450, 2007.

269. E. A. B. Abdel-Salam, and Z. I. A. Al-Muhiameed, "Analytic solutions of the space-time fractional combined KdV-mKdV equation," *Mathematical Problems in Engineering*, vol. 2015, pp. 871635-1-6, 2015.

270. Q. Zhao, S. K. Liu, and Z. T. Fu, "New soliton-like solutions for combined KdV and mKdV equation," *Communications in Theoretical Physics*, vol. 43, no. 4, pp. 615–616, 2005.

271. B. J. Hong, and D. C. Lu, "New exact solutions for the VGKdV-mKdV equation with nonlinear terms of any order," *Journal of Basic and Applied Physics*, vol. 1, no. 3, pp. 73–78, 2012.

272. H. Naher, and F. Abdullah, "Some new solutions of the combined KdV-mKdV equation by using the improved (G'/G)-expansion method," *World Applied Sciences Journal*, vol. 16, no. 11, pp. 1559–1570, 2012.

273. A. Bekir, "On traveling wave solutions to combined KdV-mKdV equation and modified Burgers-KdV equation," *Communications in Nonlinear Science and Numerical Simulation*, vol. 14, no. 4, pp. 1038–1042, 2009.

274. H. Triki, T. R. Taha, and A. M. Wazwaz, "Solitary wave solutions for a generalized KdV-mKdV equation with variable coefficients," *Mathematics and Computers in Simulation*, vol. 80, no. 9, pp. 1867–1873, 2010.

275. C. A. Gómez Sierra, M. Molati, and M. P. Ramollo, "Exact solutions of a generalized KdV-mKdV equation," *International Journal of Nonlinear Sciences and Numerical Simulation*, vol. 13, no. 1, pp. 94–98, 2012.

276. W. M. Zhang, and L. X. Tian, "Generalized solitary solution and periodic solution of the combined KdV-mKdV equation with variable coefficients using the exp-function method," *International Journal of Nonlinear Sciences and Numerical Simulation*, vol. 10, no. 6, pp. 711–716, 2009.

277. E. V. Krishnan, and Y. Z. Peng, "Exact solutions to the combined KdV–mKdV equation by the extended mapping method," *Physica Scripta*, vol. 73, no. 4, pp. 405–409, 2006.

278. X. L. Yang, and J. S. Tang, "New travelling wave solutions for combined KdV-mKdV equation and (2 + 1)-dimensional Broer-Kaup-Kupershmidt system," *Chinese Physics*, vol. 16, no. 2, pp. 310–317, 2007.

279. D. C. Lu, and Q. Shi, "New solitary wave solutions for the combined KdV-mKdV equation," *Journal of Information & Computational Science*, vol. 7, no. 8, pp. 1733–1737, 2010.

280. A. K. Golmankhaneh, A. K. Golmankhaneh, and D. Baleanu, "Homotopy perturbation method for solving a system of Schrödinger-korteweg-de vries equations," *Romanian Reports in Physics*, vol. 63, no. 3, pp. 609–623, 2011.

281. S. Saha Ray, "On the soliton solution and Jacobi doubly periodic solution of the fractional coupled Schrödinger–KdV equation by a novel approach," *International Journal of Nonlinear Sciences and Numerical Simulation*, vol. 16, no. 2, pp. 79–95, 2015.

282. M. Lakshmanan, and P. Kaliappan, "Lie transformations, nonlinear evolution equations, and Painlevé forms," *Journal of Mathematical Physics*, vol. 24, no. 4, pp. 795–806, 1983.

283. N. N. Rao, "Nonlinear wave modulations in plasmas," *Pramana Journal of Physics*, vol. 49, no. 1, pp. 109–127, 1997.

284. S. V. Singh, N. N. Rao, and P. K. Shukla, "Nonlinearly coupled Langmuir and dust-acoustic waves in a dusty plasma," *Journal of Plasma Physics*, vol. 60, no. 3, pp. 551–567, 1998.

285. E. Fan, "Multiple travelling wave solutions of nonlinear evolution equations using a unified algebraic method," *Journal of Physics A: Mathematical and General*, vol. 35, pp. 6853–6872, 2002.

286. S. S. Nourazar, A. Nazari-Golshan, and M. Nourazar, "On the closed form solutions of linear and nonlinear cauchy reaction-diffusion equations using the hybrid of fourier transform and variational iterational method," *Physics International*, vol. 2, no. 1, pp. 8–20, 2011.

287. A. Doosthoseini, "Variational iteration method for solving coupled Schrödinger-KdV equation," *Applied Mathematical Sciences*, vol. 4, no. 17, pp. 823–837, 2010.

288. L. Y. Huang, Y. D. Jiao, and D. M. Liang, "Multi-symplectic scheme for the coupled Schrödinger-Boussinesq equations," *Chinese Physics B*, vol. 22, no. 7, pp. 070201-1-5, 2013.

289. D. Bai, and J. Wang, "The time-splitting Fourier spectral method for the coupled Schrödinger–Boussinesq equations," *Communications in Nonlinear Science and Numerical Simulation*, vol. 17, no. 3, pp. 1201–1210, 2012.

290. N. N. Rao, "Near-magnetosonic envelope upper-hybrid waves," *Journal of Plasma Physics*, vol. 39, no. 3, pp. 385–405, 1988.

291. N. Yajima, and J. Satsuma, "Soliton solutions in a diatomic lattice system," *Progress of Theoretical Physics*, vol. 62, no. 2, pp. 370–378, 1979.

292. B. Guo, "The global solution of the system of equations for a complex Schrödinger field coupled with a Boussinesq-type self-consistent field," *Acta Mathematica Sinica*, vol. 26, pp. 295–306, 1983.

293. V. G. Makhankov, "On stationary solutions of the Schrödinger equation with a self-consistent potential satisfying boussinesq's equation," *Physics Letters A*, vol. 50, no. 1, pp. 42–44, 1974.

294. V. G. Makhankov, "Dynamics of classical solitons (in non-integrable systems)," *Physics Reports*, vol. 35, no. 1, pp. 1–128, 1978.

295. V. E. Zakharov, "1972, "Collapse of langmuir waves," *Soviet Physics JETP*, vol. 35, no. 5, pp. 908–914, 1972.

296. L. M. Zhang, D. M. Bai, and S. S. Wang, "Numerical analysis for a conservative difference scheme to solve the Schrödinger–Boussinesq equation," *Journal of Computational and Applied Mathematics*, vol. 235, no. 17, pp. 4899–4915, 2011.

297. X. Huang, "The investigation of solutions to the coupled Schrödinger-Boussinesq equations," *Abstract and Applied Analysis*, vol. 2013, pp. 170372-1-5, 2013.

298. S. Bilige, T. Chaolu, and X. Wang, "Application of the extended simplest equation method to the coupled Schrödinger–Boussinesq equation," *Applied Mathematics and Computation*, vol. 224, pp. 517–523, 2013.

299. H. M. Jaradat, S. Al-Shar'a, Q. J. A. Khan, M. Alquran, and K. Al-Khaled, "Analytical solution of time-fractional Drinfeld-Sokolov-Wilson system using residual power series method," *International Journal of Applied Mathematics*, vol. 46, no. 1, pp. 1–7, 2016.

300. P. K. Singh, K. Vishal, and T. Som, "Solution of fractional Drinfeld-Sokolov-Wilson equation using homotopy perturbation transform method," *Applications and Applied Mathematics: An International Journal (AAM)*, vol. 10, no. 1, pp. 460–472, 2015.

301. V. G. Drinfel'd, and V. V. Sokolov, "Equations of Korteweg-de Vries type and simple Lie algebras," *Soviet Mathematics, Doklady*, vol. 23, pp. 457–462, 1981.

302. V. G. Drinfel'd, and V. V. Sokolov, "Lie algebras and equations of Korteweg-de Vries type," *Journal of Soviet Mathematics*, vol. 30, no. 2, pp. 1975–2036, 1985.

303. G. Wilson, "The affine lie algebra $C^{(1)}{}_2$ and an equation of Hirota and Satsuma," *Physics Letters A*, vol. 89, no. 7, pp. 332–334, 1982.

304. M. Santillana, and C. Dawson, "A numerical approach to study the properties of solutions of the diffusive wave approximation of the shallow water equations," *Computational Geosciences*, vol. 14, no. 1, pp. 31–53, 2009.

305. Y. H. He, Y. Long, and S. L. Li, "Exact solutions of the Drinfel'd-Sokolov-Wilson equation using the F-expansion method combined with exp-function method," *International Mathematical Forum*, vol. 5, no. 65, pp. 3231–3242, 2010.

306. Z. Xue-Qin, and Z. Hong-Yan, "An improved F-Expansion method and its application to coupled Drinfel'd–Sokolov–Wilson equation," *Communications in Theoretical Physics*, vol. 50, no. 2, pp. 309–314, 2008.

References 305

307. K. Khan, M. A. Akbar, and M. Nur Alam, "Traveling wave solutions of the nonlinear Drinfel'd-Sokolov-Wilson equation and modified Benjamin-Bona-Mahony equations," *Journal of the Egyptian Mathematical Society*, vol. 21, no. 3, pp. 233–240, 2013.

308. C. Matjila, B. Muatjetjeja, and C. M. Khalique, "Exact solutions and conservation laws of the Drinfel'd-Sokolov-Wilson system," *Abstract and Applied Analysis*, vol. 2014, pp. 271960-1-6, 2014.

309. A. Atangana, "Exact solution of the time-fractional groundwater flow equation within a leaky aquifer equation," *Journal of Vibration and Control*, vol. 22, no. 7, pp. 1749–1756, 2014.

310. S. Sahoo, G. Garai, and S. Saha Ray, "Lie symmetry analysis for similarity reduction and exact solutions of modified KdV–Zakharov–Kuznetsov equation," *Nonlinear Dynamics*, vol. 87, no. 3, pp. 1995–2000, 2017.

311. A. M. Wazwaz, and S. A. El-Tantawy, "A new (3 + 1)-dimensional generalized Kadomtsev–Petviashvili equation," *Nonlinear Dynamics*, vol. 84, no. 2, pp. 1107–1112, 2015.

312. Q. Zhou, L. Liu, Y. Liu, H. Yu, P. Yao, C. Wei, and H. Zhang, "Exact optical solitons in metamaterials with cubic–quintic nonlinearity and third-order dispersion," *Nonlinear Dynamics*, vol. 80, no. 3, pp. 1365–1371, 2015.

313. C. F. M. Coimbra, "Mechanics with variable-order differential operators," *Annalen der Physik*, vol. 12, pp. 692–703, 2003.

314. D. Ingman, and J. Suzdalnitsky, "Control of damping oscillations by fractional differential operator with time-dependent order," *Computer Methods in Applied Mechanics and Engineering*, vol. 193, no. 52, pp. 5585–5595, 2004.

315. C. M. Soon, "Dynamics with variable-order operators," Hawai'i, University of Hawai'i, 2005. http://scholarspace.manoa.hawaii.edu/bitstream/10125/10494/1/uhm_ms_3967_r.pdf.

316. C. M. Soon, C. F. M. Coimbra, and M. H. Kobayashi, "The variable viscoelasticity oscillator," *Annalen der Physik*, vol. 14, no. 6, pp. 378–389, 2005.

317. S. G. Samko, and B. Ross, "Integration and differentiation to a variable fractional order," *Integral Transforms and Special Functions*, vol. 1, no. 4, pp. 277–300, 1993.

318. C. F. Lorenzo, and T. T. Hartley, *Initialization, Conceptualization, and Application in the Generalized Fractional Calculus*, NASA TP-1998-208415, Lewis Research Centre, NASA, Cleveland, OH, 1998. doi:10.1615/critrevbiomedeng.v35.i6.10.

319. C. F. Lorenzo, and T. T. Hartley, "Variable order and distributed order fractional operators," *Nonlinear Dynamics*, vol. 29, pp. 57–98, 2002.

320. D. Ingman, and J. Suzdalnitsky, "Application of differential operator with servo-order function in model of viscoelastic deformation process," *Journal of Engineering Mechanics*, vol. 131, no. 7, pp. 763–767, 2005.

321. L. E. S. Ramirez, and C. F. M. Coimbra, "On the selection and meaning of variable order operators for dynamic modeling," *International Journal of Differential Equations*, vol. 2010, pp. 846107-1-16, 2010.

322. D. Ingman, and J. Suzdalnitsky, "Response of viscoelastic plate to impact," *Journal of Vibration and Acoustics*, vol. 130, no. 1, pp. 011010-1-8, 2008.

323. S. Das, *Functional Fractional Calculus for System Identification and Controls*, Springer, New York, 2008.

324. S. Das, *Functional Fractional Calculus*, Springer, New York, 2011.

325. H. Zhang, F. Liu, M. S. Phanikumar, and M. M. Meerschaert, "A novel numerical method for the time variable fractional order mobile–immobile advection–dispersion model," *Computers & Mathematics with Applications*, vol. 66, no. 5, pp. 693–701, 2013.

326. R. Almeida, and D. F. M. Torres, "An expansion formula with higher-order derivatives for fractional operators of variable order," *The Scientific World Journal*, vol. 2013, pp. 915437-1-11, 2013.

327. H. G. Sun, Y. Zhang, W. Chen, and D. M. Reeves, "Use of a variable-index fractional-derivative model to capture transient dispersion in heterogeneous media," *Journal of Contaminant Hydrology*, vol. 157, pp. 47–58, 2014.

328. P. Zhuang, F. Liu, V. Anh, and I. Turner, "Numerical methods for the variable-order fractional advection-diffusion equation with a nonlinear source term," *SIAM Journal on Numerical Analysis*, vol. 47, no. 3, pp. 1760–1781, 2009.

329. B. J. West, "Colloquium: Fractional calculus view of complexity: A tutorial," *Reviews of Modern Physics*, vol. 86, no. 4, pp. 1169–1186, 2014.
330. G. W. S. Blair, B. C. Veinoglou, and J. E. Caffyn, "Limitations of the Newtonian time scale in relation to non-equilibrium rheological states and a theory of quasi-Properties," *Proceedings of The Royal Society A: Mathematical, Physical and Engineering Sciences*, vol. 189, no. 1016, pp. 69–87, 1947.
331. W. G. Glöckle, and T. F. Nonnenmacher, "Fractional integral operators and Fox functions in the theory of viscoelasticity," *Macromolecules*, vol. 24, no. 24, pp. 6426–6434, 1991.
332. W. G. Glöckle, and T. F. Nonnenmacher, "Fox function representation of non-debye relaxation processes," *Journal of Statistical Physics*, vol. 71, no. 3–4, pp. 741–757, 1993.
333. W. G. Glöckle, and T. F. Nonnenmacher, "A fractional calculus approach to self-similar protein dynamics," *Biophysical Journal*, vol. 68, no. 1, pp. 46–53, 1995.
334. Y. Khan, S. P. A. Beik, K. Sayevand, and A. Shayganmanesh, "A numerical scheme for solving differential equations with space and time-fractional coordinate derivatives," *Quaestiones Mathematicae*, vol. 38, no. 1, pp. 41–55, 2015.
335. Y. Khan, M. Fardi, K. Sayevand, and M. Ghasemi, "Solution of nonlinear fractional differential equations using an efficient approach," *Neural Computing and Applications*, vol. 24, pp. 187–192, 2014.
336. A. W. Wharmby, and R. L. Bagley, "Generalization of a theoretical basis for the application of fractional calculus to viscoelasticity," *Journal of Rheology*, vol. 57, no. 5, pp. 1429–1440, 2013.
337. K. L. Shen, and T. T. Soong, "Modeling of viscoelastic dampers for structural applications," *Journal of Engineering Mechanics*, vol. 121, no. 6, pp. 694–701, 1995.
338. M. Enelund, L. Mähler, K. Runesson, and B. L. Josefson, "Formulation and integration of the standard linear viscoelastic solid with fractional order rate laws," *International Journal of Solids and Structures*, vol. 36, no. 16, pp. 2417–2442, 1999.
339. R. C. Koeller, "Applications of fractional calculus to the theory of viscoelasticity," *Journal of Applied Mechanics*, vol. 51, no. 2, pp. 299–307, 1984.
340. D. Ingman, and J. Suzdalnitsky, "Iteration method for equation of viscoelastic motion with fractional differential operator of damping," *Computer Methods in Applied Mechanics and Engineering*, vol. 190, no. 37–38, pp. 5027–5036, 2001.
341. Y. A. Rossikhin, and M. V. Shitikova, "Application of fractional operators to analysis of damped vibrations of viscoelastic single-mass systems," *Journal of Sound and Vibration*, vol. 199, no. 4, pp. 567–586, 1997.
342. M. Enelund, and B. L. Josefson, "Time-domain finite element analysis of viscoelastic structures with fractional derivatives constitutive relations," *AIAA Journal*, vol. 35, pp. 1630–1637, 1997.
343. E. F. Elshehawey, E. M. E. Elbarbary, N. A. S. Afifi, and M. El-Shahed, "On the solution of the endolymph equation using fractional calculus," *Applied Mathematics and Computation*, vol. 124, no. 3, pp. 337–341, 2001.
344. O. P. Agrawal, "Stochastic analysis of dynamic system containing fractional derivatives," *Journal of Sound and Vibration*, vol. 247, no. 5, pp. 927–938, 2001.
345. K. S. Miller, "The Mittag-Leffler and related functions," *Integral Transforms and Special Functions*, vol. 1, no. 1, pp. 41–49, 1993.
346. D. P. Hong, Y. M. Kim, and J. Z. Wang, "A new approach for the analysis solution of dynamic systems containing fractional derivative," *Journal of Mechanical Science and Technology*, vol. 20, no. 5, pp. 658–667, 2006.
347. J. F. Gómez-Aguilar, J. J. Rosales-Garcia, J. J. Bernal-Alvarado, T. Cordova-Fraga, and R. Guzman-Cabrera, "Fractional mechanical oscillators," *Revista Mexicana de Fisica*, vol. 58, pp. 348–352, 2012.
348. N. Makris, and M. C. Constantinou, "Fractional-derivative maxwell model for viscous dampers," *Journal of Structural Engineering*, vol. 117, no. 9, pp. 2708–2724, 1991.
349. M. D. Choudhury, S. Chandra, S. Nag, S. Das, and S. Tarafdar, "Forced spreading and rheology of starch gel: Viscoelastic modeling with fractional calculus," *Colloids and Surfaces A: Physicochemical and Engineering Aspects*, vol. 407, pp. 64–70, 2012.

350. S. Sahoo, S. Saha Ray, S. Das, and R. K. Bera, "The formation of dynamic variable order fractional differential equation," *International Journal of Modern Physics C*, vol. 27, no. 7, pp. 1650074-1-12, 2016.

351. L. Gaul, P. Klein, and S. Kemple, "Impulse response function of an oscillator with fractional derivative in damping description," *Mechanics Research Communications*, vol. 16, no. 5, pp. 297–305, 1989.

352. L. Gaul, P. Klein, and S. Kemple, "Damping description involving fractional operators," *Mechanical Systems and Signal Processing*, vol. 5, no. 2, pp. 81–88, 1991.

353. A. Shokooh, and L. E. Suarez, "A comparison of numerical methods applied to a fractional model of damping materials," *Journal of Vibration and Control*, vol. 5, no. 3, pp. 331–354, 1999.

354. R. L. Bagley, and J. Torvik, "Fractional calculus-A different approach to the analysis of viscoelastically damped structures," *AIAA Journal*, vol. 21, no. 5, pp. 741–748, 1983.

355. L. E. Suarez, A. Shokooh, and J. Arroyo, "Finite element analysis of beams with constrained damping treatment modeled via fractional derivatives," *Applied Mechanics Reviews*, vol. 50, no. 11, pp. S216–S224, 1997.

356. L. Suarez, and A. Shokooh, "Response of systems with damping materials modeled using fractional calculus," *Applied Mechanics Reviews*, vol. 48, no. 11S, pp. S118–S127, 1995.

357. S. Saha Ray, and R. K. Bera, "Analytical solution of a dynamic system containing fractional derivative of order one-half by Adomian decomposition method," *Journal of Applied Mechanics*, vol. 72, no. 2, pp. 290–295, 2005.

358. L. E. Suarez, and A. Shokooh, "An eigenvector expansion method for the solution of motion containing fractional derivatives," *Journal of Applied Mechanics*, vol. 64, no. 3, pp. 629–635, 1997.

359. M. Naber, "Linear fractionally damped oscillator," *International Journal of Differential Equations*, vol. 2010, pp. 197020-1-12, 2010.

360. A. A. Stanislavsky, "Fractional oscillator," *Physical Review E*, vol. 70, no. 5, pp. 051103-1-6, 2004.

361. S. Saha Ray, S. Sahoo, and S. Das, "Formulation and solutions of fractional continuously variable order mass–spring–damper systems controlled by viscoelastic and viscous–viscoelastic dampers," *Advances in Mechanical Engineering*, vol. 8, no. 5, pp. 1–13, 2016.

362. S. Sakakibara, "Properties of vibration with fractional derivative damping of order 1/2," *JSME International Journal Series C*, vol. 40, no. 3, pp. 393–399, 1997.

363. R. Ayala, A. Tuesta, and M. Sen, "Introduction to the concepts and applications of fractional and variable orderdifferential calculus," 2007. http://www3.nd.edu/~msen/Teaching/UnderRes/FracCalc2.pdf.

364. P. G. Nutting, "A new general law of deformation," *Journal of the Franklin Institute*, vol. 191, no. 5, pp. 679–685, 1921.

365. M. Caputo, *Elasticita e Dissipazione*, Zanichelli, Bologna, Italy, 1969.

366. M. D. Choudhury, S. Das, and S. Tarafdar, "Effect of loading history on visco-elastic potato starch gel," *Colloids and Surfaces A: Physicochemical and Engineering Aspects*, vol. 492, pp. 47–53, 2016.

367. G. L. Slonimsky, "Laws of mechanical relaxation processes in polymers," *Journal of Polymer Science Part C: Polymer Symposia*, vol. 16, no. 3, pp. 1667–1672, 1967.

368. P. J. Torvik, and R. L. Bagley, "On the appearance of the fractional derivative in the behavior of real materials," *Journal of Applied Mechanics*, vol. 51, no. 2, pp. 294–298, 1984.

Index

Note: Page numbers in bold and italics refer to tables and figures, respectively.

absolute errors: four term HPM and HPTM solutions **79**; graphical comparison of *77–78*; for MHAM and HPTM solutions **93–96**, **105–106**

Adomian decomposition method 249

Adomian like polynomials 84

Adomian's polynomials 24–25, 61, 111

advection–dispersion equation 109

analytical approximate solutions: Camassa–Holm (CH) equation 110–111; Riesz fractional advection–dispersion equation 109–110; Riesz fractional diffusion equation 109–110

analytical exact solutions 125

analytical function 4

analytical methods 232

angular frequency 284

antikink-solitary wave 213, *213*

approximate analytical solutions for dynamical systems 47

approximate solution of RFADE 115–117

auxiliary linear operator 112, 119

auxiliary parameters 116

Bäcklund transformation 147

backward differentiation formulas 110

beta function 4

Boltzmann distribution 126

B-spline collocation method 210

Camassa–Holm (CH) equation 110–111

Cantor space 19–20

capillary-gravity waves 145, 146

Caputo fractional derivative 11–12

Caputo fractional differential operators 97

Caputo time-fractional derivative 110

classical order: antikink-solitary wave *213*; 3-D double periodic solution *239, 240, 241, 242, 244*; differential operators 247; kink-solitary wave solution *216*; solitary wave graph for *222*

continuously variable order 264

continuous population models 47

continuous variable fractional order frictional damping force 258–259

convolution theorem 254

coupled Schrödinger–Boussinesq equations 211

coupled sine-Gordon equations 51

critical damped system 289

critical surface-tension model 146

C-type sub-hierarchy of the Kadomtsev–Petviashvili (CKP) 147

damping ratio 284, 285

dashpot 259

3-D double periodic solution *239, 240, 241, 242, 244*

differential operator 247

diffusion equation 109

(3 + 1)-dimensional space-time fractional modified KdV–ZK 126–127

(3 + 1)-dimensional time-fractional KdV–ZK equation 126

Dirac delta function 115, 265, 267, 284

displacement-time graph *286*

dust acoustic waves (DAWs) 126

electron plasma 126

error function 6

Euler–Darboux equation 146

Euler–Lagrange equation 149

Euler psi function 3

exact solitary wave solution: K–G–S equations 49; time-fractional fifth-order mS–K equation *see* time-fractional fifth-order Sawada–Kotera (S–K) equation

F-expansion method 232

Fibonacci hyperbolic solutions 208

fifth-order approximations *54, 57, 63*

forced oscillation (damping system) 280–284; spring-mass viscoelastic-viscous damping system 280–284; spring-mass viscous-viscoelastic damping system 276–280; with viscoelastic-viscous damping 266–267; with viscous-viscoelastic/viscoelastic-viscous damping 265–266

309

Fourier transform 27, 109, 111; transform parameter 111, 115
fourth-order approximate solutions 99
fourth-order MHAM solutions 102
fractal media 47, 150, 210
fractional advection–dispersion equations 110
fractional calculus 247; Caputo derivative 11–12; defined 1; Grünwald–Letnikov fractional derivative 12–13; local fractional derivative *see* local fractional derivative; mathematical analysis studies 1; mathematical functions *see* mathematical functions; Riemann–Liouville integral and derivative 10–11; Riesz fraction *see* Riesz fraction
fractional complex transform 208, 231
fractional continuously variable-order mass-spring damper system: for forced oscillation with viscoelastic-viscous damping 288, *288*; for forced oscillation with viscous-viscoelastic damping 286–287, *287*; for free oscillation with viscoelastic damping 285, *285*; for free oscillation with viscous-viscoelastic damping 286, *286*, *287*; physical interpretations *285*, *286*, *287*, *288*, 288–289; successive recursion method for 267–284
fractional derivatives 109
fractional differential equations (FDEs) 81, 210, 231; proposed tanh method 36–7
fractional diffusion equations 110
fractional fifth-order modified Sawada–Kotera equation 183–184
fractional maxwell model 248
fractional modified Zakharov–Kuznetsov (mZK) equations 125
fractional order: antikink-solitary wave solutions for *214*; 3-D double periodic solution *240*, *241*, *243*, *245*; derivative on solitary wave solutions of fractional mZK equation 135; derivative on solitary wave solutions of fractional ZK equation 131–132; kink-solitary wave solution *217*; solitary wave graph *223*
fractional partial differential equations (FPDEs) 47, 231
fractional partial differential equations with Riesz space fractional derivatives (FPDEs-RSFDs) 109
fractional Riccati equation 127, 136
fractional sine-Gordon equations 96

fractional sub-equation method 29–31, 125, 135, 138
fractional sub-equation method for solution of space-time fractional ZK equation 127–130; fractional order derivative on solitary wave solutions 131–132; numerical simulations 130–131
fractional sub-equation method to space-time fractional modified mKdV–ZK equation 140–142; numerical simulations 142–143
fractional sub-equation method to space-time fractional mZK equation 132–133; fractional order derivative on solitary wave solutions 135; numerical simulations 134–135
fractional sub-equation method to time-fractional KdV–ZK equation 135–138; numerical simulations 138–140
free oscillation (damping system) 268–272; mass-spring viscoelastic 268–272; mass-spring viscous-viscoelastic *263*, 272–276; with viscoelastic 260–262; with viscous-viscoelastic/viscoelastic-viscous 262–265
Frenkel–Kontorova dislocation model 51

Galerkin method 232
gamma function 2–3
Gauss hypergeometric functions 7–8
generalized functional order 247
generalized hypergeometric functions 9
(G'/G)-expansion method 32–33; fractional mKdV equation *see* time-fractional modified KdV equations; parameters and arbitrary constants 181; time-fractional coupled Jaulent–Miodek equations *see* time-fractional coupled JM equations; time-fractional Kaup–Kupershmidt equations *see* time-fractional KK equations; time-fractional modified Kawahara equations 145–146; *see also* time-fractional fifth order modified Kawahara equations
graphical comparison of absolute errors *77–78*
green function approach 248
groundwater hydrology model 110
Grünwald–Letnikov fractional derivative 12–13
Grünwald–Letnikov (GL) method 258

HAM *see* homotopy analysis method (HAM)
Hamiltonian system 110

Index 311

$\hbar$-curve 50, 90, 102, *103, 113,* 117, 121; for partial derivatives *117, 121*

heaviside distribution 253

H-function 9–10

$\hbar$ graph and numerical simulations: Riesz fractional advection–dispersion equation (RFADE) 117–118; Riesz fractional diffusion equation (RFDE) 114–115; Riesz time-fractional Camassa–Holm equation 118–121

hodograph transformation 146

homotopy: defined 24; series 117; third order series 121

homotopy analysis method (HAM) 26, 50, 109, 111

homotopy perturbation method (HPM) 47, 47, 52, 55, 69, 80; adomian polynomial 24–25; exact solutions 78–79; non-linear differential equation 24; numerical simulations 72–77; numerical simulations for absolute errors in HPM and HPTM solutions 77–78; overview 23

homotopy perturbation transform method (HPTM) 25–26, 51

hot isothermal electrons 126

HPM *see* homotopy perturbation method (HPM)

HPM and HPTM for the solution of fractional coupled K–G–S equations 65; solution of fractional coupled K–G–S equations 65–72

HPM methods for solutions of fractional predator–prey model 51; analytical approximate solutions for time-fractional Lotka–Volterra competition model with limit cycle periodic behavior 59–65; analytical approximate solutions for time-fractional Lotka–Volterra model 51–55; analytical approximate solutions for time-fractional simple two-species Lotka–Volterra competition model 55–59

HPTM *see* homotopy perturbation transform method (HPTM)

HPTM and MHAM methods for the solutions of fractional coupled K–G–Z equations 79; HPTM method for the solutions of fractional coupled K–G–Z equations 80–83; MHAM for approximate solutions to K–G–Z equations 83–87; numerical results 87–96

HPTM for approximate solutions of fractional coupled S-G Equations 100–102

hyperbolic function solutions 129

hypergeometric functions: Gauss 7–8; generalized 9; Kummer 8

improved (G'/G)-expansion, mKdV equation: algebraic equations 173–174; classical order 176–179; exact solutions 174–176; exponential function 174, 175; fractional order 177–180; hyperbolic function 175, 176; non-linear ODE 173; numerical simulation 176–180; partial differential equations 181; set of coefficients 174; solitary wave solution 176–180; trigonometric function 174–176; value of n determination 173

improved method: fractional sub-equation 31–32; (G'/G)-expansion 33–35

incomplete beta function 4

integral bifurcation method 110

inverse Fourier transform 113, 116

inverse Laplace transform 80, 100

inverse scattering transform (IST) 150

isothermal multicomponent magnetized plasma 126

Jacobian matrix 54, 55, 58

Jacobi elliptic function expansion method 49

Jacobi elliptic function method 41–3, 231–232

kinetics of chaotic dynamics 110

kink-solitary wave solutions *216, 217*

Klein–Gordon–Schrödinger (K–G–S) equations 47

Klein–Gordon–Zakharov (K–G–Z) equations 47, 50

Korteweg–de Vries equation 126

Kummer hypergeometric function 8

Lagrangian time-fractional mKdV equation 148

Laplace transform 100

Laplace transformation 68, 69

Laplace transform method 50

Laplace transform of variable-order operator 254–255

Lax pair 146

local fractional calculus theory 210

local fractional continuous of order 18

local fractional derivative: Cantor space 19–20; continuous 18–19; definitions of 19

Lotka–Volterra system 48

magnet-acoustic waves 146

mass-spring damper system, formulation for 259–267

mass-spring oscillator: sliding on continuous order guide *266, 267*; under viscous-viscoelastic *263*

mathematical functions: beta function 4; error function 6; gamma function 2–3; *H*-function 9–10; hypergeometric functions *see* hypergeometric functions; incomplete gamma function 3; Mellin–Ross function 6; Mittag–Leffler function 4–5; Wright function 6

Mellin–Ross function 6

MHAM-Fourier transform (FT) 111

MHAM-FT for approximate solution: of Riesz fractional advection–dispersion equation (RFADE) 115–117; of Riesz fractional diffusion equation (RFDE) 111–112

MHAM-FT solutions 113–115, *114, 115, 118*

Mittag–Leffler function 4–5, 248, 272, 280

Miura transformations 184

modified decomposition method (MDM) 51

modified homotopy analysis method (MHAM) 26–27, 47, 111; absolute errors for MHAM and HPTM solutions 93–96, 105–106; analogize the solutions of K–G–Z equations 93; for approximate solution, of Riesz time-fractional Camassa–Holm equation 118–121; for approximate solutions of fractional coupled S-G Equations 96–99; with the Fourier transform method (MHAM-FT) 27–29; $\hbar$ graph and the numerical simulations for $u(x, t)$ and $v(x, t)$ 90–93, 102–103; numerical simulations for 87–90, 103–105; travelling wave solution *122*

modified Kudryashov method 37–38

modified Kudryashov method, fractional mS–K: classical order 197, 198; exact solitary wave solutions 194–196; fractional order 196, 197; numerical simulations 196–198

modified simple equation (MSE) method 232

modified Zakharov–Kuznetsov (mZK) equations 125

*m*th-order deformation equation 112, 116, 120

natural frequency 284

newtonian friction force 258

non-differentiable function of exponent 18

non-linear equations 61

non-linear fractional evolution 209

non-linear ion-acoustic waves 126

non-linear Klein–Gordon equation 49

non-linear Klein–Gordon–Schrödinger (K–G–S) equations 47

non-linear operator 112

non-linear partial differential equations (NPDEs) 47

non-linear systems 125

non-linear term 119

non-linear wave solutions 110

novel analytical method 209

N-soliton solutions 147

numerical simulation, mKdV equation: classical order 170–173; fractional order 170–173; solitary wave solution 170–173

numerical simulations 134; space-time fractional mKdV–ZK equation 142–143; space-time fractional mZK equation 134–135; space-time fractional ZK equation 130–131; time-fractional KdV–ZK Equation 138–140

numercal simulations, modified Kawahara equations: classical order 158, 159; fractional order 158, 159; three-dimensional solitary wave 158, 159; two-dimensional solitary wave 158, 159

ordinary differential equations (ODEs) 127, 231

over damped system 289

peakon wave 110

pole 38, 40, 43

prey–predator models 47, 48

quantum mechanical behavior 209

Riccati equation 40, 127, 136

Riemann–Liouville fractional derivative 120, 149

Riemann–Liouville integral 52, 55, 59, 66

Riemann–Liouville integrals: and derivative 10–11; operators 98

Riemann–Liouville modified derivative 18

Riemann–Liouville space fractional derivatives 110

Riesz fraction: derivative 13–14, 17; Feller derivative 17; integral 13–16

Riesz fractional advection–dispersion equation (RFADE) 109–110; $\hbar$ graph and numerical simulations 117–118; MHAM-FT for approximate solution of 115–117

Riesz fractional derivatives 109, 110–111

Index

313

Riesz fractional diffusion equation (RFDE) 109–110; $\hbar$ graph and numerical simulations 113–115; MHAM-FT for approximate solution of 111–112

Riesz-fractional time derivative 110–111

Riesz time-fractional Camassa–Holm equation: $\hbar$ graph and numerical simulations 121–122; MHAM for approximate solution of 118–121

Riesz time-fractional Camassa–Holm equation 111

right-derivatives 120

Routh–Hurwitz criterion 64

Schrödinger–Boussinesq (SB) equations 209

Schrödinger equations 49

Schrödinger–KdV (SK) equation 211

second-order diffusion equations 110

shallow water flow models 232

sine-Gordon equations 96

solitary wave graph for classical order 222

solitary wave solution 110, *130, 131, 134*; and Camassa–Holm equation 110; fractional parameter 190; fractional ZK equation 131–132; (G'/G)-expansion 170–173, 174; improved (G'/G)-expansion 176–180; Jacobi elliptic function 49; mZK equation 135; tanh method 188–189

soliton theory 150

space-fractional advection–diffusion equation 110

space-time fractional advection–diffusion equation 110

space-time fractional modified KdV–Zakharov–Kuznetsov (mKdV–ZK) equations 125

space-time fractional Zakharov–Kuznetsov (ZK) equations 125, 126, 132, 134

space-time fractional ZK equation 125–126

spring damper system, viscoelastic oscillator analysis 255–258

sub-ordinary differential equation (sub-ODE) method 110

successive recursion method 43–44, 268

Sumudu transform homotopy perturbation method (STHPM) 146

tanh method 36–37

tanh method, Sawada–Kotera (S–K) equation: algebraic equations 186; fractional complex transform 185; non-linear ODE 185; numerical simulations 187–190; sets of coefficients 186; value of n determination 185

tanh method, time-fractional coupled JM equations: exact solutions 203–206; numerical simulations 206–207

tanh method, time-fractional fifth-order modified Sawada–Kotera (mS–K): algebraic equations 191; K–K equation 184; Miura transformations 184; modified Kudryashov method 194–198; numerical simulations 192–194; ODE form 190; value of n determination 190

tanh method, time-fractional K–S equation: exact solutions 198–202; numerical simulation 202–203

Taylor series 83, 96, 115, 119

theory of derivatives and integrals 231

third-order approximate solutions 53

third order MHAM and VIM solutions **123**

third-order MHAM solutions 93

time-fractional combined KdV–mKdV equation 210; exact solutions for 214–216; numerical simulations for 216

time-fractional coupled Drinfeld–Sokolov–Wilson (DSW) Equations: by case II of set I *242, 243*; by case I of set I *239, 240, 241*; by case V of set I *244, 245, 246*; exact solutions for 232–239

time-fractional coupled JM equations: algebraic equations 160–162; exact solutions 163–164; fractional complex transform 160; (G'/G)-expansion 147; hyperbolic function 163; inverse scattering transform 146; Lax pair 146; non-linear ODE 160; sets of coefficients 162–163; STHPM 146; trigonometric function 163

time-fractional coupled non-linear K–G–S equations 49–50

time-fractional coupled non-linear K–G–Z equations 47, 50

time-fractional coupled non-linear Klein–Gordon–Schrödinger equations 47

time-fractional coupled SB equations 211; exact solutions for 223–227; numerical simulations for 227, *228, 229*

time-fractional coupled schrödinger–KdV equations 210–211; exact solutions for 217–221; numerical simulations for 221, *222, 223*

time-fractional fifth-order Kaup–Kupershmidt (K–K) equation 184

314   *Index*

time-fractional fifth order modified Kawahara equations: algebraic equations 151; capillary-gravity waves 145, 146; critical surface-tension model 146; (G'/G)-expansion 146; hyperbolic function 152–154; $\lambda = 0$, $\mu = 0$ 154–155; $\lambda = \lambda$ (arbitrary) 152–154; $\mu = 0$, sets of solutions 156–157; numerical simulations 158–159; ODE 151; trigonometric function 153–154

time-fractional fifth-order Sawada–Kotera (S–K) equation: KdV equations 184; tanh method *see* tanh method, Sawada–Kotera (S–K) equation; unidirectional nonlinear evolution 184

time-fractional KK equations: algebraic equations 165; Bäcklund transformation 147; exact solutions 165; exponential function 165, 167; (G'/G)-expansion 147; hyperbolic function 166, 167; non-linear ODE 164; N-soliton solutions 147; set of coefficients 165; trigonometric function 166, 167; value of n determination 164

time-fractional Korteweg–de Vries (KdV)–Burgers equation 209–210; exact solutions for 211–213; numerical simulations for 213

time-fractional Korteweg–de Vries (KdV)–Zakharov–Kuznetsov (KdV–ZK) equations 125

time-fractional modified KdV equations: algebraic equations 168; conservation laws and Lax pair 150; dimensionless 149, 150; Euler–Lagrange equation 149; exact solutions 168–170; field variable 148; (G'/G)-expansion 151; hyperbolic function 168, 169; IST 150; Lagrangian multipliers 148; non-linear equation 147; non-linear ODE 167; numerical simulation 170–173; physical models 149, 150; Riemann–Liouville fractional derivative 149; set of coefficients 168; soliton theory 150;

trigonometric function 169; value of n determination 167

time-fractional non-linear-coupled sine-Gordon equations 47

time-fractional order non-linear-coupled sine-Gordon equations 51

time-fractional predator–prey models 48; time-fractional Lotka–Volterra competition model 49; time-fractional Lotka–Volterra model 48; time-fractional simple two-species Lotka–Volterra competition model 48

traveling wave analysis 126

un-damped system 288–289
under-damped system 284, 285

variable order 247
variable order (VO) 247; integral operator *255*; mass-spring oscillator sliding on *257*
variable-order differential equations (VODEs) 247
variable-order differential operators (VODOs) 249–253; analysis 253–254
variable-order fractional calculus theory 249–255
variable-order model 249
variable-order operator, Laplace transform 254–255
variable viscoelastic oscillator *261*
variational approximations 110
variational iteration method (VIM) 111, 126
velocity-time graph *287*
viscoelastic material 248
viscoelastic-viscous damping 259
viscoelastic-viscous/viscous-viscoelastic 257, *263, 266*

Wright function 6

Zakharov–Kuznetsov (ZK) equations 125
zeroth-order deformation equation 85, 97, 112, 116, 119